Topics in Applied Physics   Volume 36

Topics in Applied Physics Founded by Helmut K. V. Lotsch

Amorphous Semiconductors

Edited by M. H. Brodsky

With Contributions by

M. H. Brodsky D. Carlson G. A. N. Connell
E. A. Davis R. Fischer T. M. Hayes B. Kramer
P. G. LeComber G. Lucovsky P. Nagels
I. Solomon W. E. Spear D. L. Weaire C. R. Wronski

With 181 Figures

Springer-Verlag Berlin Heidelberg New York 1979

Marc H. Brodsky

IBM Thomas J. Watson Research Center
Yorktown Heights, NY 10598, USA

ISBN 3-540-09496-2 Springer-Verlag Berlin Heidelberg New York
ISBN 0-387-09496-2 Springer-Verlag New York Heidelberg Berlin

Library of Congress Cataloging in Publication Data. Main entry under title: Amorphous semiconductors. (Topics in applied physics; v. 36). Includes bibliographical references and index. 1. Amorphous semiconductors. I. Brodsky, Marc Herbert, 1938–. QC611.8.A5A45   537.6'22   79-16148.

Monophoto typesetting, offset printing and bookbinding: Brühlsche Universitätsdruckerei, Giessen
2153/3130-543210

To My Family and Colleagues

Preface

This book addresses the physics of electronic phenomena in amorphous semiconductors with emphasis on those phenomena principally dependent upon energy levels in and near the semiconductor band gap. An attempt is made to convey to the reader those concepts which are novel to the amorphous state of semiconductors as well as those which are most easily described by analogy to the well-established base of knowledge about crystalline semiconductors. After a series of three introductory chapters, the material is arranged by phenomena: in turn, optical absorption, electronic transport, luminescence, spin effects, probes of short-range order, doping, and solar cells are covered. The chapters were written during 1978 and were reviewed for timeliness at the end of that year. For amorphous semiconductors, this is a time of intense activity and rapid advance. Therefore, each chapter first presents the underlying physics which it is hoped will be relevant both to future developments and to the current level of understanding. Where understanding is lacking or incomplete, an attempt has been made to give some cohesion to a phenomenological summary of existing facts and speculations.

The editor's personal research leading up to the preparation of this compendium has benefited greatly from many interactions with his colleagues at the IBM Thomas J. Watson Research Center over the past decade. In addition productive and enjoyable research stays were spent at laboratories in Paris, Dundee, and Stuttgart. Particular gratitude is expressed to the late Reuben Title with whom the discovery of dangling bond spins in amorphous Si and Ge was made and with whom the then seemingly fruitless search for dangling bonds in amorphous chalcogenides commenced. From those, and related observations elsewhere, several of the main themes of this book follow.

Yorktown Heights
July, 1979

Marc H. Brodsky

Contents

Contributors

Brodsky, Marc H.
  IBM Thomas J. Watson Research Center, Yorktown Heights, NY 10598,
  USA

Carlson, David E.
  RCA Laboratories, Princeton, NJ 08540, USA

Connell, G. A. Neville
  Xerox Palo Alto Research Center, 3333 Coyote Hill Road,
  Palo Alto, CA 94304, USA

Davis, Edward A.
  Cavendish Laboratory, Cambridge University, Madingley Road,
  Cambridge CB3 OHE, United Kingdom

Fischer, Roland
  Fachbereich Physik der Philipps-Universität, D-3550 Marburg, Fed. Rep.
  of Germany

Hayes, Timothy M.
  Xerox Palo Alto Research Center, 3333 Coyote Hill Road,
  Palo Alto, CA 94304, USA

Kramer, Bernhard
  Department of Physics, Heriot Watt University, Riccarton, Currie,
  Edinburgh EH14 4AS, Scotland
  Present address: Physikalisch-Technische Bundesanstalt,
  D-3300 Braunschweig, Fed. Rep. of Germany

LeComber, Peter G.
  Carnegie Laboratory of Physics, The University of Dundee,
  Dundee DD1 4HN, Scotland

Lucovsky, Gerald
  Xerox Palo Alto Research Center, 3333 Coyote Hill Road,
  Palo Alto, CA 94304, USA

Nagels, Piet
S.C.K./C.E.N., Department of Materials Sciences, B-2400 Mol, Belgium

Solomon, Ionel
Laboratoire de Physique de la Matière Condensee. Ecole Polytechnique, 91128 Palaiseau Cedex, France

Spear, Walter E.
Carnegie Laboratory of Physics, The University of Dundee, Dundee DD1 4HN, Scotland

Weaire, Denis L.
Department of Physics, Heriot-Watt University, Riccarton, Currie, Edinburgh EH14 4AS, Scotland

Wronski, Christopher R.
EXXON Research and Engineering Comapny, P.O. Box 8, Linden, NJ 07036, USA

1. Introduction

M. H. Brodsky

We live in an electronic era of technological advances based to a large extent on crystalline semiconductor devices. The marvels of modern semiconductor technology follow from sound physical knowledge and metallurgical control of single crystal materials. While semiconductor phenomena are not unique to crystals, active devices almost always involve single crystals. Amorphous semiconductors, with the well-established crystalline foundations as a departure point, offer new frontiers for research and, hopefully, promise for technological developments.

It is the purpose of this book to provide a cohesive review of the understanding of electronic phenomena in amorphous semiconductors. For each topic the authors first present a conceptual framework to place in perspective the trends and results of current research. A recurrent theme is the distinction between those phenomena that are amenable to analysis, or at least description, by reference to already understood crystalline analogies and those effects which are unique, or at least relatively novel, to the amorphous state.

The most obvious examples of these two distinctive extreme cases are first, the existence, at least in a gross qualitative sense, of band gaps in amorphous semiconductors analogous, and close in energy, to the regions forbidden to electron occupation in crystalline semiconductors; and second, the lack of long-range order in the amorphous state which introduces the search to replace the elegant simplification that periodicity permits in the formal representation of energy levels in crystals and the allowable transitions between them. Other phenomena are less clearly distinguishable as being especially analogous or novel. For example, a conceptually ideal random network of atoms often serves as a framework for modeling the amorphous state. Then, the question is: Can the deviations from this ideal be treated as localized defects analogous to point defects in crystals or are they something new and different?

1.1 Definition of Amorphous Semiconductors

It is easier to define the amorphous state by saying what it is not than by precisely specifying what it is. Amorphous semiconductors are noncrystalline. They lack long-range periodic ordering of their constituent atoms. That is not to say that amorphous semiconductors are completely disordered on the atomic scale. Local chemistry provides almost rigorous bond-length, and to a lesser extent, bond-angle constraints on the nearest-neighbor environment. Unlike amorphous metals, amorphous semiconductors do not consist of close-packed atoms, but rather they contain covalently bonded atoms arranged in an open network with correlations in ordering up to the third or fourth nearest neighbors. The short-range order is directly responsible for observable semiconductor properties such as optical absorption edges and activated electrical conductivities.

A distinction should be made between amorphous and polycrystalline materials. Polycrystalline semiconductors are composed of grains with each grain containing a periodic array of atoms surrounded by a layer of interconnective or boundary atoms. For smaller and smaller grains, i.e., microcrystallites, the surface layer of each grain contains a larger and larger number of atoms relative to the periodically arrayed interior atoms. Eventually for small enough grains, the distinction between the interior and surface is lost, and the concept of microcrystallites with a definable periodic region loses its meaning. While attempts have been made to model amorphous semiconductors by microcrystallites, it is now generally accepted that network models are more applicable. However, microclusters of noncrystalline material may exist for some semiconductors.

The terms "glassy" or "vitreous" are often used synonymously for "amorphous" or "noncrystalline." However, in some fields of study, glassy or vitreous connotes the technical preciseness of a definable thermodynamic phase. The existence of a glass state, with its glass transition temperature, has been documented for some chalcogenides but not for the tetrahedrally bonded amorphous semiconductors. For chalcogenide glasses, this is reflected in the ability to prepare them from a semiconductor melt by rapid cooling (quenching) to temperatures below the glass transition temperature. For silicon and the like, quenching from the melt, which in this case is metallic with a different short-range order than the desired semiconductor, generally cannot be done rapidly enough to freeze in an amorphous atomic arrangement. Polycrystallinity is the more common result.

Amorphous semiconductors that cannot be prepared directly from the melt are usually fabricated in the form of thin films by an atomic deposition procedure such as evaporation, sputtering, chemical vapor deposition, plasma decomposition of gases, or electroplating. Sometimes ion bombardment of crystals is used to leave an amorphous layer in the collision trail of the ions.

Once prepared, a material is empirically defined to be amorphous if its diffraction pattern consists of diffuse rings (halos) rather than sharply defined Bragg rings or spots characteristic of polycrystalline or single crystal solids. While mixtures of amorphous and crystalline materials can exist, there does not seem to be a continuous transition from one to the other. Where measured, conversion from amorphous to crystalline takes place by nucleation and growth rather than by homogeneous atomic rearrangement.

1.2 Current and Future Trends

Amorphous semiconductors, while forming a single area of study with some unifying ideas, often is divided into two subfields, the tetrahedrally coordinated siliconlike materials and the chalcogenide glasses. Arsenic, from column V of the periodic table, forms a bridge between the two classes of materials. This book deals with the common methods and theories that today place particular emphasis on defects in materials from either class. In this section, a perspective is given on how the present situation has developed.

1.2.1 Amorphous Silicon Defects and Passivation

Amorphous silicon when nominally pure is permeated with dangling bonds as evidenced by a detectable electron spin resonance (ESR) signal from about 5×10^{19} spins cm^{-3}. While the origin and properties of the dangling bond spins are interesting in themselves, they generally obscure many phenomena of interest to semiconductor physicists. For example, the electron energy levels of the dangling bonds lie in between the valence (bonding) states and conduction (antibonding) states of the fully paired, bonded electrons. These states in the gap contribute to optical absorption and electrical conduction processes, masking the measurement of an energy gap by either process. Further, there is a large density of gap states which act as fast nonradiative recombination centers, with the result that photoconductivity or photoluminescence is uninterestingly small in pure amorphous silicon. Finally, the Fermi level, which is essentially pinned by the gap states, does not move significantly when trace impurities of the conventional shallow donor or acceptor types (e.g., phosphorus or boron, respectively) are incorporated. Since normal annealing procedures reduce the ESR signal, the electrical conductivity and the tailing of the optical absorption edge, it was originally, and optimistically, assumed that "ideal" amorphous silicon might be approached either by annealing or by slowly depositing onto substrates at elevated temperatures. The assumption here was that the annealing effects were due to the healing of dangling bonds by reconstruction and rearrangement of the amorphous network and not merely due to contamination. More recently it was established that under clean conditions the annealing effects are minimal and that amorphous silicon will crystallize before all the dangling bonds are removed. Thus, ideal, pure amorphous silicon seems to exist in concept only.

If dangling silicon bonds cannot be paired to each other, the opportunity still exists to pair them with other atomic orbitals. Hydrogen appears to be suitable for this purpose. Other atoms, e.g., oxygen, nitrogen, fluorine, etc., may also aid the passivation process. Following the success of the glow discharge plasma decomposition of silane, several other procedures have evolved for incorporating hydrogen into amorphous silicon. These include reactive sputtering with a hydrogen-containing gas, proton implantation, and postdeposition treatment in a hydrogen plasma. In all cases it is presumed that hydrogen passivates the dangling bonds thereby revealing the interesting properties of intrinsic amorphous silicon, somewhat mediated by alloying effects.

Many intrinsic properties of hydrogenated amorphous silicon have been studied to date. These include the thermally activated conductivity, the optical absorption edge, photoluminescence, electroluminescence, and photoconductivity. In addition, the standard shallow donors and acceptors have been shown to dope hydrogenated amorphous silicon to be n type or p type, respectively. The phenomena and configurations of interest and import to single crystal silicon, now become amenable to study with amorphous silicon. Extrinsic conductivity, thermopower, Hall effect, and photoconductivity, amongst others, are all dopant-dependent phenomena with crystalline analogs that have been measured for amorphous silicon. Often the results are interpretable in terms of processes previously described for crystals. Sometimes – the sign of the Hall effect has been the most notable example – the crystalline analogy does not work.

Schottky barrier and pn junction diodes are two devicelike configurations that have been constructed with amorphous silicon. When operated as solar cells these diodes behave to some extent like the corresponding crystal devices. Even complicated transistorlike structures have been proposed, though practical realizations are not yet here.

The simple thesis of passivated dangling bonds is appealing, but it is not the whole story. Hydrogen or other additives might, besides passivating gap states, play active roles by, for example, enlarging the band gap (alloying with Si), sensitizing photoexcited processes, changing the lattice–electron coupling, etc. How much hydrogenation is needed to reveal the most interesting phenomena is still an open question. It appears that more than just the paramagnetic centers, e. g., weak or otherwise reconstructed bonds, pick up hydrogen. Efforts continue to quantify the process.

In summary, amorphous silicon as a semiconductor with interesting semiconductor properties has been revealed by the hydrogenation of dangling and weak bonds. What the future holds is the study of the revealed host, the revelation process, doping and associated effects, devicelike configurations, and, eventually, useful devices.

1.2.2 Chalcogenide Glasses and Proper Valence

In contrast to amorphous silicon and the other group IV tetrahedrally bonded semiconductors, the amorphous chalcogens and chalcogenides do not exhibit a dangling bond spin resonance signal in their clean, as-prepared state. Pervasive diamagnetism rather than paramagnetism is the general rule. Chalcogens with their twofold coordination and lone-pair orbitals offer the network the dual channels of structural and electronic flexibility for removal of dangling bonds. As a consequence, a typical chalcogenide has a relatively sharp optical absorption edge, a single electrical activation energy, and efficient photoexcited conductivity and luminescence. All these properties are characteristic of a well-defined and clean forbidden gap. Nevertheless, field-effect and doping experiments indicate that the Fermi level of a chalcogenide glass is nearly pinned. Luminescent emission occurs at subband gap energies. Paramagnetism can be photoexcited at low temperatures. All of these latter effects are normally characteristic of states in the gap, and until recently posed an unsolved paradox. The resolution of the paradox was found in the concept of negative correlation energies of gap states, thereby leading to pairing of electrons with opposite spins. In essence, distortions of the host network are possible so that bonded, paired electrons are favored over uncorrelated, gap-filling, single electron states. Competitive models exist for the microscopics of paired states. Theoretical and experimental activity now focuses on how to identify the microscopic pairings, be they charged dangling bonds, valence alternation pairs, or other arrangements yet to be proposed. The path for understanding of intrinsic semiconductor phenomena in chalcogenides now appears here within the framework of flexible structural coordination and available lone pair electrons.

1.2.3 Intermediate Cases

Between the extremes of the column IV tetrahedrally coordinated amorphous Si, Ge, and SiC and the twofold coordinated column VI chalcogens in elemental and compound or alloy glasses lie two other classes of amorphous semiconductors with some properties characteristic each of the two extremes. These are the amorphous group V semiconductors, e.g., arsenic, and III–V semiconductors, e.g., GaAs. These pnictide-containing amorphous semiconductors do not have dense concentrations of paramagnetic dangling bonds, but often do have a small residual dark ESR signal ($\lesssim 10^{16}$ spins cm^{-3}), particularly when prepared by quanching to cryogenic temperatures. Network rigidity is higher than for chalcogenides because of higher nearest-neighbor coordinations, but less than for group IV semiconductors because of possible lower coordination and lone pair orbitals. At present, investigations of semiconductors of the III–V and V type have not progressed as far as for the prototypical siliconlike or chalcogenide amorphous semiconductors. Undoubtedly there is room here for future study and testing of the hypotheses put forth for the conceptually simpler extreme cases.

1.3 Organization of the Book

In this chapter, we have given an introduction to previous developments on the topics of most recent interest, tracing them to emphasize the reasons for the current activity on defect states and passivation techniques, doping, and solar cells. Analogies to crystalline phenomena were drawn and those effects that are almost unique to amorphous semiconductors were pointed out. The next two chapters continue to provide an introductory framework to the book.

Chapter 2 places in perspective the various theoretical approaches relating energy eigenstates to the spatial structure of amorphous systems. All types of approaches, from brute force numerical methods to sophisticated formalisms are reviewed with respect to how well the different methods can hope to answer the most relevant questions about states near the gap, namely, localization and transitions to extended states.

Chapter 3 focuses on states in the gap and what are called "defects" in fully coordinated network models for amorphous semiconductors. Of principal concern are the consequences of electron pairing which gives diamagnetic rather than paramagnetic gap states in chalcogenide glasses.

Next follows a series of chapters, each dealing with a particular class of phenomena. In Chap. 4, a framework is put forth for understanding the optical absorption from defects in the gap, band tails, and interband transitions. Examples of all three types of observations are given.

Chapter 5 uses band models to sort electrical conduction in amorphous semiconductors into three competitive channels analogous to the three regimes of optical absorption in the previous chapter: localized states at the Fermi energy, localized states in band tails near the mobility edge, and extended state conduction. In addition to the temperature-dependent conductivity, experimental results for thermoelectric power, ac conductivity, transient conductivity, and photoconductivity are reviewed for chalcogenides and interpreted in terms of band models or the alternative small-polaron picture.

Chapter 6 reviews photoluminescent phenomena in both siliconlike and chalcogenide amorphous semiconductors. After sifting through the influence of preparation conditions, the author puts forth models for the underlying physical mechanisms.

Chapter 7 introduces the concept of hard and soft centers as the origin of the electron spin resonance signals in pure and hydrogenated amorphous silicon, respectively. Then experimentally observed spin-dependent recombination effects are described and interpreted.

Chapter 8 identifies those elements of the short-range order that are best studied by X-ray absorption fine structure (EXAFS) or infrared and Raman vibrational spectroscopy, two techniques of current interest. A survey of the vibrational spectra in various classes of elemental and compound amorphous semiconductors is given. This is followed by recent EXAFS and vibration results on local environments, bonding defects, and wrong bonds.

Chapter 9 covers the preparation and properties of doped amorphous silicon. The emphasis is on a self-consistent picture in terms of the density-of-states distribution deduced from field-effect measurements. The special aspects of the potential barrier in amorphous semiconductor junctions are presented.

Finally Chap. 10 deals with those properties of doped and undoped amorphous silicon that are relevant to photovoltaic solar cells. State-of-the-art photovoltaic characteristics are discussed within the context of preparation parameters, and a model analysis of cell operation is given.

Although a multi-author volume such as this will naturally have a variety of styles and individual approaches, all of the authors have tried to incorporate some common structure to their expositions. First the conceptual framework is set, then the calculations or experimental results are examined within this conceptual context. Further, recurring reference is made to crystalline analogies or novel aspects of the amorphous state, as is appropriate.

2. Theory of Electronic States in Amorphous Semiconductors[1]

B. Kramer and D. Weaire

With 12 Figures

Our subject is, as the cliché goes, still in its infancy. Nevertheless it has matured considerably over the last ten years. Most theorists now realize that *structure* necessarily lies at the heart of the problem of the theory of electronic properties of amorphous semiconductors. As a corollary of this, we must recognize that the questions which it poses are ill defined, for our knowledge of the structure of these materials is rather limited. They are also intrinsically difficult. There is no longer such a tendency to conceal this by such formal devices as the Gubanov Transformation (which transforms the theory of crystals into the theory of amorphous solids, and, presumably, pumpkins into stagecoaches).

While the best work in the area of band structure of crystals emphasizes quantitative results, one can hardly hope to emulate this for amorphous solids. *Qualitative* understanding is still to be achieved in this area.

Many authors have tried to classify and to idealize the varieties of disorder in solids, and particularly amorphous solids. Typical categories are *structural, compositional, cellular, quantitative, topological*, ... Such schemes are usually tied to a particular theoretical framework, such as the tight-binding method.

The complexity of the problem is such that the study of idealized model systems, in which disorder is more easily identified, quantified, and controlled, provides a very useful first step in the understanding of the properties of a real amorphous semiconductor. Thus, for instance, the "many-impurity system" [2.1] provides insight into the behavior of heavily doped semiconductors [2.2] and the dependence of band-tailing upon disorder. The model of "purely topological disorder", as represented by the Hamiltonian used by *Thorpe* and *Weaire* (Sect. 2.2.1) allows a discussion of the behavior of bands and band gaps in tetrahedrally bonded amorphous semiconductors. The Anderson Hamiltonian (Sect. 2.3.1) represents "pure compositional disorder" and is the key model for the understanding of electron localization in disordered systems.

Such idealizations are useful as the skeletal basis of an understanding of electronic properties. Sooner or later, less idealized models must be analyzed, but without such a basis they can hardly be intelligently treated.

The primary quantity of interest is the density of electronic energy levels. *What is the connection between the distribution of energy eigenvalues and the spatial structure of a disordered system? In particular, how do band gaps survive in the absence of long-range order? What is the shape of the density of states*

---

1 Work supported by the Science Research Council.

near the band edges? What is the role of the short-range order in determining the shape of the bands? Are there any significant features related to defects and dangling bonds?

Of equal importance is the nature of the electronic wave functions. They may be localized or extended, depending on the system parameters and on the corresponding energy. Localization usually appears in the band tails and strongly affects the transport properties of a material. Although the concept of localization is now quite a familiar one, many questions, particularly dealing with quantitative aspects of the problem, are still open. *What are the conditions for electron localization? What are the positions of the "mobility edges" for a given system? What is the nature of the transition from localized to extended states? How does the electrical conductivity behave at this "Anderson transition"?*

The variety of responses to these questions provides a study in contrasting styles, ranging from crude number crunching to elegant formalism. Our intention in this chapter is to gather together rather more of these separate threads than has been done in the past, although whether we succeed in weaving them into a pattern is another matter! Our initial emphasis will be on the techniques themselves (Sect. 2.1) and the reader who is uninterested in their finer points may find it more profitable to proceed to their applications and results (Sects. 2.2, 3).

Correlation effects [2.3] may in some cases be crucial in the interpretation of experimental data [2.4]. The theory of these is still in a very qualitative stage [2.5–15] and goes beyond the scope of this article.

2.1 Theoretical Methods

The most systematic and elegant work on disordered systems has been devoted to the alloy problem. Although the relevance of this to amorphous solids is indirect and there are excellent reviews [2.16, 17], we shall include it within the scope of this section, since it is the source of many key ideas.

In the case of Anderson localization there is little choice – almost all work to date has been on "alloy" Hamiltonians.

2.1.1 General Theorems

We begin with some examples of general theorems which can be invoked in this area.

Starting from the Schrödinger equation of the tight-binding type,

$$\sum_j H_{ij}C_j^v = E_v C_i^v , \qquad (2.1)$$

the methods and theorems of matrix analysis can be applied. In (2.1), H_{ij} are the matrix elements of the Hamiltonian in an orthonormal set of basis functions $|i\rangle$, each attached to an atom site i, C_i^v are the coefficients of the vth eigenstates, and E_v is the corresponding eigenenergy. Following the theorems of Peron, Frobenius, and Gerschgorin (see, e.g., [2.18–20]), the eigenvalue spectrum of H_{ij} is bounded above and below:

$$-E_{max} \leqq E_v \leqq E_{max}. \tag{2.2}$$

The maximum eigenvalue E_{max} is bounded by the condition

$$\min_{\{i\}} \sum_j H_{ij} \leqq E_{max} \leqq \max_{\{i\}} \sum_j H_{ij}. \tag{2.3}$$

As an example, we consider a simple, topologically disordered system as described by the Hamiltonian

$$H = V \sum_{\substack{i,j \\ \text{nearest} \\ \text{neighbors}}} |i\rangle\langle j| \quad (V > 0). \tag{2.4}$$

Let Z be the number of nearest neighbors connected to a given atom by the transfer matrix element V. According to the relations (2.2) and (2.3), the spectrum of eigenvalues is bounded above and below by $\pm ZV$, respectively, since each row and column of H_{ij} contains exactly Z nonzero matrix elements V. The bandwidth is finite and cannot be greater than $2ZV$. There are many such general results for the spectra of topologically and quantitatively disordered systems using more realistic forms of the tight-binding Hamiltonian (2.1), including intra- and interatomic interactions. These include the Lifshitz bounds for a binary alloy [2.1] as sketched in Fig. 2.1 and the bounds established for the sp^3 Hamiltonian by *Weaire* and *Thorpe* [2.21] which we shall discuss in more detail in Sect. 2.2.1.

2.1.2 Approximation Techniques

In the theory of periodic systems the use of Green's functions often seems to obscure what would otherwise be obvious. However, many of the most significant developments in the theory of disordered systems have been framed in terms of Green's functions and some facility in dealing with them is indispensable to a proper understanding of the subject.

a) Green's Functions and the Density of States

Using the energy-dependent one-electron Green's function defined by $G(z) = (z - H)^{-1}$, the density of states may be written as

$$n(E) = -\frac{1}{\pi} \lim_{\eta \to 0} \text{Tr}\{\text{Im}\{G(E + i\eta)\}\}. \tag{2.5}$$

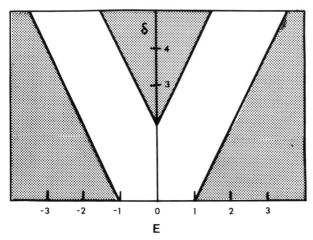

Fig. 2.1. Lifshitz bounds for a binary alloy, described by a Hamiltonian of the form (2.4) together with diagonal elements of magnitude $\pm \delta/2$ associated with A and B sites. (Here $ZV=1$ for convenience)

This may be proved by using the spectral representation

$$G(z) = \sum_{v} \frac{|v\rangle \langle v|}{z - E_v}.$$ (2.6)

$|v\rangle$ and E_v are the eigenstates and eigenvalues of H, respectively. Formally, we can calculate $G(z)$ by starting from the Hamiltonian

$$H = H_0 + U,$$ (2.7)

and expanding $(z - H_0 - U)^{-1}$ in an infinite series:

$$G(z) = G_0(z) \sum_{n=0}^{\infty} [UG_0(z)]^n.$$ (2.8)

$G_0(z) = (z - H_0)^{-1}$ is the Green's function corresponding to the disorder-independent part of H, H_0, and U is an abbreviation for the disorder-dependent part.

The actual decomposition of the Hamiltonian into H_0 and U is rather arbitrary and depends on the model system under consideration and the method used in the calculation. For an alloy, for instance, U may be identified with the (fluctuating) diagonal part of the tight-binding Hamiltonian of (2.1), and H_0 with the nondiagonal part. For the many-impurity problem, H_0 may represent the unperturbed crystal and U the superposition of the impurity potentials. For a structurally disordered system, H_0 is the kinetic energy of the electron and U is the superposition of atomic potentials.

The series expansion (2.8) converges for $\|z\| > E_{max}$. Once the summation is performed, one can use the method of analytical continuation to obtain $G(z)$ in the whole complex energy plane.

For a macroscopic system, the density of states has to be averaged over all possible configurations. Therefore, we are interested in the configurational average of $G(z)$,

$$\langle G(z) \rangle = \int ... \int d1...dN \, P(1...N) \, G(z; 1...N), \tag{2.9}$$

where $P(1...N)$ denotes the probability distribution for the configuration $\{1...N\}$. $G(z)$ is related to the scattering matrix of the system, $T(z)$, by

$$G(z) = G_0(z) + G_0(z) \, T(z) \, G_0(z), \tag{2.10}$$

$$T(z) = U \sum_{n=0}^{\infty} [G_0 U]^n. \tag{2.11}$$

Using the decomposition of U into the contributions of the individual macroscopic scattering centers, v_j, $T(z)$ can be written as an infinite sum of contributions from finite clusters of atoms [2.22]. Since the scattering centers are correlated, in general, the multiple scattering expansion of $\langle G(z) \rangle$ can only be treated approximately, using partial summation techniques and special decompositions of the probability distribution. There exist already a number of reviews of these methods [2.16, 17, 23, 24]. The idea is to treat multiple scattering processes up to infinite order correctly only for certain finite clusters of atoms.

These methods also form the basis of Anderson's analysis of localization due to disorder discussed later in this section.

In the "coherent potential approximation" [2.25, 26], the rest of the system is described by an effective potential, which may be represented formally by H_0. It can be determined, for instance, by requiring that *on the average* the cluster of atoms does not scatter relative to the effective medium, i.e.,

$$\langle T_{cluster}([H_0]) \rangle = 0. \tag{2.12}$$

If this condition is fulfilled, then

$$\langle G(z) \rangle \approx G_0(z). \tag{2.13}$$

In general, H_0, as determined by (2.12), turns out to be energy dependent and non-Hermitian. If we start with the whole system as the atom cluster, then (2.13) becomes exact, as can be seen easily from (2.10). In this case the self-consistently determined "coherent potential" $H_0(z)$ represents simply the self-energy of the electron moving in the random potential U. In the "single-site approximation", multiple scattering at only one atom is treated correctly up to infinite order and (2.13) is only a relatively crude approximation. It can be expected that the accuracy of the whole procedure improves systematically upon increasing the size of the (correctly treated) atom cluster.

We offer Fig. 2.2 as a guide to further reading in this area.

b) Green's Functions and Localization of Electron States

From the spectral representation (2.6) we conclude that the one-electron Green's function contains information concerning both the energy levels and eigenstates of a system. The dependence on the eigenstates may be used to investigate electron localization in disordered systems.

Following *Anderson*, we may use the criterion for the existence of only localized states in the system the "absence of diffusion" from a given site. The probability amplitude for an electron to move from the site j to the site j' during the time interval t is given by the retarded Green's function,

$$G_{jj'}(t) = -i\langle 0|C_j(t)\,C_{j'}^+(0)|0\rangle \qquad (t>0) \tag{2.14}$$

where $C_{j'}^+(0)$ and $C_j(t)$ are electronic creation and annihilation operators associated with the sites j' and j and the times 0 and t, respectively. $|0\rangle$ is the electronic vacuum state. $G_{jj'}(t)$ is determined by the differential equation

$$\frac{1}{i}\frac{\partial G_{jj'}(t)}{\partial t} = \sum_{j''} H_{jj''}\,G_{j''j'}(t) + \delta_{jj'}\delta(t), \tag{2.15}$$

and some initial condition $G_{jj'}(0)$.

Thus,

$$p_{00}(t) = |G_{00}(t)|^2 \tag{2.16}$$

is the probability of an electron being found on the initial site after a time interval t. If

$$\lim_{t\to\infty} p_{00}(t) \neq 0, \tag{2.17}$$

we may regard the electron as being confined to a region near 0 and therefore to be localized.

In energy space, this localization criterion reads

$$\lim_{\eta\to 0}\frac{\eta}{\pi}\int_{-\infty}^{+\infty} dE\,G_{00}(E+i\eta)\,G_{00}(E-i\eta) \neq 0, \tag{2.18}$$

where $G(z)$ is the Fourier transform of $G(t)$, and is identical to the Green's function in (2.5–13). It can be shown that, in the thermodynamic limit, branch cuts in $G_{00}(z)$ along the real energy axis correspond to extended states, whereas localized states are associated with a dense distribution of poles [2.64].

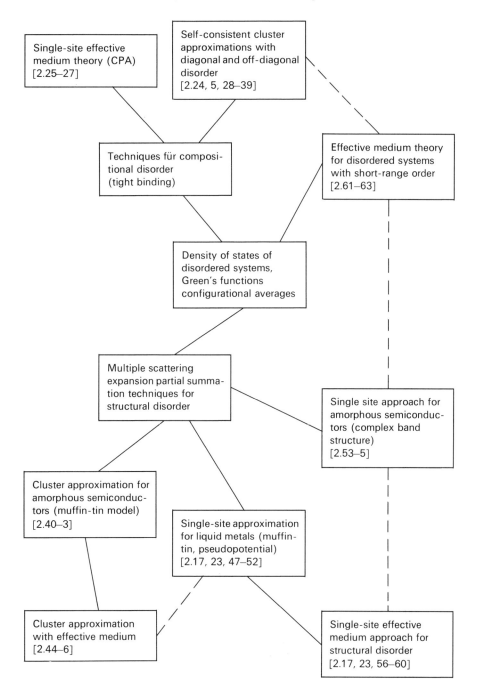

Fig. 2.2. Approximative techniques based on effective medium theories and/or multiple scattering theory

Intuitively, this is immediately clear from the case of a single atom. Equation (2.18), together with the multiple scattering formalism and partial summation techniques, has been used extensively [2.65] to estimate the position of the mobility edge and the critical value of the disorder for the Anderson transition (Sect. 2.3.3a). The quantity defined by (2.18) may also be related to the *average* participation ratio (Sect. 2.3.2) which is often used as a measure of localization [2.66, 67]. Inserting for $G_{00}(E \pm i\eta)$ the spectral representation and averaging over all sites, one obtains

$$\frac{1}{N} \sum_{j=1}^{N} \lim_{t \to \infty} p_{jj}(t) = \frac{1}{N} \sum_{j=1}^{N} \sum_{v} |\langle j|v\rangle|^4 \equiv \mathscr{P}. \tag{2.19}$$

For an ideal crystal $|\langle j|v\rangle|^2 = N^{-1}$ and the right-hand side of (2.19) is proportional to N^{-1}, whereas it is unity for an infinitely disordered system, with all the eigenstates extremely localized near the sites j. Hence, $\mathscr{P}$ may be considered to be to the inverse of the number of sites contributing to the eigenstates of the system.

c) Fluctuation States

A number of techniques are particularly useful when considering the distribution and the nature of electron states well within the tails of the band. It is evident that none of the multiple scattering methods starting from finite clusters of atoms provide an adequate description. In the simplest case, the problem is that of an impurity band in the high density limit. In this case we associate states in the band tails with fluctuations of the density of impurity atoms.

One possibility, in dealing with such low-lying states, is to apply the variational principle [2.68, 69]. Assuming, that at a given energy almost all of the localized eigenfunctions have the same shape,

$$\psi_v(\mathbf{r}) \approx f(\mathbf{r} - \mathbf{r}_v), \tag{2.20}$$

the energy corresponding to such a state,

$$E(\mathbf{r}_v) = \langle \psi_v|H|\psi_v\rangle = \langle f_v|H_0|f_v\rangle + \langle f_v|V|f_v\rangle, \tag{2.21}$$

will be very low at sites $\mathbf{r}_0$ corresponding to large negative fluctuations in the potential. Consequently, these are the places where we expect the eigenstates to be localized. Since the variational principle overestimates the ground-state energy of a system, $E(\mathbf{r}_0)$ will be larger than E_v, the true eigenenergy. A lower limit for the density of states in the low-energy tail may therefore be estimated as

$$n_f(E) = \frac{1}{\Omega} \frac{1}{\Delta E} \tag{2.22}$$

[number of local minima in $E(\mathbf{r}_0)$ such that $E \leq E(\mathbf{r}_0) \leq E + \Delta E$].

The best estimate for $n(E)$ is obtained by

$$n(E) = \max_{\{f\}} n_f(E). \tag{2.23}$$

The density of states in the band tail has been calculated for various models of fluctuation potentials. For a one-dimensional "white noise" potential one obtains, for instance,

$$n(E) \sim |E| \exp(-aE^{3/2}). \tag{2.24}$$

The above considerations can be cast into a more formal language by using functional integrals [2.70].

d) Path Integrals

Tail states may be treated more rigorously by using path integrals. Integrating the equation of motion (2.15) for the retarded Green's function in the local space representation yields [2.71–73]

$$G(r, r', t) = \int_{r(0)=r'}^{r(t)=r} \mathcal{D}[r(\tau)] \exp\left(i \int_t^0 d\tau \left\{\frac{m}{2}\dot{r}(\tau)^2 - V[r(\tau)]\right\}\right). \tag{2.25}$$

Here, $r(\tau)$ is a continuous path from $r(0)$ to $r(t)$, and $\mathcal{D}[r(\tau)]$ is the Feynman measure in configuration space [2.74]. $G(r, r', t)$ is the probability amplitude for an electron to hop from r' to r within the time interval t. Therefore, (2.25) can be interpreted as decomposing the total probability amplitude into partial contributions corresponding to the classical paths $r(\tau)$.

Decomposing $V(r)$ into statistically independent random atomic contributions, the configurational average of (2.25) can be calculated explicitly and the density of states may be obtained analytically for various potential models in the limit $|E| \to \infty$ [2.75–79].

The path integral representation may also be used to study electron localization [2.80–82].

e) Analogies with Phase Transitions and Renormalization Group Analysis

Certain aspects of our subject are strongly reminiscent of the theory of phase transitions. This is particularly true in the case of the problem of the critical behavior associated with the Anderson transition (Sect. 4.3). Various authors have speculated on the possibility of applying the renormalization group technique [2.83] in close analogy with phase transition theory (see, for example, [2.84–87]). The papers [2.86, 87] contain specific predictions of critical exponents based on the identification of the Anderson transition with a particular phase transition. However, it is by no means clear (at least to these authors) upon what rigorous basis this identification rests.

Much the same can be said of the recent work of *Wegner* [2.88] which analyses Anderson localization using the renormalization group method. While this technique is borrowed from phase transition theory, no analogy is made with particular phase transitions. For further discussion of the results of this work, see Sect. 2.3.

2.1.3 Numerical Techniques

a) Numerical Diagonalization

It is fairly easy, by means of modern computational techniques to diagonalize large matrices [2.89]. In the following we describe briefly one of them which is particularly advantageous in the study of simple tight-binding Hamiltonians.

Most diagonalization methods start from the fact that the spectrum of eigenvalues of a matrix A remains unchanged if a similarity transformation, T, is applied:

$$B = TAT^{-1}. \tag{2.26}$$

Particularly useful is a transformation, which yields a tridiagonal matrix the eigenvalue problem of which may readily be solved. One means of constructing such a transformation is the Lanczos method in which T is constructed successively from the relations

$$\gamma_n t_{n+1} = At_n - \alpha_n t_n - \beta_{n-1} t_{n-1},$$
$$\gamma_n^* t_{n+1}^* = A^\dagger t_n^* - \alpha_n t_n^* - \beta_{n-1}^* t_{n-1}^*, \tag{2.27}$$

where $A^\dagger$ is the transpose of A. α_n, β_n, β_n^* are chosen so that t_{n+1}^* is orthogonal to $t_1 \ldots t_n$ and t_{n+1} is orthogonal to $t_1^* \ldots t_n^*$, respectively.

The choice of γ_n and γ_n^* is somewhat arbitrary. They can be determined, for instance, by requiring t_n and t_n^* to be normalized. For $n \geq N$, the dimension of A, t_{n+1}, and t_{n+1}^* must be zero. T contains the vectors t_n as rows.

In principle, numerical techniques thus enable us to determine the eigenvalues and eigenvectors of our tight-binding Hamiltonian. In practice, for large systems much of the information contained in the detailed positions of the eigenvalues is redundant. A typical quantity of interest is the *density of eigenvalues* which, as we shall see, may be calculated more directly.

b) Equation-of-Motion Method

The equation-of-motion method, which enables electronic properties of large systems to be calculated numerically without referring to the details of eigenstates, is based on the time-dependent Schrödinger equation [2.90]. Using again the site representation

$$|\psi_t\rangle = \sum_{j=1}^{N} a_j(t)|j\rangle, \tag{2.28}$$

we have

$$\frac{1}{i}\frac{da_j(t)}{dt} = \sum_{j'=1}^{N} H_{jj'} a_{j'}(t). \tag{2.29}$$

As a linear differential equation in t, its numerical solution is trivial for finite systems. Generally, the time development $|\psi_t\rangle$ depends on the initial condition $|\psi_0\rangle$ and can be written explicitly in terms of eigenstates of H,

$$|\psi_t\rangle = \sum_v e^{-iE_v t}|v\rangle\langle v|\psi_0\rangle. \tag{2.30}$$

Thus, if $|\psi_0\rangle$ is a mixture of eigenstates, the same remains true for $|\psi_t\rangle$. Acting on $|\psi_0\rangle$ with a suitable energy-dependent operator, we can project out properties corresponding to a certain energy interval.

As an example, we discuss briefly the numerical calculation of the average inverse participation ratio (2.19) in terms of the time dependence of $|\psi_t\rangle$. Consider a finite system with, for instance, periodic boundary conditions. From (2.19), (2.28), and (2.30), it is easy to show that

$$\mathscr{P} = \frac{N}{2}\sum_{j=1}^{N} [2(\overline{|a_j|^2})^2 - \overline{|a_j|^4}], \tag{2.31}$$

where

$$\bar{g} \equiv \lim_{T\to\infty} \frac{1}{T}\int_0^T g(t)\,dt. \tag{2.32}$$

At the end of the calculation the thermodynamic limit $N\to\infty$ has to be performed. Using energetically weighted initial states, we can trace the electron localization with energy. In a similar way other quantities, such as, for example, the electrical conductivity may be expressed by using time averages (Sect. 2.3.5).

Alternatively, one can start from the time-dependent Green's function of (2.15). This approach has proved useful when calculating numerically the density of states of alloys, and properties of magnetic systems and superionic conductors [2.91–95].

c) Moments and the Recursion Method

Another numerical method for determining the density of states starts from a moment expansion [2.96, 97]. In the site representation we have from (2.8)

$$TG(z) = \sum_{p=0}^{\infty} \mu_p \frac{1}{z^{p+1}}, \tag{2.33}$$

where

$$\mu_p = \sum_{jj_1\ldots j_p} H_{jj_1}\ldots H_{j_{p-1}j_p} \tag{2.34}$$

is the pth moment of the density of states:

$$\mu_p = \int_\infty E^p n(E)\, dE. \tag{2.35}$$

Exact upper and lower bounds for the *integrated* density of states

$$N(E) = \int_{-\infty}^{E} dE'\, n(E') \tag{2.36}$$

can be obtained from a *finite* number of moments [2.98]. The calculation of $n(E)$ from the bounds on $N(E)$ is only possible if $n(E)$ is sufficiently smooth. Hence, the application of this method is rather limited. There are a number of systematic approximation procedures for $n(E)$ if a finite number of its moments is known. Essentially, one assumes a functional form for $n(E)$ containing free parameters, which can be determined from the known moments. The success of such a fitting method depends on a good guess of the functional form of $n(E)$.

The use of moment expansions has been largely supplanted by continued fraction expansions. These express the Green's function as [2.99, 100]:

$$\langle 0|G(z)|0\rangle = \frac{1}{z-\alpha_1-}\cdot\frac{\beta_1}{z-\alpha_2-}\cdot\frac{\beta_2}{z-\alpha_3}\cdots \tag{2.37}$$

The advantages of this formulation are threefold: firstly, the continued fraction yields a complex function of z with cuts only at the realy energy axis, even if truncated at any finite order. Secondly, $\mathrm{Im}\{G_{00}(z)\} > 0$ for $\mathrm{Im}\{z\} < 0$ and vice versa, guaranteeing a positive density of states. Thirdly, α_n and β_n may be constructed systematically using the Lanczos method [2.100].

The method converges rather rapidly, if the bandwidth is finite. For simple (periodic) Hamiltonians, it can be shown that α_n converges to the center of the band and β_n to half the bandwidth. Generally, β_n as a function of n can be related to the form of the singularities within the band [2.98, 101].

The recursion method has been applied to a great number of problems including tetrahedrally bonded amorphous semiconductors, alloys, surface states, and transition metals [2.102–107]. Recently a first attempt has been made to obtain information about electron localization (Sect. 2.3.3). The method is rather fast and is applicable to systems up to 10^5 atoms. Its main advantage is that it gives directly the local density of states, without needing such additional numerical procedures as, for instance, the Fourier transformation in the method of *Alben* et al. [2.91] or the time average in the method of

Weaire and *Williams* [2.90]. There are also no excessive storage requirements, because the coefficients α_n and β_n are calculated step by step from the Lanczos relations (2.27).

2.2 The Density of States

The broad features of the density of states of a solid may be determined by a variety of experiments: soft X-ray emission and absorption, optical uv and X-ray photoemission, optical spectroscopy, etc. (see, for instance, the reviews [2.108–112]). In addition, techniques have been developed for probing the gap region in a semiconductor, in which the density of states is very small [2.113, 114]. All of these methods have been brought to bear on amorphous Si and Ge, which are often regarded as the primary test case for theory and experiment in this area. What emerges is, at the crudest level, something of a null result – the electronic density of states is remarkably similar to that of the corresponding crystal. In particular, optical spectroscopy reveals the existence of band gaps comparable to those of crystalline Si and Ge.

2.2.1 Exact Results

The first task of theory is to understand the survival of so much of the electronic band structure in the amorphous phase. The reason is simple enough, at least in the tight-binding picture. If short-range interactions dominate, then short-range order is the chief determinant of the density of states. This may be demonstrated with particular clarity for the case of Si and Ge if we use the following simple Hamiltonian:

$$H = \sum_{jvv'} V_1 |jv\rangle \langle jv'| + \sum_{jj'v} V_2 |jv\rangle \langle j'v|, \tag{2.38}$$

where $|jv\rangle$ represent the sp^3 orbitals associated with the sites j, V_1 is the interaction between different orbitals at the same site, and V_2 describes the intersite interaction (see Fig. 2.3). V_2 is responsible for the separation of the bonding and antibonding bands, and V_1 is responsible for their finite width.

As *Weaire* and *Thorpe* showed, the following features of the electronic density of states are invariant with respect to the choice of the structure, being entirely dictated by the assumed tetrahedral coordination of nearest neighbors [2.115–117].

I) There are two bands (valence and conduction bands) of the same width as those of the diamond cubic crystal (or less) and hence a band gap of a magnitude at least of $2|V_2| - 4|V_1|$ (Fig. 2.4).

II) In each band, half the states are concentrated in a delta function consisting of purely p-like states. (This has little significance in the conduction

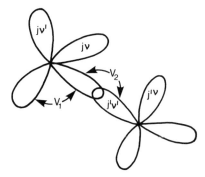

Fig. 2.3. Basis functions and interactions in the *Weaire–Thorpe* Hamiltonian

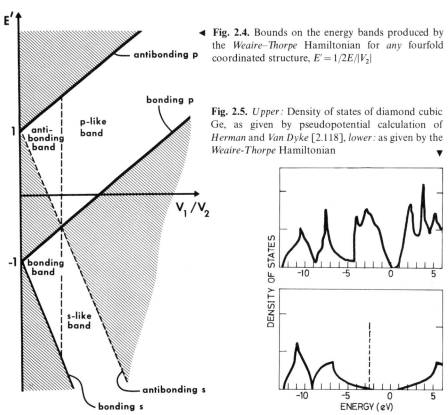

◀ **Fig. 2.4.** Bounds on the energy bands produced by the *Weaire–Thorpe* Hamiltonian for *any* fourfold coordinated structure, $E' = 1/2E/|V_2|$

Fig. 2.5. *Upper:* Density of states of diamond cubic Ge, as given by pseudopotential calculation of *Herman* and *Van Dyke* [2.118], *lower:* as given by the *Weaire–Thorpe* Hamiltonian   ▼

band, but in the valence band it corresponds to the pronounced *p*-bonding peak at the top of the band shown in Fig. 2.5.)

Whilst this model illuminates several essential features of the problem it is deficient in many respects. Specifically, interactions are confined to very short range, site orbitals are assumed to be orthogonal, and quantitative variations of the interaction parameters with small departures from perfect tetrahedral

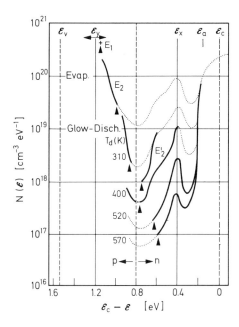

Fig. 2.6. Density of states in the gap region, for evaporated E_1, E_2 and glow discharge specimens of Si, as measured by *Spear* and co-workers [2.113] by field-effect measurements

coordination (bond stretching and bending) are neglected. To achieve a more realistic description, which is necessary if the small difference between the density of states of crystalline and amorphous phases is to be analyzed, a more realistic Hamiltonian should be used.

The type of questions which such a theory might be expected to answer are (I) *What is the significance of the asymmetry of the p-bonding peak* [2.109] *which is definitely distinct from that of the crystal?* (II) *What is the significance of the washing-out of the two-peak structure at the bottom of the valence band?* (*Thorpe* and *Weaire* predicted this on the basis of the simple model of this section, but their argument was, as later explained [2.119, 120] not entirely satisfactory.) (III) *What is the detailed shape of the density of states in the gap region shown in Fig. 2.6, and with what microscopic structural features is it associated?*

2.2.2 Tight-Binding Models

It is relatively easy to use the model Hamiltonian defined above in numerical procedures to determine the density of states.

The relationship between the local structure of a system and density-of-states features has been discussed recently using the so-called Cluster-Bethe method [2.121, 122]. Small atom clusters of a suitable structure with tetrahedral coordination are used in this method. A Bethe lattice with the same nearest neighbor configuration is attached to each dangling bond at the cluster surface, to minimize spurious features due to the surface. The density of states of the Bethe lattice is featureless. Hence all sharp structures resulting from such a

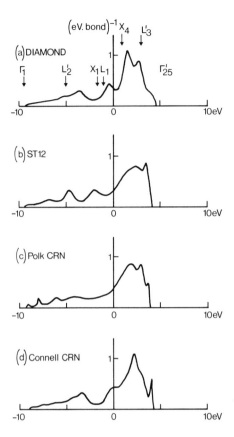

Fig. 2.7. Valence band density of states for the diamond cubic structure, the ST 12 crystal structure, and two random network models, as given by a calculation by *Bullett* and *Kelly* [2.124] using the recursion method

calculation are likely to be connected with the short-range order of the system. Detailed studies of the *s*-like valence bands were performed. It has been demonstrated that the fine structure of the bands indeed depend on atomic configurations within finite regions, in particular, that the number of odd and even rings influence critically the widths and positions of the two peaks at the bottom of the valence bands (for a recent review see [2.123]).

The relationship between ring statistics and the density of states can be analyzed also by applying the recursion method to various structural models of the tetrahedrally bonded materials [2.124]. Using random network models with and without odd-numbered rings it has been concluded that odd-numbered rings are necessary to explain the experimental data for Ge and Si.

One may also calculate the density of states within the framework of more sophisticated tight-binding theories applied to atom clusters, without the restriction of nearest-neighbor interactions. A number of cluster calculations using refined Hückel theories with Gaussian atomic orbitals have been carried out [2.125–127].

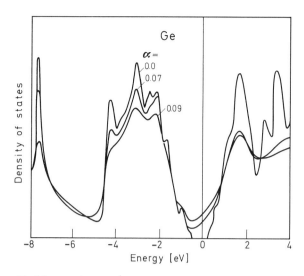

Fig.2.8. Density of states of amorphous Ge models as obtained by using the generalized pseudo-potential approximation for various degrees of disorder, α [2.130]. $\alpha = 0.09$ corresponds to a mean short-range order distance $R_0 \sim \alpha^{-1}$ of about five interatomic distances

2.2.3 Pseudopotentials and the Quasi-Crystalline Approximation

Pseudopotentials were used to study the behavior of the density of states when the long-range order of a crystalline system is relaxed [2.128, 129]. Starting from a crystal structure with the same short-range configuration as in the amorphous phase, the long-range order was destroyed by choosing a two-atom distribution function, which tends to unity for infinite separation of the two atoms, but shows peaks at small distances. Thus short-range order is represented by the model and is the same as in the crystal.

Calculations have been performed for the tetrahedrally bonded amorphous semiconductors as well as Se and Te [2.130]. The general result was that the density of states near the energy gap between the valence and conduction bands is hardly affected by the relaxation of the long-range order, i.e., it is determined only by short-range order (Fig. 2.8). This agrees with the more rigorous results from the tight-binding models described above.

The s-like valence bands do not react to the relaxation of the long-range order in this model, in contrast to experimental findings. This leads to the conclusion that fluctuations in the short-range order have to be taken into account to broaden the lower valence bands. Fluctuations in the short-range order lead to a fluctuating atomic pseudopotential. Taking into account the changes in the form factor caused by fluctuations in the dihedral angle, the shape of the s-bands can be related qualitatively to the ratio of odd and even numbered rings in the amorphous phase [2.131, 132].

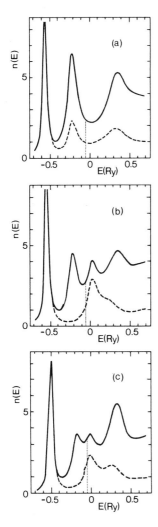

Fig. 2.9. Density of states for 5-atom clusters of Si in tetrahedral configuration without (**a**) and with (**b** and **c**) a vacancy. In (**c**) structural relaxation around the vacancy has been assumed. The dotted lines denote the Fermi energy. The vacancy is situated at one of the corners of the tetrahedron. The dashed curves are the local densities of states at the central atom [2.134]

Pseudopotential theory has also been used to calculate the density of states of the various crystalline polytypes of Si and Ge [2.123]. Results show again that short-range fluctuations in the atomic configuration are important for the understanding of the shape of the s valence bands in the amorphous phases of Si and Ge.

2.2.4 Muffin-Tin Models

The muffin-tin potential model was used to investigate the density of states of liquid metals [2.17, 23] as well as of amorphous solids with short-range order [2.53]. Cluster calculations were performed for the tetrahedrally coordinated semiconductors [2.40–43, 133] in order to obtain details of the electronic spectrum near the band gap as a function of the size of the clusters.

More sophisticated KKR techniques have been employed recently using a CPA-type cluster approximation [2.44–46, 134]. In order to obtain information about the dependence of the spectrum within the gap upon structural changes, clusters with vacancies with and without structural relaxation have been considered. The resulting density of states shows a peak well within the gap appearing upon vacancy formation. Peaks above and below the valence and the conduction band edge appear, when the atomic configuration around the vacancy is slightly relaxed (Fig. 2.9). Quantitatively the density of states within the gap region is overestimated, presumably because the cluster size was too small, thus yielding an unphysically large imaginary part of the coherent potential.

2.2.5 Summary

Let us attempt to summarize our present understanding of the electronic spectrum of amorphous Si and Ge, and draw some conclusions for further studies.

The broad features of the density of states are understandable in terms of the simple tight-binding model. The interpretation of the finer details requires a much more sophisticated theory and has generally been controversial. The most interesting area of investigation – the density of states in and around the gap – remains largely untouched. Qualitatively, one expects, according to Mott and others, exponential band tails with superimposed peaks due to states associated with dangling bonds. The band tails might be understood in terms of short range order fluctuations, using a Halperin–Lax type of theory, but no one has yet attempted such a programme. Defect states are likely to be very sensitive to the precise local environment, so that the definition of appropriate defects and the calculation of the associated states is far from straightforward (for a recent attempt see [2.135]).

2.3 Anderson Localization

In experiment, the localization of electronic states (at the Fermi level) manifests itself in the vanishing of the dc conductivity in the low-temperature limit. If the Fermi level lies in the region of localized states, conduction must therefore be activated at finite temperatures. In the case of amorphous semiconductors, *Mott* has developed a "variable-range hopping" treatment, which entails the celebrated $T^{1/4}$ behavior of the logarithm of the conductivity. This has been widely observed [2.136]. Most theoretical studies have been confined to simple model systems. Even for these there are as yet only few widely accepted results.

2.3.1 Anderson's Model

The phenomenon of Anderson localization has been studied most extensively for model Hamiltonians of the type (in the site representation)

$$H = \sum_j \varepsilon_j |j\rangle\langle j| + \sum_{\substack{jj' \\ \text{nearest} \\ \text{neighbors}}} V_{jj'} |j\rangle\langle j'| . \tag{2.39}$$

In the simplest case, $V_{jj'}$ are taken to be constant and disorder is contained only in the diagonal part, $P\varepsilon_j$, which is assumed to be randomly distributed according to some probability distribution $P(\varepsilon)$ of width W.

It seems reasonable that the band tails, which presumably correspond to rare local fluctuations of the Hamiltonian (Sect. 2.1.2), should consist of localized states. It is less obvious that, upon increasing the disorder, the energies E_c (the "mobility edges"), which are the limits of the energy range of the extended states, should eventually move into the center of the band, so that all states become localized. Nevertheless, this is thought to be the case (for dimension $d \geq 2$). By tradition, the critical strength of disorder for localization of the entire band has often been chosen as the primary point of comparison between different theoretical treatments.

Although it is a property of the simplest of independent electron Hamiltonians, Anderson localization has proved to be an extremely subtle phenomenon. *Anderson* [2.137] was the first to realise that a Hamiltonian as in (2.39), if disordered, might have *either* extended *or* localized eigenstates, depending on the strength of the disorder [width of $P(\varepsilon)$]. This is not the case in one dimension, where all states are localized for arbitrarily weak disorder (see for instance, [2.138] for an excellent review). In two dimensions and higher ($d \geq 2$), there remains some uncertainty, but the confidence of Mott and others in the existence of an "Anderson transition" between localized and extended states has received increasing support from experimental investigations and numerical calculations. Our main emphasis will be on the latter.

2.3.2 Theory of Localization of Electron States

The progress of theoretical methods in the theory of localization of electron states between 1958 and 1978 is sketched in Fig. 2.10.

The mainstream of analytical ideas is based directly on Anderson's ideas concerning the relationship of localization and the renormalized perturbation expansion, as described in Sect. 2.1.2. A detailed comparison of the various alternative approximations which may be applied to this expansion has been given by *Licciardello* and *Economou* [2.65]. Recent estimates of the critical strength of disorder for localization are quite different from Anderson's original ones.

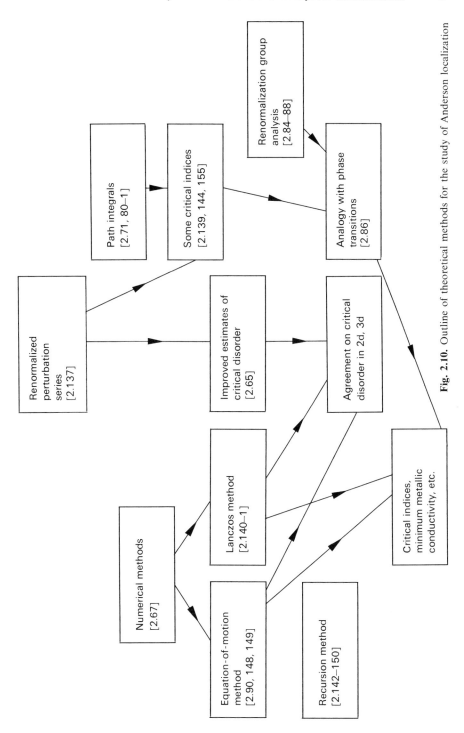

Fig. 2.10. Outline of theoretical methods for the study of Anderson localization

Table 2.1

Name	Definition $[N \to \infty]$	Localized regime	Extended regime
Inverse localization length	κ ⎫	finite	0
Prefactor index	$\left. \begin{array}{c} \\ \\ \end{array} \right\} |\psi| \sim r^{-\beta} e^{-\kappa r}$ β ⎭	finite	0
Inverse participation number	$P = \sum_{i}^{N} |\psi_i|^4$	finite	0
Participation ratio	$p = N^{-1} P^{-1}$	0	finite
Conductivity	$\sigma = [\text{KG formula}]$	0	finite

In order to test such theories, a number of numerical methods have been developed. These have included conventional methods of matrix eigenvalue and eigenvector determination which have been refined to the point where Hamiltonian matrices of dimension $10^2 \times 10^2$ or larger can be analyzed [2.139–141]. More recently, the equation-of-motion method, which avoids the calculation of individual eigenvalues and eigenvectors, has been applied to calculate average properties related to localization directly [2.90]. Another similar method, the recursion method, is being similarly adapted at present [2.142].

For the purpose of this review, we have set aside the study of Anderson localization on Bethe lattices (see, e.g., [2.143]) which has not so far been particularly helpful, despite its apparent simplicity.

As for the question of critical behavior, agreement was found at an early stage between the critical index for the localization length, as predicted by Anderson's theory, and an alternative analytical formulation due to *Edwards* ([2.144–145], Sect. 2.1.2d), but little further insight seems to have been gained until 1977, when various authors suggested that the entire critical behavior was obtainable by an analogy with an appropriate phase transition (Sect. 2.1.2e).

Table 2.1 lists the various physical quantities upon which attention is focused in the theory of localization. Firstly, if we accept the indicated exponential form for asymptotic behavior of the envelope of a localized wave function, this is characterized by two parameters β and κ, of which the more important is κ having the significance of the inverse of the localization length. The approach to the Anderson transition from the localized side is then characterized by the vanishing of κ.

Alternatively (and with less assumptions regarding the specific form of wave functions) we can choose to use the *mean fourth power* of an eigenstate as a

Table 2.2. Comparison of critical strength of disorder expressed as (W/V) for various analytical theories, applied to the case of a rectangular distribution of site energies [2.65]). The last column refers to the "G^0" approximation

	Anderson [2.137] (1958)	Ziman [2.146] (1969)	Herbert and Jones [2.147] (1971)	Licciardello and Economou [2.65] (1975)
Square lattice	28	22	—	7.2
Diamond cubic	32	22	—	8.2
Simple cubic lattice	62	32.4	40, 24, 20	14.5

measure of localization. We shall choose to call this the *inverse participation number* [cf. (2.19)].

Lastly, we may use the conductivity, for which the Kubo–Greenwood formula (Sect. 2.3.5) constitutes a sufficient definition for present purposes, to characterize extended states.

2.3.3 The Anderson Transition

a) Analytical Methods

Some idea of the wide range of analytical estimates of critical disorder strength and their evolution with the passage of time may be gained from Table 2.2. Note that Anderson's original estimate, despite its importance as the foundation of this subject and the elaborate analysis which supported it, is now thought to be much too large.

b) Matrix Diagonalization

Some appreciation of the power of recently developed matrix techniques can be gained from Fig. 2.11 which shows some results of *Yoshino* and *Okazaki* [2.139] for typical eigenstates of the Anderson Hamiltonian with rectangular disorder, defined on the two-dimensional square lattice. Samples of 10^4 sites were used. The technique used depended for its efficiency upon sampling only a limited number of eigenvectors and using "rigid wall" boundary conditions. (The latter would be a serious limitation in higher dimensions.) Figure 2.11 shows a very direct demonstration of the exponential character of localized wave functions.

A rather similar approach is to be found in the recent work of *Licciardello* and *Thouless* [2.140, 141] except that they have used periodic and antiperiodic boundary conditions and examined the effect of a change of boundary conditions and eigenvalues. The eigenvalue shifts should decrease exponentially with the size of the system if states are localized, so that the localization can be diagnosed without the calculation of eigenvectors (see also Table 2.2).

Both of these methods give a critical value of W/ZV just over 1.5 for the square lattice. *Licciardello* and *Thouless* give results for other two-dimensional

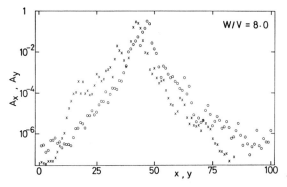

Fig. 2.11. The exponential character of localized wave functions is exemplified by this calculation by *Yoshino* and *Okazaki* [2.139]. $A_i = \sum_j |a_{ij}|^2 \qquad (i=x,y)$

lattices as well and all are remarkably consistent with the analytical predictions of *Licciardello* and *Economou* [2.65] as is the further estimate of the critical value of W/ZV of 2.0 for the diamond cubic lattice by *Licciardello* and *Thouless* [2.140].

Admittedly we have simplified matters somewhat here, since the latest of these calculations [2.141] has shown some seeds of doubt in their practitioners. It is however too early to say whether this is because the methods of *Thouless* are *better* (i.e., discern greater detail) than the rival methods or *worse* (i.e., based in part on invalid assumptions).

c) Results of the Equation-of-Motion Method

We have already introduced the equation-of-motion method in Sect. 2.1.3a. In adapting it to the calculation of localization properties, the motivation of *Weaire* and *Williams* [2.90] was identical to that of *Alben* et al. [2.91] who originated it, in the context of the calculation of the density of states.

The method evaluates the average value of the inverse participation number at a given energy [2.148, 149]. Figure 2.12 shows results for "cubic" lattices in dimensions 1–5. The estimate of the critical value of W/ZV for the two-dimensional square lattice agrees with that of the direct numerical methods of Sect. 2.3.3b. The trend with dimensions is in accord with the analytic theories of Sect. 2.3.3a. For a further discussion of dimensions $d \geq 4$, see Sect. 2.3.4.

d) The Recursion Method

The recursion method (Sect. 2.1.3c) may be used to give a *local* density of states (*not* the total density of states or the ensemble-averaged local density of states). Thus, it is clear that localization properties can, in principle, be derived from this approach. But how?

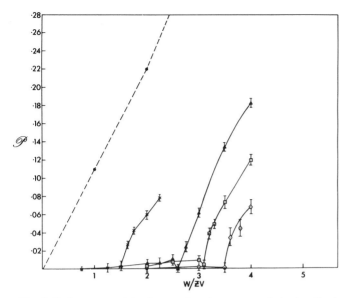

Fig. 2.12. Inverse participation number (a measure of the localization of eigenstates) at the center of the band for the Anderson Hamiltonian with rectangular disorder of width W, defined on lattices of dimension 1 to 5. Linear, chain; $\times$, square; $\triangle$, cubic; $\square$, 4 d hypercubic; $\diamond$, 5 d hypercubic. The calculations were performed by the equation-of-motion method [2.149]

Haydock [2.142] has suggested that the answer is very simple – that the asymptotic values of the continued fraction coefficients define the position of the mobility edge, according to

$$E_c = \alpha_\infty \pm 2\beta_\infty , \tag{2.40}$$

where α_∞ and β_∞ are the asymptotic limits (assuming that these exist) of the recursion coefficients α_n, β_n.

Weaire and *Hodges* [2.150] disagree with this interpretation. According to them, the coefficients do *not* converge to asymptotic values and the fluctuations which persist in that limit are a measure of localization.

2.3.4 Critical Behavior

Most attempts at rigorous derivations of critical properties have dealt with the localized side of the transition.

We may define the index α for the divergence of the localized length by

$$\kappa \sim (E - E_c)^\alpha . \tag{2.41}$$

Anderson [2.151] obtained a value of 3/5 for this index in three dimensions from an examination of dominant terms in the renormalized perturbation

Table 2.3. Critical indices for the localized side of the Anderson transition, as predicted by *Sadovskii* [2.86] and extended by the use of the relations given by *Licciardello* and *Thouless* [2.140–141] and results quoted by *Fisher* [2.152]

Physical quantity	Critical index		
	$d=2$	$d=3$	$d\geq 4$
Inverse localization length κ	$\frac{3}{4}$	$\frac{3}{5}$	$\frac{1}{2}$
Prefactor index β	0.22	1.03	$d-2$
Inverse participation number	$\frac{3}{2}$	1.12	0

series. *Abram* and *Edwards* [2.144, 145] obtained the same value from the path integral method and *Abram* gives 3/4 for the same index in two dimensions [2.152].

These values are supported by the analogy with a phase transition for an $n=0$ Landau–Ginzburg Hamiltonian, used in a number of recent papers [2.86, 87]. In addition, this approach suggests $\alpha=\frac{1}{2}$ for $d\geq 4$. The "marginal dimension" is thus found to be $d=4$, that is to say, the analytic properties associated with critical behavior should be the same for four dimensions and higher.

Using the results of *Fisher* [2.153] one can also estimate the prefactor index β, as predicted by this analogy. The resultant values together with those for other critical indices are collected in Table 2.3.

It is difficult to resist the temptation to try to extract critical indices from the numerical results which we have described in previous sections. Indeed, the temptation has not always been resisted, but it seems to us that this is a most unreliable procedure and no conclusion can yet be drawn from it. Equally, there seems no convenient experiment which yields these critical indices. *Mott* [2.152] has suggested that they imply certain inequalities for critical indices on the other side of the transition which is experimentally more accessible, but this argument has not found much support.

Unfortunately, analytical predictions of critical behavior on the other side of the transition seem much more difficult, although *Wegner* [2.88] has pioneered the direct application of renormalization group methods to the transition. Again, numerical estimates of the critical behavior of the conductivity are hazardous. Nevertheless, one might hope that something as dramatic as a discontinuous drop to zero, as suggested by *Mott*, might be clearly discernible, so the matter is worthy of more scrutiny. In addition, there are some special arguments concerning the behavior of the conductivity in two dimensions. Hence we devote the next section to this topic.

2.3.5 Conductivity

Assuming a simple form for the matrix elements of the momentum operator in the site representation, the Kubo–Greenwood formula for the conductivity may be expressed in terms of eigenstates, labelled by α and β, as

$$\sigma = 2\pi e^2 h^{-1} V^2 \Omega^{-1} \sum_{jj'\Delta\Delta} (\hat{e}\cdot\hat{\Delta})(\hat{e}\cdot\hat{\Delta}') \sum_{\alpha\neq\beta} \langle j|\alpha\rangle\langle\alpha|j'\rangle$$

$$\cdot\langle j'\Delta'|\beta\rangle\langle\beta|j\Delta\rangle\delta(E_\alpha - E_F)\delta(E_\beta - E_F). \tag{2.42}$$

Here the volume of the system is denoted by Ω, $\hat{e}$ is a unit vector in the direction of the applied field, and V is the off-diagonal matrix element in (2.39).

Given certain assumptions, the shift in eigenvalues when boundary conditions are changed may be related to the conductivity, so the method used by *Licciardello* and *Thouless*, which is discussed in Sect. 2.3.3b, may be used to calculate it.

Furthermore, a scaling argument suggests that if the conductivity drops discontinuously at the mobility edge, its limiting value (the "minimum metallic conductivity") should be a universal constant in the case of two dimensions [2.140, 141]. The best experimental system for its investigation has been the inversion layer. *Pepper* [2.154] obtained the value $0.12e^2/h$ for the minimum metallic conductivity in two dimensions for a variety of lattices and degrees of disorder. However the analysis contained some subjective elements – one cannot easily escape the essential difficulty of extracting critical behavior from noisy numerical data. Moreover, the more recent work of the above authors [2.141] for larger systems has proved somewhat inconsistent with the earlier results and difficult to interpret. To add to the present confusion, *Prelovsek* [2.155] has recently claimed to show that the minimum metallic conductivity is *zero* in two dimensions.

Kramer and *Weaire* [2.156] have shown how the equation-of-motion method may be adapted to calculate conductivities. This method is more closely tied to the Kubo–Greenwood formula than that of *Licciardello* and *Thouless* and it remains to be seen if it gives more easily interpreted results.

In the meantime, the whole question of the minimum metallic conductivity remains in a rather unsettled state. One straw in the wind which may be important is the statement of *Wegner* [2.88] that his application of renormalization group theory suggests the existence of a minimum metallic conductivity in two dimensions but not in three.

2.4 Conclusion

As we have repeatedly seen, it is not necessary (and indeed may be misleading) to regard an amorphous solid as a defective crystal in attempting to set up a theory of its electronic properties.

Instead one can attack the problem directly on three fronts – by rigorous theory, by systematic approximations, and by numerical calculations. Each of these is so limited in power and applicability that anything less than a balanced combination of all three is likely to be unproductive.

Acknowledgement. One of us (B.K.) would like to thank the members of the Physics Department of the Heriot-Watt University for their generous hospitality during the time when this work was performed.

References

2.1   I. M. Lifshitz: Sov. Phys.-Usp. **7**, 549 (1965)
2.2   V. L. Bonch-Bruyevich: "The electron theory of heavily doped semiconductors," (Elsevier, New York 1966)
2.3   P. W. Anderson: Phys. Rev. Lett. **34**, 953 (1975)
2.4   R. A. Street, N. F. Mott: Phys. Rev. Lett. **35**, 1293 (1975)
2.5   N. F. Mott, E. A. Davis, R. A. Street: Philos. Mag. **32**, 961 (1975)
2.6   S. R. Nagel, J. Tauc: Solid State Commun. **21**, 129 (1977)
2.7   R. A. Street: Solid State Commun. **24**, 363 (1977)
2.8   M. Kastner, D. Adler, H. Fritzsche: Phys. Rev. Lett. **37**, 1504 (1976)
2.9   N. F. Mott: "Silicon Dioxide and Chalcogenide Semiconductors Compared and Contrasted," in *Amorphous and Liquid Semiconductors*, ed. by W. Spear (Centre for Industrial Consultancy and Liaison, University of Edinburgh 1977) pp. 497–503
2.10 M. Kastner: "Radiative and Nonradiative Processes at Valence Alteration and Other Defects in Lone-Pair Semiconductors," in *Amorphous and Liquid Semiconductors*, ed. by W. Spear (Centre for Industrial Consultancy and Liaison, University of Edinburgh 1977) pp. 504–508
2.11 R. A. Street: "Non-Radiative Recombination, Photostructural Changes and Urbach's Rule in Chalcogenides," in *Amorphous and Liquid Semiconductors*, ed. by W. Spear (Centre for Industrial Consultancy and Liaison, University of Edinburgh 1977) pp. 509–513
2.12 K. Weiser: "Comments on the Non-Radiative Recombination of Photocarriers in Amorphous Chalcogenides," in *Amorphous and Liquid Semiconductors*, ed. by W. Spear (Centre for Industrial Consultancy and Liaison, University of Edinburgh 1977) pp. 514–518
2.13 K. A. Chao, F. A. Oliveira, N. Majlis: Solid State Commun. **21**, 845 (1977)
2.14 R. S. Tripati, L. C. Mosely: Solid State Commun. **21**, 655 (1977)
2.15 E. N. Economou, K. L. Ngai, T. L. Reinecke: "Localised Electronic States in the Gap of Amorphous Semiconductors: A New Mechanism," in *Amorphous and Liquid Semiconductors*, ed. by W. Spear (Centre for Industrial Consultancy and Liaison, University of Edinburgh 1977) pp. 296–300
2.16 R. J. Eliott, J. A. Krumhansl, P. L. Leath: Rev. Mod. Phys. **46**, 465 (1974)
2.17 H. Ehrenreich, L. M. Schwartz: Solid State Phys. **31**, 150 (1976)
2.18 R. Bellman: *Introduction to Matrix Analysis* (McGraw-Hill, New York 1960)
2.19 F. R. Gantmacher: *Matrizenrechnung Bd. II*, Hochschulbücher für Mathematik Bd. 37 (VEB Deutscher Verlag der Wissenschaften, Berlin 1970/71)
2.20 E. Kreyszig: *Advanced Engineering Mathematics* (Wiley, New York 1972)
2.21 D. Weaire: "Some theorems relating to densities of states," in *Physics of Structurally Disordered Solids*, ed. by S. S. Mitra (Plenum, New York 1976) pp. 351–367
2.22 K. M. Watson: Phys. Rev. **105**, 1388 (1975)
2.23 L. Schwartz, H. Ehrenreich: Ann. Phys. (NY) **64**, 100 (1971)
2.24 F. Yonezawa, K. Morigaki: Prog. Theor. Phys. Suppl. **53**, 1 (1973)
2.25 P. Soven: Phys. Rev. **156**, 809 (1967)

2.26   D.W.Taylor: Phys. Rev. **156**, 1017 (1967)
2.27   B.Velicky, S.Kirkpatrick, H.Ehrenreich: Phys. Rev. **175**, 747 (1968)
2.28   J.Mertsching: Phys. Status Solidi (b) **63**, 241 (1974)
2.29   J.Zittarz: Z. Phys. **267**, 243 (1974)
2.30   T.Kaplan, M.Mostroller: Phys. Rev. **139**, 1783 (1974)
2.31   A.R.Bishop, A.Mookerjee: J. Phys. C**7**, 2165 (1974)
2.32   A.Mookerjee: J. Phys. C**8**, 29 (1975)
2.33   F.Ducastelle: J. Phys. C**8**, 3297 (1975)
2.34   H.Fukuyama, H.Krakauer, L.Schwartz: Phys. Rev. B**10**, 1173 (1975)
2.35   D.M.Esterling: Phys. Rev. B**12**, 1596 (1975)
2.36   R.Bass: Phys. Rev. B**12**, 2234 (1975)
2.37   P.R.Best, P.Lloyd: J. Phys. C**8**, 2219 (1975)
2.38   V.Kumar, D.Kumar, S.K.Joshi: J. Phys. C**9**, 2733 (1976)
2.39   A.Gonis, J.W.Garland: Phys. Rev. B**16**, 2424 (1977)
2.40   J.Keller: J. Phys. C**4**, 3143 (1971)
2.41   T.C.McGill, J.Klima: J. Phys. C**5**, L163 (1972)
2.42   J.Keller, J.M.Ziman: J. Non-Cryst. Solids **8–10**, 111 (1972)
2.43   T.C.McGill, J.Klima: Phys. Rev. B**5**, 1517 (1972)
2.44   S.Yoshino, M.Okazaki, M.Inoue: Solid State Commun. **15**, 683 (1974)
2.45   M.Inoue, S.Yoshino, M.Okazaki: J. Phys. Soc. Jpn. **39**, 780 (1975)
2.46   S.Yoshino, M.Inoue, M.Okazaki: J. Phys. Soc. Jpn. **39**, 787 (1975)
2.47   J.L.Beeby: Proc. R. Soc. London A **279**, 82 (1964)
2.48   S.F.Edwards: Philos. Mag. **3**, 1020 (1958)
2.49   L.E.Ballentine: Can. J. Phys. **44**, 2533 (1966)
2.50   S.F.Edwards: Adv. Phys. **16**, 359 (1967)
2.51   P.Lloyd, P.V.Smith: Adv. Phys. **21**, 69 (1972)
2.52   P.Lloyd, J.Oglesby: J. Phys. C**9**, 4383 (1976)
2.53   K.Maschke, P.Thomas: Phys. Status Solidi (b) **39**, 453 (1970)
2.54   B.Kramer: Phys. Status Solidi (b) **41**, 649 (1970)
2.55   K.Maschke, H.Overhof, P.Thomas: Phys. Status Solidi (b) **59**, 125 (1973)
2.56   B.L.Gyorffy: Phys. Rev. B**1**, 3290 (1970)
2.57   B.L.Gyorffy: Phys. Rev. B**5**, 2382 (1972)
2.58   B.L.Gyorffy, G.M.Stocks: J. Phys. (Paris) C**4**, 75 (1974)
2.59   B.Kramer: Phys. Status Solidi (b) **69**, 429 (1975)
2.60   B.Kramer: "Mathematical Methods for Calculating the Electronic Spectrum," in *Physics of Structurally Disordered Solids*, ed. by S.S.Mitra (Plenum, New York 1976) pp. 291–349
2.61   H.J.Fischbeck: Phys. Status Solidi (b) **53**, 527 (1972)
2.62   H.J.Fischbeck: Phys. Status Solidi (b) **62**, 425 (1974)
2.63   H.J.Fischbeck: Phys. Status Solidi (b) **69**, 113 (1975)
2.64   E.N.Economou, M.H.Cohen: Phys. Rev. B**5**, 2931 (1972)
2.65   D.C.Licciardello, E.N.Economou: Phys. Rev. B**11**, 3697 (1975)
2.66   R.J.Bell, P.Dean: Discuss. Faraday Soc. **50**, 55 (1970)
2.67   D.J.Thouless: Phys. Rep. **13**C, 94 (1974)
2.68   B.I.Halperin, M.Lax: Phys. Rev. **148**, 722 (1966)
2.69   B.I.Halperin, M.Lax: Phys. Rev. **153**, 802 (1967)
2.70   J.Zittarz, J.S.Langer: Phys. Rev. **148**, 741 (1966)
2.71   E.N.Economou, M.H.Cohen, K.F.Freed, E.S.Kirkpatrick: "Electronic Structure of Disordered Materials," in *Amorphous and Liquid Semiconductors*, ed. by J.Tanc (Plenum, New York 1974) pp. 101–158
2.72   S.F.Edwards, V.V.Gulyaev: Proc. Phys. Soc. London **83**, 495 (1964)
2.73   T.Lukes: Philos. Mag. **12**, 719 (1965)
2.74   R.P.Feynman, A.R.Hibbs: *Quantum Mechanics and Path Integrals* (McGraw-Hill, New York 1965)
2.75   T.Lukes: Philos. Mag. **13**, 875 (1966)
2.76   R.Jones, T.Lukes: Proc. R. Soc. London A**309**, 457 (1969)

2.77   R.Jones: J. Phys. C **2**, 1187 (1969)
2.78   R.Friedberg, J.M.Luttinger: Phys. Rev. B **12**, 4460 (1975)
2.79   J.M.Luttinger: Phys. Rev. B **13**, 2596 (1976)
2.80   S.F.Edwards: J. Phys. C **3**, L 30 (1970)
2.81   S.F.Edwards: J. Non-Cryst. Solids **4**, 417 (1970)
2.82   K.Freed: Phys. Rev. B **5**, 4802 (1972)
2.83   S.K.Ma: *Modern Theory of Critical Phenomena* (Benjamin, New York 1976)
2.84   G.Toulouse, P.Pfeuty: C.R.Acad. Sci. t **280**, B 33 (1975)
2.85   A.Nitzan, K.F.Freed, M.H.Cohen: Phys. Rev. **15**, 4476 (1977)
2.86   M.V.Sadovskii: Zh. Eksp. Teor. Fiz. **70**, 1936 (1976)
2.87   A.Aharony, Y.Imry: J. Phys. C **10**, L 487 (1977)
2.88   F.J.Wegner: Z. Phys. B **25**, 327 (1976)
2.89   J.H.Wilkinson: *The Algebraic Eigenvalue Problem* (Clarendon Press, Oxford 1965)
2.90   D.Weaire, A.R.Williams: J. Phys. C **10**, 1239 (1977)
2.91   R.Alben, M.Blume, H.Krakauer, L.Schwartz: Phys. Rev. B **12**, 4090 (1975)
2.92   R.Alben, H.Krakauer, L.Schwartz: Phys. Rev. B **14**, 1510 (1976)
2.93   R.Alben, S.Kirkpatrick, D.Beeman: Phys. Rev. B **15**, 346 (1977)
2.94   R.Alben, G.Burns: Phys. Rev. B **16**, 3746 (1977)
2.95   R.Alben, M.Blume, Marilyn McKeown: Phys. Rev. B **16**, 3829 (1977)
2.96   N.I.Akhiezer: *The Classical Moment Problem* (Oliver and Boyd, London 1965)
2.97   J.C.Wheeler, M.G.Prais, C.Blumstein: Phys. Rev. B **10**, 2429 (1974)
2.98   J.P.Gaspard, F.Cyrot-Lackmann: J. Phys. C **6**, 3077 (1973)
2.99   R.Haydock, V.Heine, M.J.Kelly: J. Phys. C **5**, 2845 (1972)
2.100 R.Haydock, V.Heine, M.J.Kelly: J. Phys. C **8**, 259 (1975)
2.101 C.H.Hodges: J. Phys. (Paris) **38**, L 187 (1977)
2.102 M.Lannoo: J. Phys. (Paris) **34**, 869 (1973)
2.103 R.L.Jacobs: J. Phys. F **4**, 1351 (1974)
2.104 R.Haydock, V.Heine, M.J.Kelly, J.B.Pendry: Phys. Rev. Lett. **29**, 868 (1972)
2.105 M.J.Kelly: Surf. Sci. **43**, 587 (1974)
2.106 R.Haydock, M.J.Kelly: J. Phys. C **8**, L 290 (1975)
2.107 N.V.Cohan, M.Weissmann: J. Phys. C **10**, 383 (1977)
2.108 M.F.Thorpe, D.Weaire: "The Theory of the Electronic Density of States in Amorphous Semiconductors," in *Amorphous and Liquid Semiconductors*, ed. by J.Stuke, W.Brenig (Taylor and Francis, London 1974) pp. 917–937
2.109 N.Shevchik: "Photoelectron Spectroscopy of Amorphous Group IV and III–V Semiconductors," in *Tetrahedrally Bonded Amorphous Semiconductors*, ed. by M.H.Brodsky, S.Kirkpatrick, D.Weaire (A.I.P., New York 1974) pp. 72–84
2.110 W.E.Spicer: "Photoemission Spectroscopy and the Electronic Structure of Amorphous Materials – Studies of Ge and Si," in *Amorphous and Liquid Semiconductors*, ed. by J.Stuke, W.Brenig (Taylor and Francis, London 1974) pp. 499–518
2.111 J.Tauc: "Optical Properties of Amorphous Semiconductors," in *Amorphous and Liquid Semiconductors*, ed. by J.Tauc (Plenum, New York 1974) pp. 159–220
2.112 M.L.Theye: "Optical Properties of *a*-Ge, *a*-Si and *a*-III–V Compounds," in *Amorphous and Liquid Semiconductors*, ed. by J.Stuke, W.Brenig (Taylor and Francis, London 1974) pp. 479–498
2.113 W.Spear: "Localised States in Amorphous Semiconductors," in *Amorphous and Liquid Semiconductors*, ed. by J.Stuke, W.Brenig (Taylor and Francis, London 1974) pp. 1–16
2.114 W.Paul: "Photoluminescence," in *Physics of Structurally Disordered Solids*, ed. by S.S.Mitra (Plenum, New York 1976) pp. 171–197
2.115 D.Weaire, M.F.Thorpe: Phys. Rev. B **4**, 2508 (1971)
2.116 M.F.Thorpe, D.Weaire: Phys. Rev. B **4**, 3518 (1971)
2.117 D.Weaire, M.F.Thorpe: "Electronic Structure of Amorphous Semiconductors," in *Computational Methods for Large Molecules and Localised States in Solids*, ed. by F.Herman, A.D.McLean and R.K.Nesbet (Plenum, New York 1973) pp. 295–315

2.118 F.Herman, J.P.Van Dyke: Phys. Rev. Lett. **21**, 1575 (1968)
2.119 M.F.Thorpe, D.Weaire, R.Alben: Phys. Rev. B**7**, 3777 (1973)
2.120 R.Alben, D.Weaire, P.Steinhardt: J. Phys. C**6**, L384 (1973)
2.121 J.D.Joannopoulos, F.Yndurain: Phys. Rev. B**10**, 5164 (1974)
2.122 F.Yndurain, J.D.Joannopoulos: Phys. Rev. B**11**, 2957 (1975)
2.123 J.D.Joannopoulos, M.L.Cohen: Solid State Phys. **31**, 71 (1976)
2.124 D.W.Bullett, M.J.Kelly: Solid State Commun. **16**, 1379 (1975)
2.125 B.Y.Tong, T.C.Wong: Phys. Rev. B**6**, 4482 (1972)
2.126 F.C.Choo, B.Y.Tong: "Imperfections and Hydrogen Incorporation in Crystalline and Amorphous Structures," in *Amorphous and Liquid Semiconductors*, ed. by W.Spear (Centre for Industrial Consultancy and Liaison, University of Edinburgh 1977) pp. 120–124
2.127 B.Y.Tong: "On the Application of Extended Hückel Theory in Tetrahedrally Bonded Amorphous Semiconductors," in *Tetrahedrally Bonded Amorphous Semiconductors*, ed. by M.H.Brodsky, S.Kirkpatrick, D.Weaire (AIP, New York 1974) pp. 145–150
2.128 B.Kramer: Phys. Status Solidi **47**, 501 (1971)
2.129 B.Kramer, K.Maschke, P.Thomas: Phys. Status Solidi **48**, 635 (1971)
2.130 B.Kramer: "Electronic Properties of Amorphous Solids," in Advances in Solid State Physics, Vol. XII, ed. by O.Madelung (Pergamon, New York 1972) pp. 133–182
2.131 J.Treusch, B.Kramer: Solid State Commun. **14**, 169 (1974)
2.132 B.Kramer, J.Treusch: "Electronic properties of tetrahedrally bonded amorphous semiconductors," in *Physics of Semiconductors*, ed. by M.H.Pilkuhn (Teubner, Stuttgart 1974) pp. 1061–1065
2.133 J.Keller: "Cluster Scattering in Amorphous Semiconductors and Liquid Metals," in *Computational Methods for Large Molecules and Localised States in Solids*, ed. by F.Herman, A.D.McLean, R.K.Nesbet (Plenum, New York 1973) pp. 341–356
2.134 M.Okazaki, S.Yoshino: "Electronic Structure of Amorphous Si: Cluster Theory for Gap States," in Proc. of 6th Conference on *Amorphous and Liquid Semiconductors* (Leningrad, 1975) pp. 93–97
2.135 J.D.Joannopoulos: Phys. Rev. B**16**, 2764 (1977)
2.136 N.F.Mott, E.A.Davis: *Electronic Processes in Non-Crystalline Solids* (Oxford University Press, London 1971)
2.137 P.W.Anderson: Phys. Rev. **109**, 1492 (1958)
2.138 K.Ishii: Suppl. Progr. Theor. Phys. **53**, 77 (1973)
2.139 S.Yoshino, M.Okazaki: J. Phys. Soc. Jpn. **43**, 415 (1977)
2.140 D.C.Licciardello, D.J.Thouless: J. Phys. C**8**, 4157 (1975)
2.141 D.C.Licciardello, D.J.Thouless: J. Phys. C**11**, 925 (1978)
2.142 R.Haydock: Philos. Mag. B **37**, 97 (1978)
2.143 R.Abou-Chacra, P.W.Anderson, D.J.Thouless: J. Phys. C**6**, 1734 (1973)
2.144 R.A.Abram, S.F.Edwards: J. Phys. C**5**, 1183 (1972)
2.145 R.A.Abram, S.F.Edwards: J. Phys. C**5**, 1196 (1972)
2.146 J.M.Ziman: J. Phys. C**2**, 1230 (1969)
2.147 D.C.Herbert, R.Jones: J. Phys. C**4**, 1145 (1971)
2.148 D.Weaire, V.Srivastava: J. Phys. C**10**, 4309 (1977)
2.149 D.Weaire, V.Srivastava: "New Numerical Results for Anderson Localisation," in *Amorphous and Liquid Semiconductors*, ed. by W.Spear (Centre for Industrial Consultancy and Liaison, University of Edinburgh 1977) pp. 286
2.150 D.Weaire, C.H.Hodges: J. Phys. C, **11**, 685 (1978)
2.151 P.W.Anderson: Proc. Nat. Acad. Sci. USA **69**, 1097 (1972)
2.152 N.F.Mott: Commun. Phys. **1**, 203 (1976)
2.153 M.Fisher: Rev. Mod. Phys. **46**, 597 (1974)
2.154 M.Pepper: Contemp. Phys. **18**, 423 (1977)
2.155 P.Prelovsek: Phys. Rev. Lett. **40**, 1596 (1978)
2.156 B.Kramer, D.Weaire: J. Phys. C**11**, L5 (1978)

3. States in the Gap and Defects in Amorphous Semiconductors

E. A. Davis

With 16 Figures

The development of our understanding of amorphous semiconductors over the past ten years has in many ways paralleled that of crystalline semiconductors about thirty years earlier. For both classes of materials, initial studies were plagued by poor agreement between results obtained in different laboratories on ostensibly the same material – a feature that was attributed in many cases to inadequate characterization of samples in terms of structure, stoichiometry, and impurity content, and to the presence of defects of various kinds. For crystalline semiconductors it became clear, as time progressed, that many properties could be associated with such local departures from ideality in an otherwise perfect material. Subsequent careful control over conditions of growth and the development of elaborate techniques for reducing the impurity content, particularly in the case of Ge and Si, eventually allowed investigation of their "intrinsic" properties. Later, the controlled addition of impurities, coupled with an understanding of the role they played, paved the way towards the application of these materials in devices. For amorphous semiconductors a similar road has been followed and, although there is clearly still a long way to go, it seems fair to say that sufficient advances have been made to allow a separation of "intrinsic" from "extrinsic" properties – at least as far as certain simple elemental and binary materials are concerned. Just as for crystals, defects play a major and often dominating role.

3.1 General Models for States in the Gap of Amorphous Semiconductors

Early models for gap states in amorphous semiconductors were developed in somewhat of a phenomenological way without much attention being given to what aspects of the noncrystalline state – intrinsic or extrinsic – might produce them. They were proposed primarily to explain the principal features observed experimentally in thin amorphous films and glasses. These features, which in some aspects are still valid today, may be summarized as follows:

(I) *A general insensitivity of properties to the incorporation of impurities.* For example, the electrical conductivities of amorphous films of Ge, Si, and

III–V compounds, prepared by vacuum evaporation of the corresponding crystalline phases, are virtually independent of whether the starting materials are pure or whether they are doped sufficiently to make the crystals degenerate n or p type. Chalcogenide glasses prepared by supercooling the liquid are likewise relatively oblivious to the presence of several percent of an additional element added to the melt. This type of behavior led *Mott* [3.1] to propose that the normal valency of impurities incorporated into the host is satisfied, i.e., all electrons are taken up in bonds such that, say, a boron atom in amorphous Ge is threefold coordinated (rather than fourfold as in crystalline Ge) and hence is rendered electrically inactive. Although exceptions to this behavior are now known, the so-called 8-N rule (where N is the valency) is remarkably well satisfied in the majority of materials. It turns out that there are other additional reasons (having to do with defect states) why amorphous semiconductors cannot easily by doped, but we note here that, irrespective of other factors, the 8-N rule in itself is sufficient to explain the insensitivity to impurity addition and that successful doping requires a breakdown in the rule.

(II) *Pinning of the Fermi level near midgap.* This feature, when considered in respect of impurity incorporation, is of course equivalent to (I); if the Fermi level is spinned then doping cannot be achieved. Movement of the Fermi level by impurity addition is possible but only under special conditions of preparation. For the moment the emphasis is on generalities. The strong tendency for the Fermi level to be pinned near the center of the gap in amorphous semiconductors, not only in the presence of impurities but also irrespective of preparation conditions or postpreparation treatments, has in fact yet to be fully explained.

(III) *A general similarity of the overall distribution of the density of states in the valence and conduction bands to that in the corresponding crystalline phase.* The importance of short-range atomic configuration, which is normally the same in the crystalline and noncrystalline phases of a given material, is essentially responsible for this feature. In particular, for the purposes of this chapter, we stress the existence of an optical gap in amorphous semiconductors which in many cases is comparable to that in the crystalline phase.

There are other characteristic features of amorphous semiconductors which could be mentioned – for example, a breakdown of the k-conservation selection rules for optical excitation of electronic and vibrational states, a Hall effect anomalous in sign and magnitude, and an ac conductivity that increases with frequency. However, early models concentrated on attempting to explain the three enumerated above.

One of the first proposals for the effect of disorder on electronic states in the gap of amorphous semiconductors was made by *Gubanov* [3.2] who suggested that the conduction- and valence-band edges tailed into the gap and that in these tails the states were localized in space. Subsequently *Cohen* et al. [3.3] introduced the CFO model in which the tails were considered to be sufficiently extended that they overlapped near the center of the gap. On this model the Fermi level E_F is pinned in the region of overlap: states derived from the conduction band and lying below E_F are positively charged and those derived from the valence band and lying above E_F are charged negatively. In the pseudogap the states are localized and critical energies separate these states from extended states in the bands: the critical energies are called mobility edges and the energy separation between them the mobility gap. Although it seems certain that *some* tailing of the bands occurs, it is now believed that this is not as extensive as in the CFO model but rather that the Fermi level is pinned near the gap center by fairly discrete levels associated with specific defects. An early model proposed by *Davis* and *Mott* [3.4] was that the Fermi level lies in a single narrow band of levels lying near midgap, these levels being associated with dangling bonds in an otherwise fully connected network. In this model, the mobility-edge concept was retained but the band tails were assumed to be only a small fraction of the bandgap – a few tenths of an electron volt wide. Similar models followed these principles; in some of them the single band of defect levels is replaced by two or more bands situated approximately equidistant on either side of E_F (see [3.5, 6]).

Detailed measurements of drift mobility, field effect, photoconductivity, luminescence, etc., on a variety of materials have subsequently verified the existence of fairly discrete levels or, at least, structure in the distribution of the density of states throughout the gap of amorphous semiconductors. In addition, the evidence for mobility edges also seems convincing. In these respects, therefore, the early phenomenological models for states in the gap appear to have been correct. Before describing the detailed information now available and our current understanding of defects in amorphous materials, it seems appropriate to end this section with a few further remarks concerning general models.

First we mention the suggestion of *Phillips* [3.7] that the structure of an amorphous material might relax energetically so as to cause the expulsion of states from the gap. This "clean-gap" model is based on the proposition that the general tendency for electrons is to lower their energy by being paired in covalent bonding situations and thus for any states associated with, say, exceptionally long or short bonds or with defects, to lower their energy towards the valence band. Likewise, antibonding states of these configurations would be pushed up into the conduction band. Related to this idea is the more recent model due to *Anderson* [3.8] in which the gap is essentially free of one-electron states, but that a pseudogap containing a continuum of two-electron states remains. This important concept is based on a model of strong coupling between electrons and phonons leading to a negative term in the energy which more than compensates for the Coulomb repulsive force between electrons at the same site. This idea of an effective negative correlation energy for electron states has formed the basis of subsequent models for defect levels in amorphous semiconductors as will be described later.

3.2 A Distinction Between Two Classes of Materials

When one comes to consider specific defects and energy levels in amorphous semiconductors, it is convenient to distinguish two classes of materials. They will be referred to as "Ge-type" and "Se-type." There are two distinctions to be made:

I) *Network flexibility.* In accordance with the 8-N rule, the coordination numbers of Ge and Se are, as in the crystalline phases, four and two respectively. Fourfold coordination means symmetrical bonding and the formation of rigid structures; a continuous random network (CRN) with tetrahedral bonds can be constructed with negligible density deficit and very little possibility for local reorganization of atoms. On the other hand, twofold coordination is very asymmetrical and model structures must necessarily be fairly open (unless the chains are folded back and forth in a very tight ordered arrangement – in which case one has the trigonal crystalline phase) and have a greater degree of flexibility. In other words, steric hindrance to configurational changes is less in the lower-coordinated structures.

II) *The presence or otherwise of high-energy, nonbonding electrons.* In Se, but not in Ge, the uppermost valence band is formed from nonbonding, lone-pair p electrons. As stressed principally by *Ovshinsky* [7.9], lone-pair orbitals in the amorphous state lose the special orientational relationship present in the crystal and their random arrangement could, in principle, give rise to an extended valence-band tail. However, it is when one considers the nature of possible defects in these materials that lone-pair electrons can be seen to play a crucial role as will be described below.

Materials that fall into the Ge-type class are the tetrahedrally coordinated substances, Ge, Si, and III–V compounds. Materials in the Se-type class are Se, Te, S, and multicomponent systems containing a large proportion of a chalcogen element. Clearly, a mixed system such as Ge_xSe_{1-x} can fall into either class according to the value of x. Amorphous As and other pnictides, with a coordination number of three and having only low-lying, nonbonding s electrons, are also intermediate materials and have to be considered separately. Likewise, amorphous Si prepared by the glow-discharge decomposition of silane appears to have properties very different from those of films prepared by evaporation or sputtering; the presence of hydrogen in the glow-discharge films is important and it is becoming common practice to identify such films as Si–H alloys. For the most part the class in which a given material falls can be deduced from their properties which turn out to be rather distinctive. These are as follows:

a) *Paramagnetic Centers*

The Ge-type class of materials is characterized by a large density of free spins as evinced by ESR and a Curie-like susceptibility. The density of paramagnetic centers is normally found to lie between 10^{19} and 10^{20} cm^{-3} and their g factor is close to 2.01. Although temperature of deposition and/or subsequent annealing affects these numbers somewhat, the variation between samples is remarkably small. The only exception seems to be glow-discharge-deposited Si for which no spin signal is observed; the presence of hydrogen undoubtedly serves to saturate dangling bonds which would otherwise give an ESR signal. In contrast to the Ge-type class of materials, Se and chalcogenide glasses, certainly when well annealed, appear to contain no paramagnetic centers or at least a density less than the experimentally detectable limit $\sim 10^{15}\,cm^{-3}$. Exceptions are rare and have normally been accounted for in terms of trace impurities such as Fe. Chalcogenide films deposited onto low-temperature substrates and maintained at a low temperature [3.10], however, do exhibit an ESR signal.

b) *Photoinduced* ESR

Although the Se-type class of materials do not normally exhibit ESR signal in the dark equilibrium state, it is possible to photoinduce paramagnetic centers by illuminating with radiation of energy close to that of the band gap at a low temperature [3.11, 12]. The ESR signal associated with such metastable centers has a g value ~ 2.00 and corresponds to a density of about $10^{17}\,cm^{-3}$. Although the signal remains after cessation of the illumination, it disappears on heating (thermal bleaching) or on irradiation with sub-band-gap light (infrared quenching) [3.12, 13]. As will be seen, the existence of such metastable centers plays an important role in current models for defect states in these materials. An ESR signal can also be generated by irradiation with fast electrons [3.14] but the lineshape of this can be distinguished from the photoinduced signal. The Ge-type class of materials do not exhibit photoinduced ESR, with an exception again being that of hydrogenated silicon.

c) *Luminescence*

Following excitation with radiation in the wavelength range identical to that giving rise to photoinduced ESR, the Se-type class of materials luminesce with varying degrees of efficiency [3.15].. The radiation is emitted in a band whose width and position (normally close to half the band gap) can be interpreted in terms of strong electron–phonon coupling. The luminescence is consequently associated with a large Stokes shift. On the other hand luminescence has not been observed in the Ge-type class of materials (for an isolated exception, see [3.16]), if again we exclude films containing hydrogen such as glow-discharge-deposited Si. In the latter material luminescence is observed but the electron–phonon coupling appears to be much less than in chalcogenides [3.17].

d) *Variable-Range Hopping Conduction*

Particularly at low temperatures, the conductivity of the Ge-type class of materials obeys the $T^{-1/4}$ law characteristic of variable-range hopping amongst localized states at the Fermi level. In fact, only at high temperatures and after annealing can band conduction be observed [3.18, 19]. The density of states $N(E_F)$ at E_F can be evaluated from the slope of the $T^{-1/4}$ plots if an intelligent guess is made for the spatial extent (α^{-1}) of the localized wavefunction. If $T^{-1/3}$ behavior (two-dimensional, variable-range hopping) can be observed in thin films, then $N(E_F)$ and α^{-1} can be determined independently [3.20, 21], as they can also from an analysis of high-field conductivity data [3.22, 23]. The density turns out to be $\sim 10^{18} \text{cm}^{-3} \text{eV}^{-1}$ with remarkable consistency. Although there is not a one-to-one correspondence between this density and that of the ESR-active centers, a correlation does exist [3.24]. The density of states can be reduced by annealing [3.18, 19] or enhanced by irradiation [3.22]. In contrast, the conductivity of Se and chalcogenides is temperature-activated in an Arrhenius manner with an energy close to one-half the optical band gap – indicating a Fermi level pinned near midgap and conduction in the (valence) band. $T^{-1/4}$ hopping in this class of materials has been observed in low-temperature-deposited films [3.10] but this is an exceptional case. Irradiation with fast ions fails to induce the effect in films deposited at room temperature (A. P. Troup and N. Apsley, personal communication).

Armed with the general classification of materials into two types as suggested by their different chemical bonding and by their physical properties, we can now consider in more detail defects in these materials and the associated levels in the gap.

3.3 Amorphous Germanium and Silicon

3.3.1 Evaporated or Sputtered Films

Films of amorphous germanium and silicon prepared by evacuation in vacuo or by sputtering in argon always appear to have a high density of states in the gap, in particular at the energy of the Fermi level E_F. The principal evidence for this is the observation of variable-range ($T^{-1/4}$ behavior) hopping conduction and a small value of the thermopower – both indicative of transport at E_F. Figure 3.1 illustrates typical behavior for amorphous germanium. Only at high temperatures does it appear possible to excite carriers into the bands, as suggested by the tendency for the activation energy to approach a value comparable to half the optical gap. Even then a difference between the activation energies for conduction and thermopower suggests transport by hopping in band-tail states. Annealing, or deposition onto high-temperature substrates, extends the range of temperatures over which band conduction occurs, but the small variation in the slope of the $T^{-1/4}$ plots indicates that the density of states at E_F cannot be reduced by more than a factor of about five by such treatments. As already mentioned, densities of states are typically of the

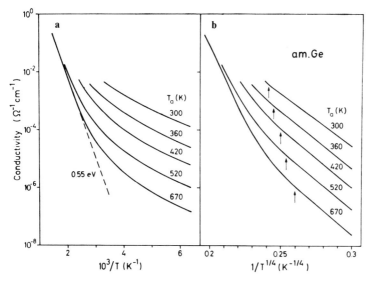

Fig. 3.1a, b. Temperature dependence of the dc conductivity of an amorphous Ge film prepared by evaporation on to a quartz substrate held at 300 K and successively annealed for 15 min at the temperatures shown. (a) Plotted versus T^{-1}, (b) plotted versus $T^{-1/4}$. Straight lines below the arrows in (b) indicate hopping conduction at the Fermi level. After the highest anneal temperature, band conduction sets in at high temperatures [0.55 eV slope in (a)]. [3.18]

order of $10^{18}\,\mathrm{cm}^{-3}\,\mathrm{eV}^{-1}$ as deduced from the various methods outlined in Sect. 3.2. Since there is virtually no information available on the *distribution* in energy of gap states in evaporated or sputtered films of amorphous germanium or silicon, a volume density of states cannot be deduced with any certainty. If it is assumed that the states are contained within a narrow band of width 0.1 eV at E_F, then the volume density is $\sim 10^{17}\,\mathrm{cm}^{-3}$ – a value considerably lower than the density of paramagnetic centers (10^{19}–$10^{20}\,\mathrm{cm}^{-3}$) observed by ESR. If the states are distributed over the whole of the bandgap (say 1 eV) then a discrepancy still exists and one is forced to conclude that the volume density of states as deduced from dc hopping conduction is lower than the density of free spins[1].

This discrepancy is not serious since there is no reason to assume that all the centers carrying a spin contribute to dc hopping conductivity. Nevertheless the two are related in that annealing, or an increase in substrate temperature during preparation, reduces both densities. Furthermore the ESR linewidth increases with temperature at a rate approximately proportional to the dc

---

1 The density of hopping states deduced from an analysis of ac conductivity data is much higher than that obtained from the dc conductivity [3.25]. However, there are problems associated with analyzing ac data; for example, the density of states determined by application of the Austin–Mott formula is so high as to render use of the formula invalid. Furthermore, the technique might be greatly affected by sample inhomogeneities.

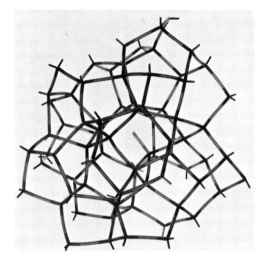

Fig. 3.2. A continuous random network (CRN) model of amorphous Ge (or Si) containing a dangling bond

hopping conductivity [3.24, 26, 27] – as expected if the spin–lattice relaxation time is decreased by motion. One might postulate that the hopping states are a subset of the paramagnetic states and are associated with the same defect. The ESR signal is devoid of hyperfine splitting or any fine structure which, if present, could be used to help in identification of the center, but it does not seem unreasonable to associate the defects with single dangling bonds which may exist in isolation (Fig. 3.2) or in more complex configurations to be described below.

3.3.2 Hydrogenated Germanium and Silicon

By introducing hydrogen into evaporated or sputtered films either during preparation or by subsequent infusion of atomic hydrogen [3.28, 29], it is possible to reduce the magnitude of both the hopping conductivity and ESR signal below the level of detectability. Preparation of silicon by glow-discharge decomposition of silane (SiH_4) has a similar effect, if the temperature of deposition is not too low. Clearly, incorporation of hydrogen can saturate dangling bonds and remove states in the gap associated with them. If a film of glow-discharge-deposited silicon is bombarded with fast ions, $T^{-1/4}$ behavior and an ESR signal reappear but these can be removed in stages by annealing [3.24].

The most comprehensive investigation of states in the gap of amorphous silicon has been made on glow-discharge-deposited (g.d.) material for which the density of states is sufficiently low to permit the distribution to be probed by the field-effect technique ([3.30, 31]; see also Chap. 9). Two peaks in the distribution, one above and one below E_F, have been identified. An early suggestion

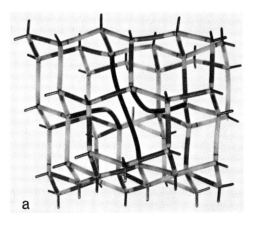
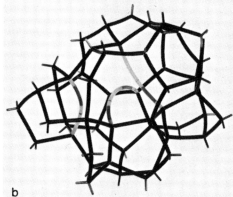

Fig. 3.3a, b. Models of a divacancy in (**a**) crystalline Ge (or Si), (**b**) amorphous Ge (or Si). The six free orbitals have been paired in each case

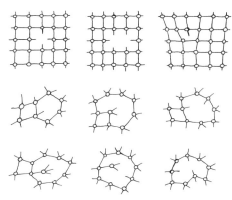

Fig. 3.4. Schematic representation of possible defects in crystalline and amorphous Ge (or Si). The three defects in the crystals (top) are a monovacancy, a divacancy, and a dislocation (linear array of dangling bonds extending perpendicular to plane of paper). The defects at the bottom correspond to one, two, and three dangling bonds in various configurations in a CRN. Note that they are all of lower order than the monovacancy which produces four dangling bonds

[3.31] was that these states are associated with a defect similar to the divacancy in crystalline silicon (Fig. 3.2a). The similarity of the energy levels to those known to arise from the divacancy in the crystal [3.32] was cited as justification, along with the knowledge that the monovacancy in crystalline silicon is rather unstable. Figure 3.3b illustrates a divacancy in a continuous-random-network (CRN) model of amorphous silicon. A possible pairing of the six dangling bonds has been made: however, it should be noted that there is no reason to expect two short and one long covalent bond as in the crystal (Fig. 3.3a), and it may be unreasonable to assume that they give rise to similar energy levels in the gap. A further warning that the divacancy model may be too simple is that the two peaks in the distribution of states probed by field effect in g.d. Si have quite different heights, the number of states below E_F being at least an order of magnitude larger than that above E_F.

Although the divacancy is the lowest-order stable defect in crystalline silicon, defects containing a smaller number of dangling bonds are clearly possible in an amorphous network. Furthermore, unlike in the crystal, the number of dangling bonds need not be even: defects which have associated with them any number of dangling bonds from one upwards are possible (see Fig. 3.4). All of the defects shown may not, of course, be favorable energetically but are presented here to illustrate what configurations are possible topologically, i.e., within the normal constraints of a CRN (bond-angle distortions less than about 20°). It seems to be generally accepted that, wherever possible, dangling bonds will pair up in such defects (even on the interior surface of a large void). Defects with an odd number of dangling bonds can give rise to ESR signals in unhydrogenated material, but when hydrogen is present, these become saturated. Hydrogen may also break up long paired bonds and then attach itself to the unpaired orbital. Bonds saturated with hydrogen will be strong and give rise to states lying deep in the valence band; however, the possibility that antibonding states associated with such bonds lie close to the bottom of the conduction band or even in the upper half of the energy gap cannot be ruled out [3.33, 34].

It remains to consider any states associated with unduly long, short, or distorted bonds in the amorphous network. Clearly calculations of the energy levels associated with some of the defect configurations shown in Fig. 3.4 would be desirable. Theoretical techniques exist for doing this – see for example the local-density-of-states calculations for states associated with dislocations in crystalline silicon and germanium [3.35, 36] – but the problem of correctly incorporating distortion energies related to local configurational changes may be too great to expect immediate success.

The distribution of states in the gap of g.d. silicon referred to above is supported by a number of experiments described in detail in Chap. 9 by Spear and LeComber. It has, however, not gone unchallenged. A reinterpretation of the field-effect data by *Singh* and *Cohen* [3.37] implied that the density of states throughout the gap is rather *lower* than first suggested but more importantly that the energies of the peaks in the distribution may have been misplaced. Evidence for the original placements from luminescence studies [3.38] has also been brought into doubt by the report [3.17] of a large ($\sim 0.5\,eV$) Stokes shift associated with electron–phonon coupling and also by the suggestion [3.39] that fine structure in the luminescence spectra might be associated with interference effects. A Stokes shift suggests that hydrogenated silicon films may have a more flexible structure than that of pure amorphous silicon and therefore possess properties more like those characteristic of chalcogenides (see Sect. 3.4) than has hitherto been believed.

One property that brings these two classes of materials into closer coincidence is that of photoinduced ESR signals, observed in films that exhibit negligible ESR in the dark. As first reported by *Knights* et al. [3.40], excitation of g.d. silicon with band-gap light results in an ESR signal the strength of which depends on the intensity of the light and, in contrast to the dark ESR signal in unhydrogenated films, of asymmetric shape. Studies as a function of microwave power reveal that the signal can be decomposed into two components with g values centered on 2.005 and 2.025, that latter being the broader of the two. Similar signals have also been observed by spin-dependent luminescence [3.41]. *Kaplan* [3.42] has suggested that the two signals arise from charged states of a weak bond, i.e., a hole or an electron trapped at one of the long orbitals present at a divacancy or similar defect. Other models have been proposed [3.43–45]. Interestingly, the density of such centers seems to be relatively independent of the amount of hydrogen incorporated into the films. This suggests that such centers might exist in all films, the corresponding ESR signals being unobservable in those films that exhibit a large dark ESR signal.

Chapter 7 discusses these spin effects further. In addition, a comprehensive review of ESR in amorphous germanium and silicon, including the effect of doping and irradiation, has been given by *Stuke* [3.24]. The reader is referred to these sources for further details.

3.4 Amorphous Chalcogenides

3.4.1 Luminescence in Chalcogenides

As mentioned earlier, defects in the Se-type class of materials are characterized by strong electron–phonon coupling. The first suggestion that this might be so was made by *Street* et al. [3.46] when attempting to understand the nature of the luminescence in these materials, particularly As_2S_3. As a fairly general rule, luminescence in chalcogenides occurs in a band peaking at an energy close to that of half the optical band gap, the integrated intensity being greatest when excitation is by photons of energy corresponding to the tail of the optical absorption edge (see Fig. 3.5). Early measurements were by *Kolomiets* et al. [3.47] and were interpreted by them and by *Davis* and *Mott* [3.4] in terms of recombination via a simple band of defect states near midgap. A problem with this model is the absence of an absorption band corresponding to excitation of carriers into or out of such states. An alternative "recombination-edge" model [3.48] also seems unsatisfactory for the same and other reasons – notably the presence of very similar luminescence behavior in crystalline chalcogenides.

The model proposed by *Street* et al. [3.15] is that the luminescence does indeed occur at defects (the nature of which will be described below) but that the associated states in the gap lie near to one or other of the bands (see Fig. 3.6). Excitation of the defect requires photons of energy comparable to the band-gap energy as illustrated by process A – creation of an electron–hole pair with subsequent rapid trapping of the hole, or by the process A' – direct excitation of an electron in the defect levels. Complementary processes can be

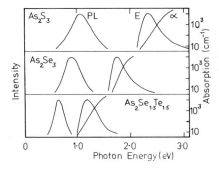

Fig. 3.5. Low-temperature photoluminescence spectra (PL), excitation spectra (E) and optical absorption edges (α) for glassy samples of three chalcogenides [3.15]. Note that the luminescence band occurs at roughly half the energy gap (which can be approximately taken as the energy corresponding to $\alpha = 10^3 \, \text{cm}^{-1}$) and that the width of the band is lower in the smaller gap materials for which the Stokes shift is less

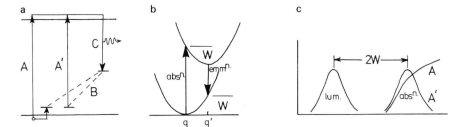

Fig. 3.6. (a) Photogeneration and luminescence processes. (A) photoexcitation across the gap followed by rapid trapping of the hole at the defect center. (A) direct excitation of the defect center. Following both events the center relaxes (B) producing a deep level. The electron then recombines with the hole (C) radiatively. (b) Configurational coordinate diagram illustrating absorption and emission processes. The distortion energy is W and the Stokes shift is $2W$. The vibrational potential energy functions for ground and excited states (with equilibrium coordinates q and q' respectively) are taken to be the same. (c) Lineshapes of luminescence and absorption for the case of direct (A') and continuum (A) excitation. The width of the luminescence band is proportional to $(2W)^{1/2}$. [3.15]

envisaged for a shallow level near the conduction band. Following excitation by either of these processes, the lattice surrounding the defect distorts, leading to a new energy level for the defect which lies near midgap (process B). Subsequent recombinations of the electron with the hole at the defect then leads to the radiative transition (process C). The difference between the absorption and luminescence energies is then to be regarded as a Stokes shift and is more adequately represented on a configurational-coordinate diagram as shown in Fig. 3.6b.

An important experimental observation is that the luminescence efficiency is greatest at low temperature. The temperature dependence is believed to arise from a competing nonradiative mechanism which occurs if the initially excited electron escapes from the vicinity of the center[2]. Should this occur the centers

---

2 In a model due to *Street* et al. [3.49] the electron is envisaged as tunneling away through weak potential fluctuations in the conduction-band edge, the probability of escape then being proportional to $\exp(T/T_0)$ where T_0 is a parameter related to the shape of the fluctuations.

are then left in a metastable state and, as long as they remain in this condition, are unavailable for the primary excitation processes A or A'. Such a situation would then be expected to give rise to "luminescence fatigue," i.e., a decrease in the luminescence intensity during prolonged photoexcitation, and also to "photoinduced absorption" with a threshold energy of roughly half that of the gap corresponding to excitation of an electron into or out of the metastable center. Both these effects are observed experimentally [3.11, 12, 15, 50, 51].

The fall-off of the photoexcitation spectrum on the high-energy side (see Fig. 3.5) – a feature which, incidentally, resulted in the early failure of many workers to find luminescence in these materials – can be explained by assuming that it is necessary either to excite the defect directly or at least to create a carrier pair within the vicinity of a defect. If excitation is elsewhere, then the low mobility of carriers in amorphous semiconductors makes it extremely unlikely for them to reach the defects before they find a nonradiative channel for recombination[3]. Since the exponential absorption edge (Urbach tail) in amorphous semiconductors is believed to be associated with internal random electric fields, then the fact that photoluminescence is most efficiently excited in these tails suggests that the defects in their ground state are charged. This point will be returned to later.

The fall-off of the photoexcitation spectrum on the low-energy side does *not* represent a decreasing quantum efficiency for photoluminescence but merely arises from a decreasing number of photons absorbed in the sample. The fall-off in fact mirrors the absorption edge, including a very low energy tail of reduced slope which is believed to correspond to direct absorption at the defects (process A' in Fig. 3.6) [3.52].

3.4.2 Defects in Chalcogenides

Turning now to the nature of the defects responsible for luminescence and other properties of chalcogenides, the original proposals of *Street* et al. [3.53, 54] will first be presented followed by the chemical-bond description given by *Kastner* et al. [3.55, 56]. Both of these sets of papers followed the important proposal by *Anderson* [3.8] mentioned in Sect. 3.1. This is that in amorphous semiconductors there is a strong tendency for electrons to be paired in bonding configurations, the Coulomb repulsion between electrons at the same site being outweighed by a negative term in the energy due to electron–phonon interaction which leads to configuration changes in the local atomic structure.

In retrospect, the raison d'être of *Anderson*'s proposal seems fairly self-evident, since the absence of ESR signals and $T^{-1/4}$ hopping conduction in chalcogenides (apart from films deposited at low temperatures [3.10]) requires that no single-spin centers be present. Early attempts to avoid the necessity of

---

3 An absence of a corresponding fall-off in the excitation spectrum of several crystalline chalcogenides (for which the luminescence process appears to be otherwise identical to that described above), supports this model because the mobility in crystals is likely to be higher.

spin pairing invoked the possibility that any ESR signal might be below the level of detectability, due either to a low concentration of centers or to a spread in g values such that the signal would be considerably broadened. However, the known pinning of the Fermi level and the difficulty of doping chalcogenides, combined with less direct evidence, suggests that a considerable density of states in the gap exists. Furthermore, although a fully connected network structure can be envisaged for As_2Se_3 for example, it is clearly not possible to construct a model for the structure of amorphous Se, using only twofold coordinated atoms, without producing chain ends, unless all the chains close up on themselves (rings), or, as mentioned earlier, an essentially infinite-length chain is folded back and forth.

Without considering specific defects, *Anderson* [3.8] considered the totality of all possible bonding configurations in an amorphous network and proposed that electrons would fill the resulting states in pairs up to the Fermi energy. The early CFO band-tail model was thus resurrected – with the important difference that the pseudogap containing one-electron states was replaced by one containing two-electron states. Thermal excitation of single electrons requires, on this model, an energy of U_{eff} — the effective negative correlation energy necessary to split the pairs.

While recognizing that a quasi-continuous distribution of energy levels provides an attractive explanation for the pinning of the Fermi level near the gap center, the evidence for fairly discrete levels in the gap is quite strong. *Street* and *Mott* [3.53], and *Mott, Davis,* and *Street* (MDS) [3.54] therefore chose to apply the *Anderson* negative U_{eff} concept to specific defects in an otherwise fully connected network. The model will be illustrated with reference to selenium.

Consider the two dangling bonds at the ends of the selenium chain illustrated in Fig. 3.7a. When they each contain a single electron the defects are neutral and will be designated D^0. Transfer of an electron from one chain end to the other will lead to the creation of two charged defects D^+ and D^-. It is proposed that the reaction

$$2D^0 \rightarrow D^+ + D^- \tag{3.1}$$

is exothermic, the necessary lowering in energy arising from local lattice distortions. On a configurational-coordinate diagram (Fig. 3.7b), the positive correlation energy U associated with the two electrons at D^- in the absence of configurational changes become negative (U_{eff}) after lattice relaxation. The "chemical" reason for the exothermic nature of the reaction is that, at D^+, an extra bond with a neighboring chain can be formed by utilizing the normally nonbonding lone-pair electrons. The coordination of Se atoms at D^+ is therefore three, in contrast to that at D^- where it is one and at a normally bonded Se atom where it is two. MDS proposed that the lattice distortion at D^- is negligible, at D^+ it is considerable and at D^0 it is intermediate.

a

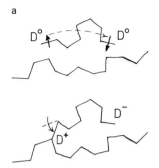

b

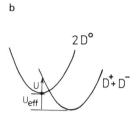

Fig. 3.7. (a) Illustration of the transfer of an electron from one chain end to another creating two charged defects D^+ and D^-. The reaction is assumed to be exothermic, the D^+ defect forming a threefold coordinated atom.

(b) The same reaction on a configurational coordinate diagram [3.74]. The positive correlation energy U associated with two electrons at one site is turned into an effective negative correlation energy U_{eff} because of the configurational changes

In a chemical-bond description of these defects given by *Kastner* et al. [3.55, 56], the charged states of the defect are designated C_1^- and C_3^+, C standing for chalcogenide and the subscript indicating the atomic coordination. The neutral center is labelled C_3^0 since these authors believe that an extra electron placed on C_3^+ is shared equally between the three bonds of the atom, which therefore remains threefold coordinated. In this respect the description of the neutral center differs from that proposed in [3.53, 54], in which it was suggested that the extra electron is located primarily on only one of the three bonds.

The structures and energies of several simple bonding configurations for a chalcogen, as given by *Kastner* et al. [3.56], are displayed in Fig. 3.8. The normal bonding configuration is C_2^0, the straight lines representing bonds (σ states) and the dots the lone-pair (LP), nonbonding electrons. In this configuration the antibonding (σ^*) states are empty and the energy, relative to the LP level, is $-2E_b$ as shown. Antibonding states are pushed up from the LP energy more than bonding states are pushed down. Thus the next configuration shown, C_3^0 – a neutral threefold coordinated atom with an extra electron placed in the antibonding orbital – has a higher energy than C_2^0 by an amount Δ. The C_3^+ configuration, with an energy of $-3E_b$, is the defect having the lowest energy. The energy of C_3^- is $-E_b + 2\Delta + U_{\sigma^*}$, the positive correlation energy term U_{σ^*} arising because *two* electrons are in the antibonding state. The extra electron at a "normal" dangling bond C_1^0 is indistinguishable from the two lone-pair electrons at the site and so the energy of this defect is $-E_b$. Finally a negatively charged dangling bond, C_1^-, has four electrons in the lone-pair state; its energy is $-E_b + U_{LP}$, the second term arising from the correlation energy in this level. The reaction corresponding to (3.1), in *Kastner*'s notation, is

$$2C_3^0 \rightarrow C_3^+ + C_1^-, \tag{3.2}$$

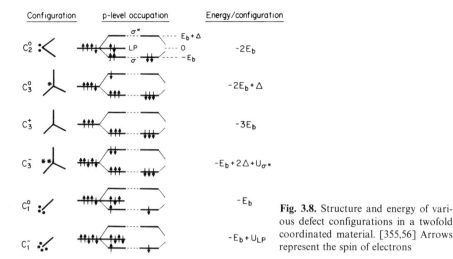

| Configuration | p-level occupation | Energy/configuration |

Fig. 3.8. Structure and energy of various defect configurations in a twofold coordinated material. [355,56] Arrows represent the spin of electrons

which can be seen to be exothermic if

$$-4E_b + 2\Delta > -3E_b - E_b + U_{LP},$$

i.e., if

$$2\Delta - U_{LP} > 0. \tag{3.3}$$

It should be stressed that whether this is so or not has not been proved, since Δ and U_{LP} have not been calculated. In addition, relaxation effects around the configuration shown in Fig. 3.8, which would change the energy of electrons on near neighbors, have not been included. Finally there is an implicit assumption in the model that the lattice is infinitely "soft," i.e., the configurations illustrated are assumed to form without steric hinderance from the surrounding network; introduction of lattice stiffness would be expected to make the reaction *less* exothermic.

The charged defects C_3^+ and C_1^- have been called by *Kastner* a valence-alternation pair (VAP). Their creation, starting from a fully bonded network in which all atoms are in the C_2^0 configuration can be described by

$$2C_2^0 \rightarrow C_3^+ + C_1^- \tag{3.4}$$

which costs an energy

$$-3E_b - E_b + U_{LP} + 4E_b = U_{LP}. \tag{3.5}$$

The concentration of VAPs present in a sample prepared by cooling a melt, assuming equilibration at the glass transition temperature T_g, is then

$N \exp(-U_{LP}/2kT_g)$ where N is the concentration of lattice sites. For films deposited onto a substrate at a temperature $T < T_g$, the concentration might be expected to be lower[4]. For $T \sim 500\,\text{K}$, the fraction of lattice sites that could be VAPs is $\sim 10^{-6}$ if U_{LP} is $\sim 1.1\,\text{eV}$.

The energy to create VAPs may be reduced if they form close to each other because of the Coulomb energy of attraction. Such bound pairs have been called by *Kastner*, intimate valence-alternation pairs (IVAPs). Although certain configurations of IVAPs are self-annihilating, others may not be [3.56, 59, 74]. The fraction of VAPs that form IVAPs is difficult to estimate, although interestingly it can be shown [3.56] that, if they form in equilibrium, the fraction is either close to zero or close to unity. Anticipating information to be presented later, it might be mentioned here that VAPs pin the Fermi energy, IVAPs do not; the concentration of VAPs is expected to be far more sensitive than that of IVAPs to the addition of dopants that form charged centers; transport properties that are affected by the presence of charged centers, e.g., drift mobilities, will be influenced more by VAPs than IVAPs since the latter are essentially neutral if the pair separation is comparable to an interatomic spacing. Luminescence could in principle arise from either VAPs or IVAPs.

Before considering the states in the gap expected from the presence of the atomic configurations described above, it is important to realise that the various under- and over-coordinated sites can be considered as different charge states of the *same* defect. Reverting back to the notation D^-, D^0, and D^+ (and thereby avoiding the question of whether D^0 is precisely threefold-coordinated – see above), removal of an electron from D^- converts it to D^0, removal of another electron converts it to D^+. Furthermore, they can be considered as true defects, rather than simply stretched or distorted bonds, since they either have a charge and no spin (D^+ and D^-) or else a spin and no charge (D^0). In the absence of any other charged centers (which may be introduced via impurity addition for example), the concentrations of D^+ and D^- are precisely equal and, assuming reaction (3.1) is exothermic, D^0 centers will only be created (in pairs) in low concentration by thermally activated transfer of an electron from D^- to D^+. We shall, however, see that, under nonequilibrium conditions, D^0 centers can be created optically in a metastable state.

Another way of looking at the negative correlation energy which makes reaction (3.1) exothermic, is as follows. Let the addition of an electron from, say, the valence band to D^+ cost an energy E_1 and the addition of an electron to the resulting D^0 an energy E_2. Then we can write

$$D^+ + e(+E_1) \rightarrow D^0 \tag{3.6}$$

$$D^0 + e(+E_2) \rightarrow D^- . \tag{3.7}$$

---

4 Somewhat paradoxically, the density of defects in deposited films is actually higher than in a glass; this is undoubtedly due to the low mobility of the adatoms which are thereby prevented from equilibrating.

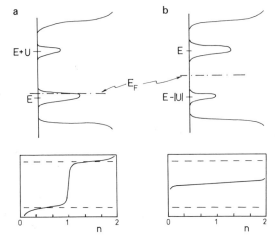

Fig. 3.9a, b. Density of states and variation of Fermi level with electron occupancy n for a semiconductor with defects having (**a**) a positive correlation energy and (**b**) an effective negative correlation energy. [3.57]

Thus

$$2D^0 \rightarrow D^+ + D^- + (E_1 - E_2). \tag{3.8}$$

If the addition of the second electron to D^+ costs less energy than the first (i.e., $E_2 < E_1$ – which is what a negative correlation energy means), then $E_1 - E_2$ is positive and the total reaction is exothermic.

3.4.3 States in the Gap of Chalcogenides

Energy levels in the gap of a semiconductor associated with various charge states of a defect under conditions of positive correlation energy U are well known to semiconductor physicists from earlier work on, say, Cu or Au impurities in crystalline germanium. Figure 3.9a illustrates two (slightly broadened) levels associated with singly and doubly occupied states of a positive U defect. As a function of increasing electron concentration (and at a low T) the Fermi level rises in the lower singly occupied level. The density of the upper levels is equal to the density of the lower levels that are filled, because a state cannot be doubly occupied until it is first singly occupied. When the concentration of electrons n (expressed as a fraction of the density of defect sites) reaches unity, the density of levels becomes equal. Further increase of N leads to the Fermi level jumping across the gap U to the upper level whereupon it moves slowly up through the level until, at $n = 2$, all states are doubly occupied.

The situation for defects that have associated with them a negative U is very different. The levels corresponding to double occupancy lie *below* those corresponding to single occupancy (Fig. 3.9b). As n is increased from zero, the electrons enter into the defect sites in pairs. The density of lower levels is equal to the number of electrons introduced and this band is therefore always full.

Just as for an intrinsic semiconductor, the Fermi level lies midway between the two levels; at a finite T, some occupancy of the upper level occurs at the expense of the lower level. This "pinning" of E_F between the two levels is maintained at all values of n between 0 and 2 except near the two extremes.

Expressions for the variation of E_F with n for the two situations at low temperatures have been given by *Adler* and *Yoffa* [3.57]. Using the notation of Fig. 3.9, they are

$$
\left.
\begin{aligned}
E_F &= E - kT \ln \left\{ \frac{2|n-1|}{1-|n-1|} \right\} \quad 0<n<1 \\
E_F &= E + U + kT \ln \left\{ \frac{2|n-1|}{1-|n-1|} \right\} \quad 1<n<2
\end{aligned}
\right\} \quad (U>0),
\tag{3.9}
$$

$$
E_F = E - |U|/2 - \tfrac{1}{2}kT \ln \left\{ (2/n) - 1 \right\} \qquad (U<0).
\tag{3.10}
$$

These are the functions plotted in the lower half of Fig. 3.9. Alternative arguments for the pinning of E_F in the negative U situation are given in [3.54, 58].

When the above considerations are applied to the system of defect levels associated with D^-, D^0, and D^+, a rather complicated energy level scheme for states in the gap is needed. This is shown in Fig. 3.10. First consider Fig. 3.10a) and the levels labelled D^- and D^+. The location of D^- is based on the idea that the two extra electrons (additional to the lone-pair electrons) lie at an energy close to that of the lone-pairs in the normal fully bonded configuration, i.e., the valence band. However D^- is charged and can capture a hole from the valence band making the analogy with a normal shallow acceptor level fairly exact. The D^+ center has a low energy (see C_3^+ in Fig. 3.8). However, the antibonding state of this configuration lies at an energy close to that of the conduction band and the shallow donor level labelled D^+ in Fig. 3.10a denotes this state. When D^- captures a hole or when D^+ captures an electron, D^0 levels are formed and reexcitation of the carriers back to their respective bands costs more than E^- or E^+. In fact the energies are $W^- + E^-$ and $W^+ + E^+$ respectively, where W^- and W^+ are the distortion energies associated with D^0 relative to D^- and D^+ respectively. According to the Franck–Condon principle, optical excitations occur in times short compared to those of lattice relaxations and optical excitation of an electron from the valence band to D^+ costs $E_g - E^+$, where E_g is the band gap; likewise optical excitation from D^- into the conduction band costs $E_g - E^-$.

The two levels associated with D^0 can be understood as follows. The upper level corresponds to the energy required to excite an electron thermally from the valence band on to D^+ (which then becomes D^0) and the lower level B' corresponds to exciting a second electron from the valence band on to the same defect (which then becomes D^-). The position of the Fermi level can be derived from this description since the energy to create one hole in the valence band is

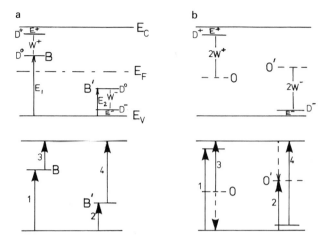

Fig. 3.10. Energy levels in the gap of a semiconductor containing D^+ and D^- defects. Thermal transitions are shown on the left and optical transitions on the right (see text).
Thermal transitions: 1. $E_1 + e_{vb} + D^+ \rightarrow D^0$, 2. $E_2 + e_{vb} + D^0 \rightarrow D^-$, 3. $(E_g - E_1) + D^0 \rightarrow D^+ + e_{cb}$, 4. $(E_g - E_2) + D^- \rightarrow D^0 + e_{cb}$.
Optical transitions: 1. $(E_1 + W^+) + e_{vb} + D^+ \rightarrow D^0$, 2. $(E_2 + W^-) + e_{vb} + D^0 \rightarrow D^-$, 3. $(E_g - E_1 + W^+) + D^0 \rightarrow D^+ + e_{cb}$, 4. $(E_g - E_2 + W^-) + D^- \rightarrow D^0 + e_{cb}$.
Downward facing arrows (dashed) correspond to radiative transitions

the average of these two energies, i.e., E_F lies between levels B and B' and, as shown above, is very effectively pinned there. An analogous argument for the creation of one electron in the conduction band is that the energy to take one electron from D^- and place it in the conduction band is $E_g - (W^- + E^-)$ but the energy to take the second electron from the same defect (which is now D^0) is $W^+ + E^+$. The energy per electron is therefore $[E_g - (W^- + E^-) + (W^+ + E^+)]/2$. This is $E_C - E_F$ and so E_F lies midway between B and B' as before.

The energy separation between B and B' is the negative U_{eff} associated with the exothermic reaction (3.8). This can be seen quite simply. Referring to Eqs. (3.6) and (3.7) and Fig. 3.10a, $E_1 = E_g - (W^+ + E^+)$ and $E_2 = (W^- + E^-)$. Thus $E_1 - E_2 = E_g - (W^+ + E^+) - (W^- + E^-) = B - B'$.

Optical transitions associated with D^0 will now be considered. These involve the two additional levels O and O' shown in Fig. 3.10b lying below B by W^+ and above B' by W^-. To excite an electron optically from the valence band on to D^0 requires more energy than to do it thermally by an amount W^-. Likewise to excite an electron optically from D^0 into the conduction band requires more energy than to do it thermally by the amount W^+. Also shown in Fig. 3.10b are the optical energies to excite an electron from the valence band onto D^+ and an electron from D^- into the conduction band. These energies, which again are larger than the corresponding thermal energies by W^+ and W^-, were referred to above. Finally in Fig. 3.10b are displayed luminescence transitions associated with D^0, corresponding to capture of an electron from the conduction band and release of an electron into the valence band (capture of a hole), respectively.

The energy difference between O' and O is the positive correlation energy associated with optical transfer of an electron from one D^0 to another, i.e., for the reaction $2D^0 \rightarrow D^+ + D^-$ *in the absence of relaxations* (see Fig. 3.7b). This can most easily be seen by performing the transfer in two stages; optical excitation from D^0 into the conduction band, followed by a downward transition from the conduction band to another D^0 (see Fig. 3.10b).

3.4.4 Current Experimental Situation for Chalcogenides

Having described in fair detail the gap states associated with various defects to be expected in Se-type materials, the current experimental situation will be discussed briefly.

The review by *Street* [3.15] explores most aspects of luminescence in chalcogenides and here we shall give mainly a summary followed by a discussion of some recent work. As was illustrated in Fig. 3.5, luminescence occurs at an energy close to midgap, the width of the emission peak being consistent with the magnitude of the Stokes shift [3.15]. A block diagram of the scheme proposed is shown in Fig. 3.11; the individual processes were outlined earlier. In principle the electronic transitions involved in the radiative transition could be either of those shown by dotted arrows in Fig. 3.10b, although in most materials the evidence is in favour of the upper of these two. Since the Fermi energy lies rather close to midgap in chalcogenides, the considerations described above suggest that the difference in energies of the two transitions is not great. Complete symmetry would result if $E^+ = E^-$ and $W^+ = W^-$ (Fig. 3.10). A model to explain such symmetry has in fact been proposed [3.59]. However, evidence from drift mobility and photoconductivity studies [3.54, 60] suggests that W^- is rather less than W^+.

Although fatigue, and the existence of photoinduced ESR and absorption are explained quite naturally on *Street*'s model in terms of the conversion of D^- (and perhaps D^+) into metastable D^0 centers, it has recently been proposed by *Kastner* and *Hudgens* [3.61] that they might be distinct processes, not closely coupled to the luminescence. In this model, fatigue occurs, not because some of the luminescence centers are "used up" during excitation but rather because nonradiative centers are created. The details of such an independent process have not, however, yet been described.

Kastner and *Hudgens* [3.61] also suggest that the luminescence centers are not charged, as in the model put forward by *Mott* et al. [3.54] and *Street* [3.15], but are in fact neutral – being the intimate valence alternation pairs referred to earlier. Experimental results on As_2Se_3 cited by these authors to be at variance with the charged-center model are:

I) A decrease in the intensity of luminescence at low excitation energies that is *not* associated with reduced absorption in the sample.

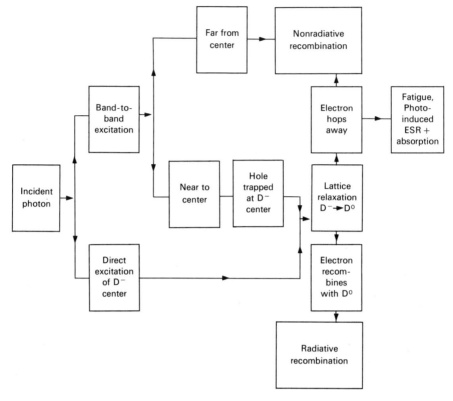

Fig. 3.11. Block diagram showing the various steps in photogeneration and recombination mechanisms for amorphous chalcogenides. In principle D^- could be exchanged for D^+ and the roles of electrons and holes reversed. For a more detailed consideration of the role of excitons and alternative paths for nonradiative recombination see [3.74]

II) An independence ($<5\%$) of the luminescence intensity on magnetic fields up to 100 kG. It is argued that if spins of the hole captured at D^- and the electron waiting to recombine with it are not correlated, then a magnetic field should decrease the rate of recombination by aligning the spins.

III) An independence ($<1\%$) of the luminescence intensity on electric fields up to 2×10^5 V cm^{-1}. One might expect that the probability for an electron to hop away from the luminescence center would be increased in a sufficiently strong field leading to a decreased intensity.

IV) A slight shift of the high-energy side of the luminescence peak to higher energies with increasing excitation intensity, suggesting energy transfer between centers.

V) The presence [3.62] of very short-lived luminescence ($\sim 10^{-8}$ s) having lower efficiency than the slower luminescence ($\sim 10^{-4}$ s).

The reader is referred to [3.61] for a discussion of these points and the case given there for believing that the luminescence centers are neutral.

Recently *Pfister* et al. [3.63] have performed an interesting set of experiments on amorphous Se to which small concentrations of Tl have been added. It was found that the hole drift mobility in such samples decreased in a manner consistent with an increase in the density of trapping centers which determine the transit time. Assuming that the trapping levels are D^- centers, then an increase in their density might be expected if the Tl atoms adopt a positive charge (Tl^+) because, by the law of mass action, the product of positively and negatively charged species should remain constant. Surprisingly, however, the luminescence, as well as the photoinduced spin density and absorption in the samples, was found to be completely independent of the Tl concentration. Although such results could be interpreted in favor of the idea of the luminescence centers being neutral (since the concentration of these would be unchanged by the addition of Tl^+), another explanation is possible. *Pfister* et al. [63] propose that each of the Tl^+ centers introduced are paired to D^- centers, these then being responsible for the decrease in the drift mobility. The density of charged defect centers and luminescence associated with them would remain unaffected. The absence, in all samples, of drift mobility signals corresponding to electron transport is also easier to understand on this model, since if the density of D^- were increased, a concomitant decrease in the density of D^+ centers would occur with the possibility of reduced trapping for electrons.

3.5 Amorphous Arsenic

3.5.1 Defects in Arsenic

As mentioned in Sect. 3.2, amorphous As and other group V elements might be expected to have properties intermediate between those of the Ge- and Se-type materials. The atomic coordination of three leads to a structure that is more flexible than Ge but less flexible than Se. In addition the electronic structure differs from either, the uppermost valence band being composed primarily of *p*-bonding electrons with lone-pair *s* electrons lying lower in energy. The question then arises as to whether the defect states in a-As exhibit positive- or negative-*U* characteristics.

For the most part, the properties of a-As parallel those of the chalcogenides. The temperature dependence of dc conductivity exhibits Arrhenius behavior with a slope approximately equal to half the optical gap. Photoluminescence [3.64, 65] occurs with emission in a band centered at close to midgap energy (Fig. 3.12) and there is a concomitant photoinduced ESR signal and optical absorption similar to those found for chalcogenides. However other properties suggest Ge-like features. Thus $T^{-1/4}$ behavior of the conductivity can be induced by application of pressure [3.66], by irradiation at low-temperatures with fast ions (N. Apsley and A. P. Troup, personal communication), or by preparing films by deposition onto substrates held at 78 K [3.10]. Furthermore the material exhibits an ESR signal in the dark and this is weakly temperature

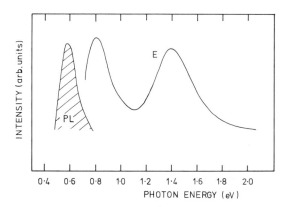

Fig. 3.12. Low-temperature photoluminescence (PL) and excitation spectra (E) for bulk amorphous As. The optical absorption edge (not shown) cuts down through the high-energy (1.4 eV) peak in the excitation spectrum as for chalcogenides (Fig. 15). The low-energy (0.8 eV) peak in the excitation spectrum is believed to correspond to direct excitation of defects. If so the Stokes shift is small ($\sim$ 0.2 eV). [3.65]

dependent [3.67]. Two other features that distinguish a-As from chalcogenides are the facts that conduction is by electrons [3.68] (negative thermopower) and that the photoluminescence excitation spectrum (at least in bulk material) has a double-peaked form [3.64, 65].

All the above properties have been incorporated into a model for defect states in amorphous arsenic described in a review article by *Greaves* et al. [3.69]. In essence the model is based on the concept of dangling bonds, many of which react to form D^+ and D^- defect states but some, because of topological constraints, do not so react but instead remain as unpaired dangling bonds. The defect configurations and energies relevant to a detailed discussion are shown in Fig. 3.13 which parallels that of Fig. 3.8 for Se. The notation of *Kastner* and *Fritzsche* [3.56] (who have also considered defects in group V materials) in labelling the As atoms P (pnictide) and using subscripts and superscripts to describe the coordination and charge state respectively, has been followed. A further subscript (either p or sp) refers to the predominant type of bonding at the defect – either p-like or sp^3-like.

The lowest energy entity is the normally bonded atom $_pP_3^0$ in which three p-like bonds are formed, the electronic energy being $-3E_p$ relative to the p level in the isolated atom. It should be emphasized at this point that pure p-like bonds require a bond angle of 90°, whereas it is known [3.69] that the average bond angle in a-As is 98° $\pm$ 10°. The degree of s–p mixing will therefore vary from site to site, in some locations being exactly p-like and in others close to sp^3 (requiring the tetrahedral angle 109.5°). A neutral twofold-coordinated atom $_pP_2^0$ has a single dangling-bond electron in a nonbonding configuration and the total electronic energy of this defect is $-2E_p$. Creation of the fourfold-coordinated defects $_{sp}P_4^0$ and $_{sp}P_4^+$ both require promotion of the two s lone-pair electrons to form sp^3 hybrids, i.e., a hybridization energy E_h. The neutral defect $_{sp}P_4^0$ has the extra electron in an antibonding orbital, the energy Δ denoting the asymmetry of the bonding–antibonding splitting. In the negatively charged center $_pP_2^-$, two electrons are in a nonbonding orbital and the energy is raised relative to $_pP_2^0$ by a positive correlation energy U_{lp}. The two other defects

Level Occupation   Configuration Energy

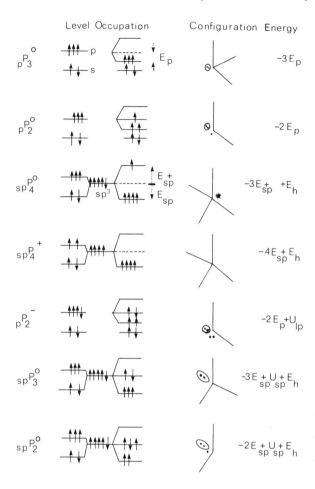

Fig. 3.13. Structure and energy of various defect configurations in a threefold-coordinated material [3.69]. Arrows represent the spin of electrons (see text)

shown are $_{sp}P_3^0$ and $_{sp}P_2^0$, i.e., threefold- and twofold-coordinated atoms with bond angles such that the bonds are sp^3 hybrids. U_{sp} is the correlation energy involved in these configurations.

The stability and possible reactions of some of the above defects will now be considered. First however it is necessary to estimate some of the energies involved. For E_p we take the value 2.5 eV, being the average energy of the p states below the Fermi level as determined by X-ray photoemission [3.70]. On a simple tight-binding picture, E_h is three-quarters of the energy separation between atomic s and p states, i.e., ~ 7.5 eV. The strength of sp^3 bonds is expected to be greater than that for p orbitals but not by a large factor; *Pauling* [3.71] gives $E_{sp}/E_p = 4/3$, but in view of the fact that the sp^3 bonds in our defects link to other atoms that have p bonds, this ratio may lie closer to unity. U_{lp} and U_{sp} are difficult to estimate but are expected to have comparable magnitudes and to be less than E_p and E_h.

The possible reaction

$$2_pP_2^0 \rightarrow {}_pP_2^- + {}_{sp}P_4^+$$

is expected to be exothermic if $E_h + U_{lp} - 2E_p - 4(E_{sp} - E_p) < 0$. If $E_{sp} - E_p$ takes the maximum value of 0.8 eV, then this will be so only if $U_{lp} < 0.7$ eV; for $E_{sp} = E_p$, the condition is certainly not satisfied. It is therefore assumed that the $_pP_2^0$ defect is most probably stable and that it is the paramagnetic center observed by ESR in the dark equilibrium state [3.67]. Nevertheless in certain configurations dissociation into the charged spinless centers might just be favored and then the weak temperature dependence of the ESR signal could then arise from the back reaction.

There is however another route to the charged centers $_pP_2^-$ and $_{sp}P_4^+$. Consider the interaction of a $_pP_2^0$ center with a neighboring atom that is forced, by topological constraints from the surrounding network, to have bond angles such that three sp^3 bonds are formed. The reaction

$$_pP_2^0 + {}_{sp}P_3^0 \rightarrow {}_pP_3^0 + {}_{sp}P_4^0$$

is exothermic if $\varDelta - E_p - U_{sp}$ is negative, which it almost certainly is. The end products of the reaction are a normally bonded atom and a fourfold-coordinated neutral center $_{sp}P_4^0$, which in turn now dissociates according to

$$2_{sp}P_4^0 \rightarrow {}_pP_2^- + {}_{sp}P_4^+ \, .$$

The reaction energy here is $2(E_{sp} - E_p) + U_{lp} - E_h - 2\varDelta$, which is negative even if the maximum value of $E_{sp} - E_p$ is used. Thus charged pairs of under- and over-coordinated centers, analogous to the D^- and D^+ centers in chalcogenides, can be created at sites where the hybridization energy has been "provided" by the network. A photograph of such a pair is shown in Fig. 3.14. Spectroscopic support for the existence of the fourfold coordinated site has been obtained by *Nemanich* et al. [3.73].

The final defect shown in Fig. 3.13, a dangling bond at an atom with two sp^3-like orbitals, is of high energy and unlikely to be created. If it were formed it would clearly dissociate into charged centers, the reaction energy being large.

3.5.2 States in the Gap of Arsenic

The energy levels in the gap of a-As arising from the defects described can be roughly located from simple considerations. Thus the $_{sp}P_4^+$ defect, being in the normal donor-like configuration is expected to lead to a shallow level just below the conduction band. The $_pP_2^-$ defect is expected to lie at an energy close to that of the atomic p states and hence near midgap. In this respect there is a distinct difference from chalcogenides (see Sect. 3.7) where the analogous D^- defect lies

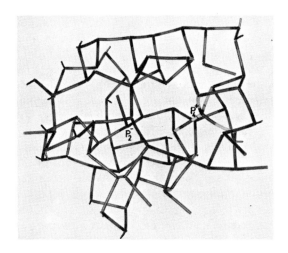

Fig. 3.14. A fourfold- and a twofold-coordinated center in a CRN model of amorphous As

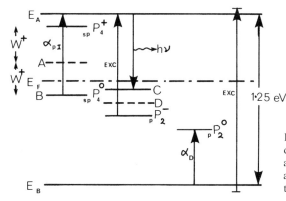

Fig. 3.15. Proposed energy level diagram for states in the gap of amorphous As. The various levels and transitions are described in the text

just above the top of the lone-pair valence band. The level associated with the stable $_pP_2^0$ defect will lie below $_pP_2^-$ by a correlation energy U_{lp}.

Figure 3.15 shows placement of the levels obtained using the above considerations and with an eye on experimental data. Apart from the $_pP_2^0$ level, the diagram is in essence the same as Fig. 3.10 but with the levels crowded towards the upper half of the gap. The Fermi level is spinned between levels A and D which makes the material n-type (as observed) with the separation $E_A - E_F$ being 0.5 eV [3.68]. The distortion energy associated with placement of an electron on $_{sp}P_4^+$ is W^+, considerably larger than that associated with placing a hole on $_pP_2^-$.

The luminescence peak at 0.55 eV (Fig. 3.12) is taken as resulting from the transition marked hv, i.e., the recombination of an electron at $_{sp}P_4^0$ after previous capture of the hole by $_pP_2^-$. This luminescence can be excited by the two transitions marked EXC. These are equivalent to the two processes envisaged for chalcogenides (Fig. 3.6) but here direct excitation of the defect

$_p\mathrm{P}_2^-$ leads to a separate peak in the excitation spectrum whereas in chalcogenides it appears only as a tail to the band-to-band excitation process. The Stokes shift is therefore only $\sim 0.2\,\mathrm{eV}$ – much less than in chalcogenides. Two additional optical absorption processes α_{pI} and α_D are marked in Fig. 3.15. These correspond to the onset of the photoinduced band at $0.6\,\mathrm{eV}$ [3.64] and a band that is reported [3.67] to accompany the dark, equilibrium ESR signal.

Although the metastable neutral center has here been identified with $_{sp}\mathrm{P}_4^0$ it seems unlikely that the defect is truly fourfold coordinated. Analysis of the shape of photoinduced ESR signal [3.12] suggests that the spin is associated with a p orbital and it therefore appears necessary to assume that one of the four orbitals is considerably more extended than the others. When the center captures an extra electron, one of the four bonds must break away to form the undistorted $_p\mathrm{P}_2^-$ center and it seems reasonable to suppose that this is the extended orbital. For chalcogenides also an asymmetric neutral center was postulated whereas *Kastner* and *Fritzsche* [3.56] favor a symmetric arrangement with the antibonding electron being shared equally between all bonds (see Sect. 3.4).

The energy level diagram of Fig. 3.15 refers to bulk material at low temperatures. In sputtered films [3.65], the lower energy peak of the photoluminescence excitation spectrum is absent and the luminescence much weaker. Perhaps such films have less opportunity to relax during formation, with a consequent preference for dangling bonds to remain in the $_p\mathrm{P}_2^0$ configuration without dissociation into the charged centers. An enhanced density of paramagnetic centers in such films (J. Knights, personal communication) supports this suggestion. Whatever the density of defect centers, one would expect the energy levels shown in Fig. 3.15 to be broadened because of a spectrum of possible bonding configurations around any particular defect – in fact the width of the luminescence band (Fig. 3.12) suggests a contribution to the broadening in addition to that expected from the electron–phonon coupling responsible for the Stokes shift.

3.5.3 Recent Calculations for Arsenic; Other Group V Materials

A few final remarks about defects in amorphous As should be made with reference to the schematic configurations shown in Fig. 3.16. Figure 3.16b shows a configuration consisting of nearest-neighbor defects $_{sp}\mathrm{P}_4^+$ and $_p\mathrm{P}_2^-$, i.e., an intimate valence-alternation pair. This defect has been considered by *Pollard* and *Joannopoulos* [3.73] as a possible candidate for photoinduced ESR. While this is possible, it should be noted that self-annihilation of the defect is quite likely. Movement of one bond (as shown), which can involve minimal atomic rearrangement, restores the fully connected threefold coordinated network. A similar self-annihilation mechanism has been suggested by *Street* [3.59, 74] for IVAPs in chalcogenides. It seems preferable, therefore, to associate the photoinduced ESR, as well as the luminescence, with well-separated charged

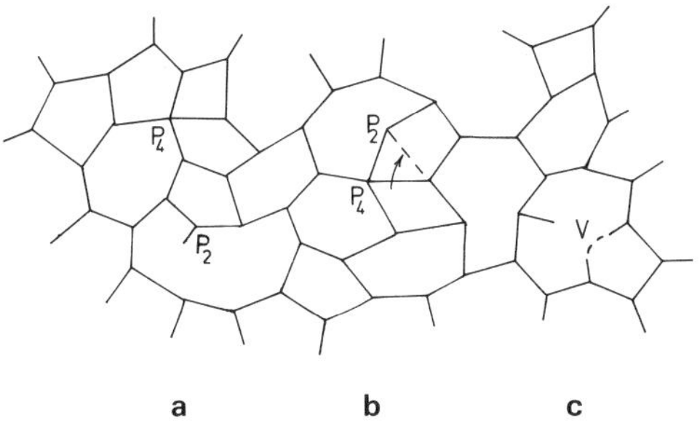

Fig. 3.16a–c. Schematic representation of defects in amorphous As. (**a**) $_{sp}P_4^0$ and $_pP_2^-$ defects (as in Fig. 3.14); (**b**) self-annihilation of intimate valence-alternation pair; (**c**) a "vacancy"

defect centers. *Pollard* and *Joannopoulos* [3.73] using a tight-binding method in fact find energy levels for the isolated twofold- and fourfold-coordinated centers which lie near midgap and in the upper half of the gap, respectively – in rough agreement with our locations. The same authors also calculate the electronic levels associated with a vacancy (Fig. 3.16c) which is the defect considered by *Taylor* et al. [3.67, 68] to be responsible for the dark ESR signal. The reason for this assignment is the similarity of the ESR signal to one that can be induced by high-energy electron bombardment [3.14]. While it is certainly plausible that vacancies can be created by bombardment, in that atoms can be knocked out into interstitial sites, the concept of a vacancy in an unbombarded film is not exactly clear cut. Amorphous films of arsenic have a density that is ~ 15 % less than the crystalline forms and the structure is therefore expected to be pervaded by voidlike caverns. Indeed it is impossible to avoid free space in constructing a threefold-coordinated CRN because of the asymmetrical nature of the bonds to each atom [3.69, 70].

How does one then define a vacancy in a CRN? Three unsatisfied orbitals pointing approximately towards the same point as in Fig. 3.16c might suffice as a definition, but then two of these orbitals will undoubtedly form a bond leaving a single orbital whose properties might then be indistinguishable from the isolated $_pP_2^0$ defect. According to *Pollard* and *Joannopoulos* [3.73] electronic states associated with the dangling bond at a vacancy lie in the upper half of the gap but their position depends on the degree of bonding at the paired orbital. Whereas *Taylor* et al. [3.67] ascribe the temperature dependence of the dark ESR signal to the breaking of this paired orbital, *Pollard* and *Joannopoulos* [3.73] suggest that the vacancy captures an electron (forming a center like V $^-$ in Si) and that this electron can make transitions between states associated with the dangling bond and the paired orbital.

States associated with defects in amorphous P and Sb have been described by *Elliott* and *Davis* [3.75]. The average bond angle in amorphous films of these materials are 102° and 96°, respectively, compared with 98° for amorphous As. Thus, relative to As, one might expect a higher degree of sp hybridization in the bonds in P and a smaller degree in Sb. In discussing defects in As, it was seen that the hybridization energy was an important factor in deciding whether neutral paramagnetic centers were stable or not. Following the same arguments, one might expect fewer paramagnetic centers in P and more in Sb, relative to the density in As. Conversely the density of charged defects and photoinduced spins as well as the efficiency of luminescence are predicted to be larger in P than in As or Sb. In essence the properties of a-P should be similar to those of the Se class of materials and the properties of a-Sb more like those of a-Ge. Further experimental work will be necessary to decide how far these predictions are correct and whether the group V elements, taken together, then provide a complete link between the two classes of material described in an earlier section of this article.

References

3.1   N. F. Mott: Adv. Phys. **16**, 49 (1967)
3.2   A. I. Gubanov: *Quantum Theory of Amorphous Conductors* (Consultants Bureau, New York 1965)
3.3   M. H. Cohen, H. Fritzsche, S. R. Ovshinsky: Phys. Rev. Lett. **22**, 1065 (1969)
3.4   E. A. Davis, N. F. Mott: Philos. Mag. **22**, 903 (1970)
3.5   N. F. Mott: J. Non-Cryst. Solids **8–10**, 1 (1972)
3.6   J. M. Marshall, A. E. Owen: Philos. Mag. **24**, 1281 (1971)
3.7   J. C. Phillips: Comments Solid State Phys. **4**, 9 (1971)
3.8   P. W. Anderson: Phys. Rev. Lett. **34**, 953 (1975)
3.9   S. R. Ovshinsky: Phys. Rev. Lett. **36**, 1469 (1976)
3.10 J. J. Hauser, F. J. Di Salvo, R. S. Hutton: Philos. Mag. **35**, 1557 (1977)
3.11 S. G. Bishop, U. Strom, P. C. Taylor: Phys. Rev. Lett. **34**, 1346 (1975); **36**, 543 (1976)
3.12 S. G. Bishop, U. Strom, P. C. Taylor: Phys. Rev. B **15**, 2278 (1977)
3.13 P. C. Taylor, U. Strom, S. G. Bishop: Philos. Mag. B **37**, 241 (1978)
3.14 P. C. Taylor, U. Strom, S. G. Bishop: Phys. Rev. B **18**, 511 (1978)
3.15 R. A. Street: Adv. Phys. **25**, 397 (1976)
3.16 D. Engemann, R. Fischer, H. Mell: In Proc. 7th Int. Conf. Amorphous and Liquid Semiconductors, ed. by W. E. Spear (Centre for Industrial Consultance and Liaison, University of Edinburgh 1977) p. 387
3.17 R. A. Street: Philos. Mag. B **37**, 35 (1978)
3.18 H. Mell: In Proc. 5th Int. Conf. Amorphous and Liquid Semiconductors, ed. by J. Stuke, W. Brenig (Taylor and Francis, London 1974) p. 203
3.19 W. Beyer, J. Stuke: In Proc. 5th Int. Conf. Amorphous and Liquid Semiconductors, ed. by J. Stuke, W. Brenig (Taylor and Francis, London 1974) p. 251
3.20 M. L. Knotek, M. Pollak, T. M. Donovan, H. Kurtzmann: Phys. Rev. Lett. **30**, 853 (1973)
      M. L. Knotek, M. Pollak, T. M. Donovan: In Proc. 5th Int. Conf. Amorphous and Liquid Semiconductors, ed. by J. Stuke, W. Brenig (Taylor and Francis, London 1974) p. 225
3.21 M. L. Knotek: Solid State Commun. **17**, 1431 (1975)
3.22 N. Apsley, E. A. Davis, A. P. Troup, A. D. Yoffe: Proc. 7th Int. Conf. Amorphous and Liquid Semiconductors, ed. by W. E. Spear (Centre for Industrial Consultancy and Liaison, University of Edinburgh 1977) p. 447
3.23 N. Apsley, H. P. Hughes: Philos. Mag. **30**, 963 (1974); **31**, 1327 (1975)
3.24 J. Stuke: "Structure and Properties of Non-Crystalline Semiconductors," in Proc. 6th Int. Conf. Amorphous and Liquid Semiconductors, ed. by B. T. Kolomiets, (Academy of Sciences of USSR, Nauka 1976) p. 193
3.25 M. Abkowitz, P. G. LeComber, W. E. Spear: Commun. Phys. **1**, 175 (1976)
3.26 J. Stuke: In Proc. 7th Int. Conf. Amorphous and Liquid Semiconductors, ed. by W. E. Spear (Centre for Industrial Consultancy and Liaison, University of Edinburgh 1977) p. 406

3.27 G.A.N.Connell, J.R.Pawlik: Phys. Rev. B **13**, 787 (1976)
3.28 I.R.Pawlik, G.A.N.Connell, D.Prober: "Electronic Phenomena in Non-Crystalline Semiconductors," in Proc. 6th Int. Conf. Amorphous and Liquid Semiconductors, ed. by B.T.Kolomiets (Academy of Sciences of USSR, Nauka 1976) p. 304
3.29 J.I.Pankove, M.A.Lampert, M.L.Tarng: Appl. Phys. Lett. **32**, 439 (1978)
      D.Kaplan, N.Sol, G.Velasco, P.A.Thomas: Appl. Phys. Lett. **33**, 440 (1978)
3.30 W.E.Spear: In Proc. 5th Int. Conf. Amorphous and Liquid Semiconductors, ed. by J.Stuke, W.Brenig (Taylor and Francis, London 1974) p. 1
3.31 W.E.Spear, P.G.LeComber: Philos. Mag. **33**, 935 (1976)
3.32 G.D.Watkins, J.W.Corbett: Phys. Rev. A, **138**, 543 (1965)
3.33 T.D.Moustakas, D.A.Anderson, W.Paul: Solid State Commun. **23**, 155 (1977)
3.34 W.Y.Ching, D.J.Lam, C.C.Lin: Phys. Rev. Lett. **42**, 805 (1979)
3.35 R.Jones: Philos. Mag. **35**, 57 (1977)
3.36 R.Jones: Philos. Mag. **36**, 677 (1977)
3.37 H.Singh, M.H.Cohen: Bull. Am. Phys. Soc. **23**, 247 (1978)
3.38 D.Engemann, R.Fischer: In Proc. 5th Int. Conf. Amorphous and Liquid Semiconductors, ed. by J.Stuke, W.Brenig (Taylor and Francis, London 1974) p. 947
3.39 J.I.Pankove, D.E.Carlson: In Proc. 7th Int. Conf. Amorphous and Liquid Semiconductors, ed. by W.E.Spear (Centre for Industrial Consultancy and Liaison, University of Edinburgh 1977) p. 402
3.40 J.C.Knights, D.K.Biegelson, I.Solomon: Solid State Commun. **22**, 133 (1977)
3.41 D.K.Biegelson, J.C.Knights, R.A.Street, C.Tsang, R.M.White: Philos. Mag. B **37**, 477 (1978)
3.42 D.Kaplan: In Proc. 14th Int. Conf. Physics of Semiconductors, ed. by B.L.H.Wilson (Institute of Physics, London 1979)
3.43 D.K.Biegelson, J.C.Knights: In Proc. 7th Int. Conf. Amorphous and Liquid Semiconductors, ed. by W.E.Spear (Centre for Industrial Consultancy and Liaison, University of Edinburgh 1977) p. 429
3.44 R.Fisch, D.C.Licciardello: Phys. Rev. Lett. **41**, 889 (1978)
3.45 S.R.Elliott: Philos. Mag. B. **38**, 325 (1978)
3.46 R.A.Street, I.G.Austin, R.M.Searle, B.A.Smith: J. Phys. C. **7**, 4185 (1974)
3.47 B.T.Kolomiets, T.N.Mamontova, V.V.Negreskul: Phys. Status Solidi **27**, K 15 (1968)
      B.T.Kolomiets, T.N.Mamontova, A.A.Babaev: J. Non-cryst. Solids **4**, 289 (1970); **8–10**, 1004 (1972)
3.48 R.Fischer, V.Heim, F.Stern, K.Wieser: Phys. Rev. Lett. **26**, 1182 (1971)
      K.Weiser: J. Non-cryst. Solids **8–10**, 922 (1972)
3.49 R.A.Street, T.M.Searle, I.G.Austin: In Proc. 5th Int. Conf. Amorphous and Liquid Semiconductors, ed. by J.Stuke, W.Brenig (Taylor and Francis, London 1974) p. 953
3.50 F.Mollot, J.Cernogora, C.Benoit à la Guillaume: Phys. Status Solidi A **21**, 281 (1974)
      J.Cernogora, F.Mollot, C.Benoit à la Guillaume: Phys. Status Solidi A **15**, 401 (1973)
3.51 R.A.Street, T.M.Searle, I.G.Austin: J. Phys. C. **6**, 1830 (1973)
3.52 R.A.Street: Philos. Mag. **32**, 431 (1975)
3.53 R.A.Street, N.F.Mott: Phys. Rev. Lett. **35**, 1293 (1975)
3.54 N.F.Mott, E.A.Davis, R.A.Street: Philos. Mag. **32**, 961 (1975)
3.55 M.Kastner, D.Adler, H.Fritzsche: Phys. Rev. Lett. **37**, 1504 (1976)
3.56 M.Kastner, H.Fritzsche: Philos. Mag. **37**, 199 (1978)
3.57 D.Adler, E.J.Yoffa: Phys. Rev. Lett. **36**, 1197 (1976)
3.58 N.F.Mott: Philos. Mag. **34**, 1101 (1976)
3.59 R.A.Street: In Proc. 14th Int. Conf. Physics of Semiconductors, ed. by B.L.H.Wilson (Institute of Physics, London 1979)
3.60 N.F.Mott, R.A.Street: Philos. Mag. **36**, 33 (1977)
3.61 M.Kastner, S.J.Hudgens: Philos. Mag. B. **37**, 665 (1978)
3.62 K.Murayama, T.Ninomiya, H.Suzuki, K.Morigaki: Solid State Commun. **24**, 197 (1977)
3.63 G.Pfister, K.Liang, M.Morgan, P.C.Taylor, E.J.Friebele, S.G.Bishop: Phys. Rev. Lett. **41**, 1318 (1978)
3.64 S.G.Bishop, U.Strom, P.C.Taylor: Solid State Commun. **18**, 573 (1976)

3.65 P.B.Kirby, E.A.Davis: In Proc. 14th Int. Conf. Physics of Semiconductors, ed. by B.L.H.Wilson (Institute of Physics, London 1979)

3.66 S.R.Elliott, E.A.Davis, G.D.Pitt: Solid State Commun. **22**, 481 (1977)

3.67 P.C.Taylor, E.J.Friebele, S.G.Bishop: Solid State Commun. **28**, 247 (1978)

3.68 E.Mytileneou, E.A.Davis: In Proc. 7th Int. Conf. Amorphous Liquid Semiconductors, ed. by W.E.Spear (Centre for Industrial Consultancy and Liaison, University of Edinburgh, 1977) p. 632

3.69 G.N.Greaves, S.R.Elliott, E.A.Davis: Adv. Phys., **28**, 49 (1979)

3.70 L.Ley, R.A.Pollak, S.P.Kowalczyk, R.McFeely, D.A.Shirley: Phys. Rev. B **8**, 641 (1973)

3.71 L.Pauling: *The Nature of the Chemical Bond* (Cornell University Press, Ithaca, N.Y. 1960)

3.72 R.J.Nemanich, G.Lucovsky, W.B.Pollard, J.D.Joannopoulos: Solid State Commun. **26**, 137 (1978)

3.73 W.B.Pollard, J.D.Joannopoulos: Phys. Rev. B **19**, 4217 (1979)

3.74 R.A.Street: Phys. Rev. B **17**, 3984 (1978)

3.75 S.R.Elliott, E.A.Davis: J. Phys. C **12**, 2577 (1979)

4. Optical Properties of Amorphous Semiconductors

G. A. N. Connell

With 29 Figures

The optical properties of amorphous semiconductors are presented in three parts. In Sect. 4.1, a general basis for interpreting optical spectra is developed qualitatively from the structural characteristics of fully coordinated amorphous materials and some parallels with the interpretation of crystalline spectra are indicated. This discussion is then put in a more quantitative form within the random phase approximation. The modifications of these results that are induced by defects, in particular by voids in tetrahedral semiconductors and by coordination variation in chalcogenides, are also reviewed and numerous features that are unique to the amorphous state become apparent here. The outcome of this discussion is that the optical absorption in amorphous semiconductors can be separated into three regions, shown schematically in Fig. 4.1. Regions B and C are created by transitions within the fully coordinated system, perturbed to some extent by defects, while region A arises from transitions involving the defect states directly. The absorption edge, spanning regions A and B, is therefore particularly complicated and has a defect-induced tail at the lowest energies, an exponential region at intermediate energies, and a power law region at the highest energies. In Sect. 4.2, the focus is placed on the experimental results for the major classes of amorphous semiconductors and attempts are made to establish the degree to which these can be understood within the general framework. Not surprisingly, the complexity of the absorption edge, just described, is mirrored by the inability in many materials to establish a unique explanation of its origin. Finally, in Sect. 4.3, some of the outstanding problems, most of which are of major current interest, are summarized.

4.1 General Features of Optical Excitation in Amorphous Semiconductors

The optical properties of amorphous and crystalline semiconductors may be deduced almost entirely from the general one-electron expression for the imaginary part of the dielectric constant [4.1],

$$\varepsilon_2(\omega) = \frac{2}{V}\left(\frac{2\pi e}{m\omega}\right)^2 \sum_i \sum_f |\langle f|P|i\rangle|^2 \delta(E_f - E_i - \hbar\omega) \tag{4.1}$$

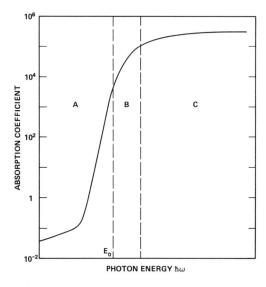

Fig. 4.1. Schematic representation of the absorption spectrum of amorphous semiconductors. The regions A, B, and C are discussed in the text

where V is the sample volume, P is the momentum operator, and the sum is over all initial valence, i, and final conduction, f, band states separated by an energy $\hbar\omega$. E_f and E_i are the eigenenergies of the eigenstates $|f\rangle$ and $|i\rangle$ respectively. It is assumed throughout this section that the sum is averaged over all angles of polarization. In a crystal, the behavior of $\varepsilon_2(\omega)$ is set by the requirement of momentum conservation in optical transitions and by the form of the electronic density of states. The first occurs because the translational symmetry of the crystal only permits nonzero momentum matrix elements when the wavevectors of the initial and final states of the transition are identical. The second then arises because for such vertical transitions, the momentum matrix elements vary slowly with wave vector $\boldsymbol{k}$. Thus

$$\varepsilon_2(\omega) = \frac{2}{V}\left(\frac{2\pi e}{m\omega}\right)^2 |P_c(\omega)|^2 \sum_i \sum_f \delta[E_f(\boldsymbol{k}) - E_i(\boldsymbol{k}) - \hbar\omega] \tag{4.2a}$$

$$= 2\left(\frac{2\pi e}{m\omega}\right)^2 |P_c(\omega)|^2 \frac{1}{(2\pi)^3} \int_{Br} d^3k \cdot \delta[E_f(\boldsymbol{k}) - E_i(\boldsymbol{k}) - \hbar\omega] \tag{4.2b}$$

where $|P_c(\omega)|^2$ is independent of $\boldsymbol{k}$ but varies with $\hbar\omega$. The integral in (4.2b) is taken over the Brillouin zone, and is called the joint density of states for interband transitions. It shows strong variations as a function of ω in the vicinity of critical points at which $V[E_i(\boldsymbol{k}) - E_f(\boldsymbol{k})] = 0$ and gives rise to the characteristic features of the optical spectra of crystals. In the amorphous phase, the structure again determines the behavior of the momentum matrix elements which in turn specify the nature of the transitions between filled and empty states. On the whole, the development parallels the crystalline case, although a number of important and unique situations arise, particularly in chalcogenide glasses, when the one-electron approximation is insufficient. The details are set out in the following sections.

4.1.1 The Role of Atomic Structure in Establishing Optical Properties

The structures of amorphous semiconductors are developed by the repetition of one or more basic molecular units, in a way that cannot be identified topologically with any known crystalline structure or indeed with any infinite periodic array. Thus for some materials, the atomic order within a molecular unit might be similar within small bond angle distortions in both the crystalline and amorphous phases, yet quite different within large volumes. For other materials, the structural differences between the crystalline and amorphous phases can involve the nature of the molecular units themselves. As one example, the short-range order as specified by the distribution and number of nearest neighbors may differ from one site to another in the amorphism, although it is fixed in the crystal. Such coordination variation is common in chalcogenide glasses, for instance [4.2]. As a second example, the short-range order as specified by the types of nearest neighbors may differ in the amorphous and crystalline phases. This commonly happens when the crystalline phase is chemically ordered and the amorphous phase is chemically disordered, for instance in SiC [4.3]. In some cases, and arsenic is the best-known example, the difference between the molecular units is less easily deciphered and yet the difference is crucial in establishing the semiconducting properties of the amorphous phase. In the particular example mentioned, arsenic has three one-bond neighbors at 2.5 Å in both phases, but mesomeric bonding between the atomic layers of the crystal creates three neighbors at 3.15 Å that are absent from the amorphism [4.4]. Therefore, the molecular units of the two phases cannot be considered as identical, a point that is well reflected by the optical properties.

It is a common practice to distinguish between a fully coordinated amorphous state and defective amorphous states. The latter occur in numerous ways. For example, although amorphous silicon, like crystalline silicon, displays fourfold coordination, most preparation techniques give material containing relatively large numbers of voids [4.5]. Other examples are the possible occurrence of like bonds in a chemically ordered system like amorphous GaAs [4.6] or the occasional occurrence of coordination variation in amorphous Se, when two Se atoms, each typically twofold coordinated, form instead one positively charged threefold coordinated atom and one negatively charged singly coordinated atom [4.2].

How then are these structural features and the optical properties to be related? In fully coordinated systems, the order within the molecular units is responsible in large part for establishing the electronic band widths, which explains the early observation of the similarity of the optical band gaps of many materials in their amorphous and crystalline phases [4.7]. Conversely, the large difference between the band gaps, of the arsenic phases, for example, in which the molecular units are different, can also be rationalized. The topological disorder is responsible for establishing the major features in the density of states [4.8]. Some features are shown to be independent of topology and these are observed in the density of states of crystalline and amorphous phases that are related by their molecular units. Others depend dramatically on ring statistics and these should lead in principle to differences in the optical spectra of the two phases. Such changes however are heavily masked because in amorphous semiconductors, even at energies well separated from the band edges, the atomic potential is strong enough that even slight distortions of the nearest-neighbor bond length and bond angle distributions, within otherwise identical molecular units, lead to strong electronic scattering and a short coherence length of the wave functions. Indeed, the coherence length is of order the lattice spacing and the resulting uncertainty in the wave vector is of order the wave vector itself. Under these circumstances, the momentum selection rule breaks down in optical transitions, creating perhaps the single most important difference between the responses of the crystalline and amorphous phases. Certainly, it leads to many of the characteristic optical properties that are described later.

Near the band edges, two additional effects occur. The first arises because fluctuations in the atomic configurations away from the average are necessarily accompanied by fluctuations in the potential acting on an electron. When the potential fluctuations are sufficiently deep or wide, the energies of the electronic states are perturbed and the band broadens. The associated energy levels depend of course on the precise details of the fluctuations. Nevertheless, the states furthest from the center of the band are associated with the widest or deepest, and thus least likely, fluctuations. The result is a band tail. The symmetry properties of the electrons and of the fluctuations of the local order are both important in determining the magnitude of the effect on the density of states. For example, the energies of s states are less sensitive to fluctuations of the local coordination that leave nearest-neighbor distances fixed but vary the orientation of nearest-neighbor atoms than are the energies of p or d states. Thus, the effect of disorder is expected to be different for the valence and conduction band edges of almost any amorphous semiconductor. The probability of optical transitions between filled and empty tail states is limited in large extent by the second effect: disorder-induced or Anderson localization of the electron states near the band edges [4.9, 10]. Transitions are therefore allowed only to the extent that there is spatial overlap of the localized wavefunctions. In contrast, transitions between localized and delocalized states beyond the mobility edges are strongly allowed.

The foregoing outline of the optical properties of fully coordinated amorphous semiconductors is modified when defective materials are considered. In situations in which voids are present, numerous effects can occur. First, the material is microscopically inhomogeneous. Light may therefore be considered to propagate by multiple scattering and in cases where the concentration of voids is smaller than the percolation threshold and the dimensions of the voids are considerably smaller than the wavelength, the Maxwell–Garnett theory should be a reasonable first approximation [4.11]. Two other effects occur, however, both of which are associated with solid-state effects at the surfaces of the voids. By analogy with the effects at surfaces of crystalline material, atomic relaxation at void surfaces is expected to produce changes both in bonding strength between atoms in close proximity to the surfaces and concomitant potential fluctuations. Parenthetically, the length scale of the latter is the intervoid separation and is clearly different from the length scale of the potential fluctuations caused by the quantitative disorder that is present in the fully coordinated system. These two surface-related effects then lead to modifications of the density of states in both the gap and the bands and to changes in the matrix elements for transitions between filled and empty states near the mobility edges. In situations in which coordination variation occurs, and this is most in-vestigated in the chalcogenide glasses, there is often an interplay between the structural configuration and the state of optical excitation. For example, highly absorbed photons can induce structural changes and attendant optical changes that can later be reversed by thermal annealing. In other examples, coordination variation may simply lead to optical band gaps that are very different from those expected from the crystalline phases. All of these phenomena will be examined in greater detail in the following sections.

4.1.2 Optical Properties in Fully Coordinated Material

a) Microscopic Formulation

Many of the results in this section were first derived by *Tauc* et al. [4.12], although the approach used here is based on *Hindley's* [4.13] work. It is assumed that the wave functions $|i\rangle$ and $|f\rangle$ of the valence and conduction bands respectively can be expanded in terms of a set of orthonormal, localized functions $|nv\rangle$ and $|nc\rangle$, centered on different atoms n. Thus

$$|i\rangle = \sum_n a_{inv}|nv\rangle \tag{4.3a}$$

$$|f\rangle = \sum_n a_{fnc}|nc\rangle \tag{4.3b}$$

In a crystal, on the one hand, the a's are plane waves and $|nv\rangle$ and $|nc\rangle$ are Wannier functions. In an amorphism, on the other hand, the phases of the a's are randomized because the uncertainty of the wave vector is of the order of the wave vector itself. Furthermore, the amplitudes of the a's may vary from atom to atom, even when $|i\rangle$ and $|f\rangle$ are delocalized, although they are presumably of the same order of magnitude in this case. When the $|i\rangle$ and $|f\rangle$ are localized, of course, one of the a's will dominate over any of the others. These are the basic assumptions of the random phase approximation and they allow the optical properties to be evaluated. It should be emphasized at the outset that the approach is not exact in view of the persistence of short-range order, at least within the molecular units, but it does provide a valuable insight into and first approximation to the origin of the optical properties of amorphous semiconductors.

The momentum matrix element, averaged over an ensemble at random systems, can now be evaluated for delocalized to delocalized transitions as

$$\langle |\langle f|P|i\rangle|^2\rangle_{\text{ensemble}} = (a^3/V)\left\langle \sum_{n'} |\langle n'c|P|nv\rangle|^2\right\rangle_{\text{ensemble}}, \tag{4.4}$$

where a is the interatomic spacing. When $\langle nc|P|nv\rangle \neq 0$, this term will dominate the sum, and

$$\langle |\langle f|P|i\rangle|^2\rangle_{\text{ensemble}} = (a^3/V)|\langle nc|P|nv\rangle|^2| \equiv (a^3/V) P_{\text{cv}}^2 \equiv |P_{\text{am}}(\omega)|^2 \tag{4.5}$$

is obtained where P_{cv} is of order $\hbar/a$. Substitution in (4.1) thus gives $\varepsilon_2(\omega)$ for an amorphous system,

$$\varepsilon_2(\omega) = \frac{2}{V}\left(\frac{2\pi e}{m\omega}\right)^2 |P_{\text{am}}(\omega)|^2 \sum_i \sum_f \delta(E_f - E_i - \hbar\omega) \tag{4.6a}$$

$$= 2\left(\frac{2\pi e}{m\omega}\right)^2 a^3 P_{\text{cv}}^2 \int_0^{\hbar\omega} dE \, g_i(-E) g_f(\hbar\omega - E), \tag{4.6b}$$

when the one-electron density of states

$$g(E) = (1/V)\sum_n \delta(E - E_n) \tag{4.7}$$

is introduced.

It is instructive to compare the expressions for $\varepsilon_2(\omega)$ in the amorphous and crystalline cases, given in (4.6) and (4.2), respectively. Clearly, energy is conserved in the transitions in both cases but the critical points associated with $\mathbf{k}$ conservation in the crystalline case are absent in the amorphous case. In fact the joint density of states is replaced by a convolution of the density of states in the valence and conduction bands for which energy is conserved. An interesting feature of both results is that the average momentum matrix element is taken to depend only on energy and it might be guessed that they are related in amorphous and crystalline phases built from the same molecular units. This may in fact be demonstrated in the approximation that the Wannier functions of the crystalline state may be identified with the localized functions $|nv\rangle$ and $|nc\rangle$ of the amorphous state. In this case

$$|P_{\text{am}}(\omega)|^2 \approx \langle |P_{\text{cr}}(\omega)|^2\rangle \tag{4.8a}$$

where the average is for all $E_i(\mathbf{k})$ and $E_f(\mathbf{k})$ such that $E_i(\mathbf{k}) - E_f(\mathbf{k}) = \hbar\omega$. The averaging has the effect that the enhancement of transitions by umklapp processes in the crystal is not present in $|P_{am}(\omega)|$ [4.14]. Of course, large discrepancies must be expected when indirect edges are involved in the crystal.

Transitions between localized states in the band tails and delocalized states are also allowed within the random phase approximation with the same matrix element, given in (7.5), as for transitions between delocalized states. This intuitively odd result occurs because normalization requires that the amplitude of a delocalized state be smaller than that of a localized state by a factor of N, where N is the number of atoms. Thus for localized-to-delocalized transitions,

$$|P_{am}(\omega)|^2 = (1/N)\, P_{cv}^2 = (a^3/V)\, P_{cv}^2 \qquad (4.8b)$$

as before for delocalized to delocalized transitions, and no break will appear in $\varepsilon_2(\omega)$ at the mobility gap. This result of course must be treated with some caution, since electron correlation and electron–phonon coupling are likely to become important in localized states and the validity of the random phase approximation under these conditions must be questioned.

b) Excitation Above the Absorption Edge

Theoretical calculations tend to suggest that the conduction band density of states of amorphous semiconductors should be smooth and featureless [4.15]. This has not been confirmed by direct measurement in many cases, but it seems reasonable nevertheless to take the conduction band density of states as a step $g_f(E_1)$ at an energy E_1 above the valence band mobility edge. With this approximation, an expression relating the valence band density of states to $\varepsilon_2(\omega)$ may readily be derived from (4.6):

$$g_i(E_1 - \hbar\omega) = \text{const} \cdot d[\omega^2 \varepsilon_2(\omega)/|P_{am}(\omega)|^2]/d\omega . \qquad (4.9)$$

As explained earlier, $|P_{am}(\omega)|^2$ may be approximated by $\langle|P_{cr}(\omega)|^2\rangle$ if it is known, or, with less accuracy and only for energies above the *Penn* gap, $\hbar\omega_g$, by the matrix element of the *Penn* model of a semiconductor [4.16],

$$|P_{Penn}(\omega)|^2 = \hbar k_f (\omega_g/\omega)^2 , \qquad (4.10)$$

where k_f is the magnitude of the wave vector at the Fermi level [4.17]. This method, it must be emphasized, cannot give an exact representation of the valence band density of states, but like photoemission, it will exhibit the major features. To date, it has proved valuable in understanding the optical properties of amorphous tetrahedral and chalcogenide semiconductors.

c) The Absorption Edge

Power Law Absorption Region

It is normally assumed that the densities of states just beyond the mobility edges can be represented by power laws:

$$g_i(-E) = \text{const} \cdot E^p, \tag{4.11a}$$

$$g_f(E) = \text{const} \cdot (E - E_0)^q, \tag{4.11b}$$

where energies are measured from the valence band mobility edge, and the conduction band mobility edge is at E_0. Then using (4.6),

$$\omega^2 \varepsilon_2(\omega) = \text{const} \cdot (\hbar\omega - E_0)^{p+q+1}. \tag{4.12}$$

When the band edges are both parabolic, this reduces to

$$\omega^2 \varepsilon_2(\omega) = \text{const} \cdot (\hbar\omega - E_0)^2, \tag{4.13}$$

a result that is obtained in many materials.

This interpretation of the origin of power law absorption and the meaning of E_0 must be treated with caution. Since localized-to-delocalized transitions are also permitted in the random phase approximation, absorption with the form of (4.13) also arises when transitions occur, for example, from the mobility edge at the top of the valence band to a conduction band tail of the form $g_f(E) \alpha (E - E_0)$, as was argued by *Davis* and *Mott* [4.18]. In this case, E_0 is the energy between the mobility edge in the valence band and the edge of the conduction band tail. In principle these opposing explanations can be separated by considerations of the magnitude of the absorption and of the energy range over which the behavior is seen, but even then, the possibility that the densities of states have other than parabolic form has not been disproved. Nevertheless, the experimental measurement of E_0 provides a useful yardstick with which to compare different materials.

Exponential Absorption Region

At photon energies below E_0, the absorption in almost all materials varies exponentially with energy as

$$\alpha(\omega) = (\omega/nc)\varepsilon_2(\omega) = \text{const} \cdot \exp[-\beta(E_e - \hbar\omega)] \tag{4.14}$$

where n is the refractive index, and E_e is an energy of order E_0. The values of β at 300 K range from about 10–25 eV^{-1}, and are constant up to near the glass transition temperature T_g. The absorption is thus reminiscent of the exponential edge first observed by *Urbach* [4.19] in alkali halides although in those materials β is temperature dependent near 300 K and given by $\beta \approx 0.8/kT$. It has been

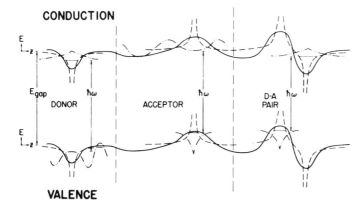

Fig. 4.2. Valence and conduction band edges as a function of position (solid line), illustrating regions of effective donor, effective acceptor, and effective donor-acceptor pairs in the potential fluctuations. The effective impurity potentials are represented by dashed lines, and wave functions of typical states by dashed-dot lines. Note that the electron wave functions are exponentially attenuated in the regions where the hole wavefunctions are large and vice versa [4.21]

suggested that β in amorphous semiconductors corresponds to a high effective temperature of 400–700 K, related to T_g. Certainly, in the small number of cases investigated, β approaches $1/kT$ at temperatures above T_g [4.20]. Thus, the implication is that the phenomenon or phenomena of interest are frozen-in at T_g.

Despite this universal behavior, the precise origin of the exponential absorption in any particular material is uncertain. This happens because the behavior of transitions involving tail states and the behavior of a small part of the oscillator strength of the dominant transitions above E_0 need to be known. The many facets of the problem will become clear in what follows.

The relation between the absorption and the density of states when transitions occur between localized and delocalized states was discussed in Sect. 4.12a. Thus from (4.6), it is evident that an exponential edge arises when the density of states in the more extensive of the tails has the form

$$g(E) = \text{const} \cdot \exp(\beta E). \tag{4.15}$$

It is most unlikely however that this explanation, requiring as it does a relatively invariant value for β, could be the explanation in all materials.

Another suggestion for the origin of the phenomenon is based on the effects of the disorder induced potential fluctuations sketched in Fig. 4.2. In principle, both symmetric (or electrostatic) and antisymmetric (or elastic) fluctuations occur, but small angle X-ray scattering measurements indicate that the latter are small and cannot lead to the observed magnitude of the exponential tailing. The former may be modeled by effective donors and acceptors [4.21] for which the electron wave functions are exponentially attenuated in the regions where the hole wave functions are large and vice versa. Thus, the lowest energy optical transitions between valence states near the top of potential maxima and conduction states in potential minima at effective donor–acceptor pairs have tiny oscillator strength because of the spatial separation of the highly localized initial and final states. Clearly the oscillator strength increases as the energy of the transitions grows until the transition involves the delocalized state at one of the mobility edges. Hence, the spatial correlations produce an energy-dependent matrix element. *Tauc* [4.22] gives an estimate of the effect by assuming that the amplitudes of the localized wave functions decrease as $\exp(-\text{const}\cdot r)$ at distance r from the center. Then the transition probability will depend on the distance R between two centers as $\exp(-\text{const}\cdot R)$. Of course, the energy of such a transition differs from E_0 by

$$\Delta E = eFR \tag{4.16}$$

where F is the average electric field between the centers. Therefore, the matrix element just below E_0 is dominated by a factor $\exp[-\text{const}\cdot(E_0-\hbar\omega)/F]$. If the distribution of F decreases fast enough about the average value F_{av}, then F may be assumed equal to F_{av}. Hence, the spatial correlations cause the absorption tail to be exponentially smaller than the energy dependence expected from the convolution of the densities of tail states in (4.6) and as such the effect is independent of the exact form of them. β is therefore proportional to F_{av} and the universal behavior of the absorption depends on a universal origin for the potential fluctuations.

Dow and *Redfield* [4.23] suggest that the effect of the potential fluctuations on the center of mass and the relative motions of an exciton can also give rise to an exponential absorption edge. Their calculations indicate that in uniform electric fields F, which are strong enough to shift the exciton energy by an amount equal to the binding energy, the Stark broadening is much larger than the Stark shift and the absorption is given by

$$\alpha(\hbar\omega, F) = \text{const}\cdot\exp[-\text{const}\cdot(E_e-\hbar\omega)/F], \tag{4.17}$$

where E_e is now the exciton energy in zero field and differs from E_0 by the exciton binding energy. The potential fluctuations are then assumed to lead to a distribution of microfields $P(F)$ and hence the absorption in an amorphous semiconductor can be obtained numerically from

$$\alpha(\hbar\omega) = \int_0^\infty P(F)\alpha(\hbar\omega, F)\,dF. \tag{4.18}$$

When a Gaussian distribution of electric fields is used, the exponential dependence of $\alpha(\hbar\omega, F)$ survives the averaging, at least over experimentally relevant regions of the spectrum, with β proportional to F_{av} in the expression for $\alpha(\hbar\omega)$. The persistence of the exponential form after averaging is an important feature of their work since it indicates that the exponential shape is somewhat insensitive to the details of the microfield distribution. Therefore, the universal

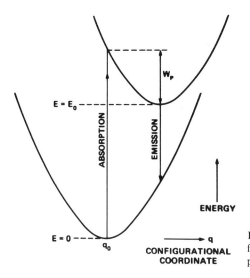

Fig. 4.3. Configurational coordinate diagram for a system exhibiting strong electron-phonon interaction [4.24]

behavior of the absorption must again be sought in the origin of the potential fluctuations but in this case there appear to be fewer restrictions on their exact form.

Tauc [4.22] has made an attractive suggestion for the origin of potential fluctuations. He considers the disorder induced fluctuations as deriving from longitudinal optical phonons, frozen-in at the glass transition temperature. In that case, the "frozen-in" phonons have the Gaussian distribution typical of T_g. The electric fields so produced are of order $10^6 \, \mathrm{V \cdot cm^{-1}}$ and appear to be sufficient to explain the observed values of β in compound amorphous semiconductors. The explanation also accounts neatly for the temperature dependence of β above T_g. Of course, it cannot account for the exponential edges of elemental materials in which the electric dipole moment is close to zero. This problem will be taken up again in Sect. 4.1.2a.

Strong electron–phonon interaction has also been suggested as a possible explanation of the exponential absorption and the simplest version of the model [4.24] is shown schematically in Fig. 4.3. The curves represent the total energy of the ground and excited states as a function of a normal coordinate q. The equilibrium configuration of the lattice is represented by q_0, while the potential well centered on q_0 defines the local vibrational frequency ω_0. The minimum of the excited state is displaced from q_0 because of the electron–phonon interaction, which is assumed to be linear in q. Thus, the interaction gives a local atomic distortion that lowers the energy of the excited state by W_p compared to its value at q_0. W_p is therefore a measure of the interaction strength.

The optical properties of the model are determined by calculating the matrix elements between the various vibrational levels of the ground and excited states. When the phonon interaction is weak, the zero-phonon transition is strongest, but as the interaction strength increases, higher vibrational levels dominate the optical transition. When $W_p \gg \hbar\omega_0$, the line shape for optical absorption is given by [4.25]

$$\alpha(\hbar\omega) = \text{const} \cdot \exp[(-E_a - \hbar\omega)^2/\gamma^2], \qquad (4.19)$$

with

$$\gamma^2 = 2W_p \hbar\omega_0, \qquad (4.20)$$

and

$$E_a = E_p + W_p. \qquad (4.21)$$

E_p is of course the zero-phonon energy of the transition.

The absorption in (4.19) is Gaussian rather than exponential. This is not necessarily a problem since, in many cases, the Gaussian distribution resembles the experimental "exponential" behavior in the regions measured. Furthermore, it has been shown that the introduction of quadratic terms to the electron–phonon coupling modifies the Gaussian distribution and, under admittedly restrictive conditions [4.25], can lead to exact exponential absorption. The major prediction of the model, and consequently a point for test, is that the luminescence intensity is

$$I_e(\hbar\omega) = \text{const} \cdot \exp[-(E_e - \hbar\omega)^2/\gamma^2] \qquad (4.22)$$

where

$$E_e = E_p - W_p. \qquad (4.23)$$

It thus has exactly the same dependence on photon energy as the absorption.

All of these mechanisms appear a priori as reasonable possibilities and it is likely that each of them may contribute to some extent in any particular material. At the present time, it is an experimental challenge to determine which, if any of them, dominate the others.

4.1.3 Effects of Defects on Optical Properties

a) Effects of Voids in Tetrahedral Materials

Voids are the single most important defect in tetrahedral amorphous semiconductors. Recently, *Webman* et al. [4.26] developed an effective medium theory appropriate to this situation which improves on the Maxwell–Garnett approach. Their result, for light propagating by multiple scattering through a heterogeneous system characterized by media of complex dielectric constants $\tilde{\varepsilon}_i(\omega)$, is

$$\left\langle \frac{\tilde{\varepsilon}_i(\omega) - \tilde{\varepsilon}(\omega)}{\tilde{\varepsilon}_i(\omega) + 2\tilde{\varepsilon}(\omega)} \right\rangle = 0, \tag{4.24}$$

where $\tilde{\varepsilon}(\omega)$ is the effective complex dielectric constant of the system. For the binary case, of interest when voids occur, (4.24) simplifies to

$$(1 - x_v) \frac{\tilde{\varepsilon}_m(\omega) - \tilde{\varepsilon}(\omega)}{\tilde{\varepsilon}_m(\omega) + 2\tilde{\varepsilon}(\omega)} + x_v \frac{\tilde{\varepsilon}_v(\omega) - \tilde{\varepsilon}(\omega)}{\tilde{\varepsilon}_v(\omega) + 2\tilde{\varepsilon}(\omega)} = 0 \tag{4.25}$$

where x_v is the fractional volume of voids, and $\tilde{\varepsilon}_m(\omega)$ and $\tilde{\varepsilon}_v(\omega)$ are the complex dielectric constants of the medium and the voids respectively. The theory is expected to provide a good approximation for $\tilde{\varepsilon}(\omega)$ whenever

$$0.05 < |\tilde{\varepsilon}_m(\omega)/\tilde{\varepsilon}_v(\omega)| < 20 \tag{4.26}$$

and the characteristic dimensions of the voids are considerably smaller than the wavelength of light. Of course, the void dimensions may be much larger than the interatomic spacings, and in this case it is probably a good approximation to take $\tilde{\varepsilon}(\omega)$ equal to 1. Unfortunately, small angle X-ray scattering experiments on a number of tetrahedral semiconductors show that a large proportion of the voids are typically 5 Å in diameter [4.27, 28], and no simple approximation for $\tilde{\varepsilon}(\omega)$ can be given.

A different approach [4.29, 30] is to assume that small voids are in fact accessible parts of an approximately homogeneous network. Then their gross effect is to reduce the average bond strength and plasma frequency of the material and the optical properties may be calculated within the framework of the Penn model. Specifically, *Phillips* [4.29] argues that the Penn gap $\hbar\omega_g$ will decrease as the square of the average coordination C, whereas the plasma energy $\hbar\omega_p$ will decrease as the square root of the macroscopic atomic density ϱ. Thus

$$\tilde{\varepsilon}(0) = 1 + (2/3) \left(\frac{\omega_p^0}{\omega_g^0}\right)^2 (\varrho/\varrho^0) (c/c^0)^{-4}, \tag{4.27}$$

where the superscript 0 indicates a property of fully coordinated material. If a void is considered as a cluster of atoms removed from a fully coordinated network, a fraction f of its bonds occur on its surface. In this case, it can be shown [4.27] that

$$d \ln C / d \ln \varrho = f \tag{4.28}$$

for $0 < f < 1$, and (4.27) can be written as

$$\tilde{\varepsilon}(0) = 1 + (2/3) (\omega_p^0/\omega_g^0)^2 (\varrho/\varrho^0)^{1-4f}. \tag{4.29}$$

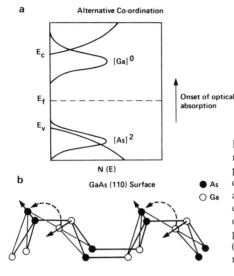

a Alternative Co-ordination

E_c

$[Ga]^0$

E_f — — — — — — — — — — — — Onset of optical absorption

E_v

$[As]^2$

N (E)

b GaAs (110) Surface   ● As   ○ Ga

Fig. 4.4. (a) Schematic representation of atomic relaxation on a void surface of a III–V amorphous semiconductor. The solid arrows show the direction of atom movement during the relaxation and the broken arrows show the direction of electron transfer. Similar models can be applied equally well to void surfaces of group IV amorphous semiconductors [4.32]
(b) The electronic states associated with the relaxed surface of amorphous GaAs [4.32]

When small voids are involved, containing fewer than two coordination shells, f is always greater than 0.53 and the interesting result that $\tilde{\varepsilon}(0)$ is increased over its value in fully coordinated material is obtained from (4.29). This contrasts with the effects of larger voids for which f may be less than 0.25 and $\tilde{\varepsilon}(0)$ is correspondingly decreased. It is worth noting that a similar result is also expected from the effective medium theory when $\tilde{\varepsilon}(\omega)$ is set equal to 1, but for different reasons. There the voids are not accessible parts of the network, and consequently have no effect on the plasma energy or Penn gap.

To understand the effects of voids on the absorption edge requires a microscopic model of the electronic states associated with the void surfaces and general, yet specific, statements are consequently more difficult to make. In sputtered amorphous germanium and silicon, the striking result that only one spin may be associated with every 10–100 electrons on the void surfaces has been obtained [4.31] and this is a useful starting point for a model. There are, it appears, at least two possible explanations for the observation: first, by a surface reconstruction, in which almost all the dangling bonds link to form weak molecular bonds and the signal arises from the small number of unpaired electrons that remain; second, by bipolaron formation, in which almost half the dangling bond sites have two antiparallel spins and the others have none. This situation is energetically favorable when structural relaxation allows the filled dangling bond states to become more s-like and the empty ones to become more p-like [4.32], as demonstrated schematically in Fig. 4.4, and the electron repulsion energy is more than compensated. It is perhaps worth noting that this behavior may occur not only for electron transfer between dangling bonds on a void surface but also for electron transfer between isolated dangling bonds, a situation that is unique to the amorphous state.

The defect states with spin are expected to act both as deep donors and deep acceptors, the acceptor state lying above the donor state by a repulsion energy in the range 0.1–1 eV. In contrast, bipolaron formation leads to a reversal of these two levels, with the doubly occupied state lying at lower energy [the separation being above 0.5 eV for the similar situation on the (111)-surface of crystalline silicon]. Finally, reconstructed molecular bonds on a void surface should produce bonding and antibonding states near the edges of the valence and conduction bands, respectively. The gap states should therefore be distributed over a large part of the gap.

Their effect on the optical properties arises both from transitions in which they are either initial or final states and from the influence of potential fluctuations, induced by the charged gap states, on transitions within the fully coordinated part of the network. Therefore, the voids potentially provide possible explanations for absorption tails and for the origin of the potential fluctuations that can induce exponential absorption.

Similar arguments may be applied to amorphous III–V compounds. For example, Fig. 4.4 can also be used to demonstrate bipolaron formation at void surfaces in amorphous gallium arsenide [4.32]. Thus, the more s-like and filled dangling bond state on arsenic lies closer to the valence band, and the more p-like and empty dangling bond on gallium lies closer to the conduction band.

b) Effects of Coordination Variation in Chalcogenides

Coordination variation (see Chap. 3) appears to explain many of the observed optical properties of the chalcogenides. The model, developed by *Street* and *Mott* [4.33], assumes that the electron–phonon interaction makes electron pairing energetically favorable at defects, which when neutral (D^0) have orbitals containing one electron. Thus the reaction

$$2D^0 \rightarrow D^+ + D^- \tag{4.30}$$

is exothermic. Here, D^+ and D^- represent the defect when empty and when containing two electrons respectively. The chemical origin of these states in chalcogenide glasses has been explained by *Kastner* et al. [4.2]. A twofold coordinated chalcogen contains four p electrons, of which two are in bonding orbitals and two are a lone pair. Thus, D^- is formed when one bond is broken and an extra electron completes a second lone pair. It might also appear that D^+ is formed when two electrons are removed from D^- to leave an empty lone-pair orbital. However, energy is released if D^+ forms a strong bond to a neighboring chalcogenide, making D^+ threefold coordinated. Thus the release of energy by the bonding change between D^+ and D^- is the origin of the electron–phonon interaction that overcomes the electron repulsion energy in D^- and makes electron pairing energetically favorable. The bonding configurations of the two states are shown schematically in Fig. 4.5 together with the alternative notation, C_1^- and C_3^+, of *Kastner* et al. Taken together, the number of bonding and lone

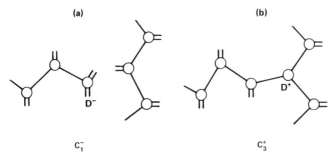

Fig. 4.5a, b. Bonding configurations of D^+ and D^- defects in a twofold coordinated chalcogenide glass, showing lone-pair and bonding p electrons [4.2]

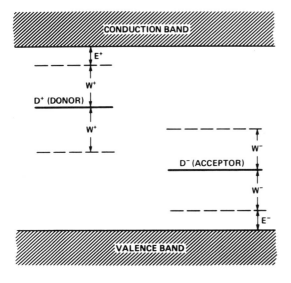

Fig. 4.6. Energy level for the model of charged defects. Solid lines show the donor (D^+) and acceptor (D^-) level. Dashed lines indicate energies of the maximum oscillator strength of optical transitions to and from these defects [4.33]

pair electrons is the same as in the ideal network and therefore the energy to form a D^+ and D^- pair is small compared with that to break a bond.

The electronic states associated with the defects are shown in Fig. 4.6. Since D^+ makes bonds to three other atoms, an extra electron placed on it must occupy a conduction band antibonding state. The center, therefore, acts as a donor with Coulombic binding energy E^+. However, when D^0 is formed, the electron–phonon interaction lowers the donor energy further by W^+, so that the donor level has depth $E^+ + W^+$. Similarly, D^- can be argued to be an acceptor, binding a hole with energy $E^- + W^-$. The relative sizes of W^+ and W^- depend on how much of the bonding energy is released when the first electron is added compared to the second. Clearly, the defects are amphoteric, since D^0 is both a neutral donor and a neutral acceptor and the glass acts as a perfectly compensated semiconductor.

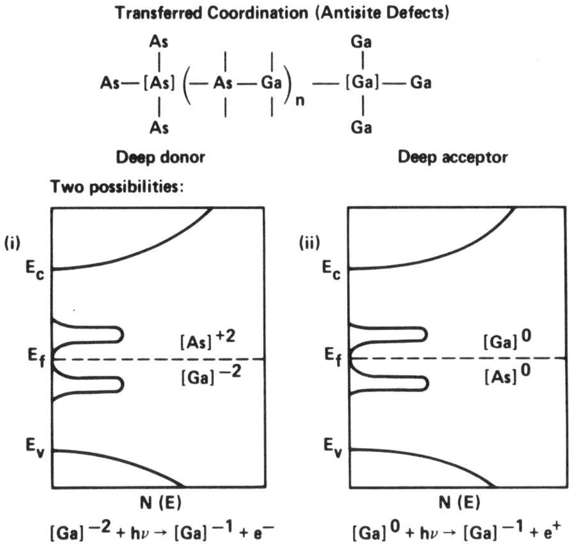

Transferred Coordination (Antisite Defects)

$$\text{As} \quad\quad\quad\quad\quad\quad\quad \text{Ga}$$
$$|\quad\quad\quad\quad\quad\quad\quad\quad |$$
$$\text{As—[As]} \left(- \text{As} - \text{Ga} \right)_n - \text{[Ga]} - \text{Ga}$$
$$|\quad\quad\quad | \quad | \quad\quad\quad\quad |$$
$$\text{As} \quad\quad\quad\quad\quad\quad\quad \text{Ga}$$

Deep donor   Deep acceptor

Two possibilities:

(i)

E_c

E_f - - - - - - - - - - - $[\text{As}]^{+2}$
                          $[\text{Ga}]^{-2}$

E_v

N (E)

$[\text{Ga}]^{-2} + h\nu \rightarrow [\text{Ga}]^{-1} + e^-$

(ii)

E_c

E_f - - - - - - - - - - - $[\text{Ga}]^{0}$
                          $[\text{As}]^{0}$

E_v

N (E)

$[\text{Ga}]^{0} + h\nu \rightarrow [\text{Ga}]^{-1} + e^+$

Fig. 4.7.. Two possible models for the electronic levels associated with isolated like-bonds in amorphous III–V semiconductors [4.32]

Optical excitation of such states is described of course by the configurational coordinate diagram of Fig. 4.3 when W_p takes the values W^+ and W^- for D^+ and D^- respectively. Thus, contributions to the absorption of the form given in (4.19) arise and an absorption with intensity proportional to the concentration of the defects occurs. In Fig. 4.6, the details of these optical (and thermal) transitions are given. The solid lines represent the deep donor and acceptor energies, and the broken lines are the energies of the maximum transition probability for absorption and emission. Of course, in addition, these charged defects influence the matrix elements for transitions between delocalized states near the mobility edges by generating potential fluctuations.

c) The Effects of Like-Bonds in Compounds

The possibility of like-bonds must always be considered in compound semiconductors, particularly when deposited from the vapor. Their effects in amorphous III–V compounds are probably best seen by considering the two types of like-bonds as the two possible antisite defects [4.32]. This situation is shown in Fig. 4.7 for GaAs. It is likely that the deposition conditions lead to isolated like-bonds so that each may be considered separately. Thus, arsenic like-bonds act as deep donors and gallium like-bonds act as deep acceptors. Of course, the occupancy and charge state depend on the relative energies of the levels and two possibilities are shown in Fig. 4.7. In either case, however, the effect is to produce an absorption tail extending well below the power-law edge E_0 and potential fluctuations with their attendant effects on the transitions of the chemically ordered network.

The effect of like-bonds in chalcogenides is qualitatively different. In the III–V's, chemical disorder produces large changes in potential through the Coulombic interaction because of the large ionic contribution to the bonding. In the chalcogenides, the bonding is basically covalent and chemical disorder leads only to small changes in the local potential. Thus, provided the atoms keep their normal coordination, the binding potential is expected to be weak and a shallow level is the result. Their effects on the optical properties are consequently less dramatic than those of like-bonds in the III–V's but nevertheless are significant in virgin vapor-deposited films.

4.2 Optical Properties of Amorphous Semiconductors

Arguments based on tight binding models and chemical bonding requirements have been responsible for the classification of amorphous semiconductors into two categories: those elements and compounds, often from groups III, IV, and V, in which the valence and conduction bands are formed from bonding and antibonding states respectively, and those systems containing group VI elements, in which the lower valence and the conduction band are again formed from bonding and antibonding states, but the upper valence band, in contrast, is derived from nonbonding lone-pair states [4.34]. This fundamental difference in the origin of the uppermost valence band is of crucial importance in establishing the characteristic and sometimes contrasting optical properties of the two classes of material, and is used here as the basis for their presentation. Of course, limitations on space necessarily demand that the discussion be restricted to prototypical materials in each group, and some tetrahedral materials in the first group and selenium and arsenic sulphides and selenides in the second group have been chosen for review.

4.2.1 Tetrahedral Materials

Almost all tetrahedrally bonded materials must be prepared by vapor deposition techniques since the liquid of the same chemical composition does not generally have tetrahedral coordination. This inability to prepare these systems as glasses makes the interpretation of their optical properties difficult, particularly in the region of the absorption edge, because the effects of voids are dominant. Only relatively recently have deposition methods been found to overcome this problem in large part, yet this, not by elimination of voids through subtle choice of deposition conditions, but rather by elimination of the contribution of voids to the states in the band gap by hydrogen coverage of their surfaces [4.31].

a) Excitation Above the Absorption Edge

In Fig. 4.8, the complex dielectric function of amorphous germanium [4.35, 36] and amorphous gallium arsenide [4.37] are shown, superposed on their

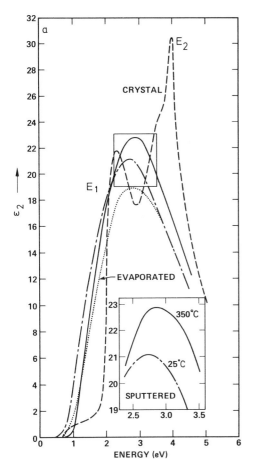

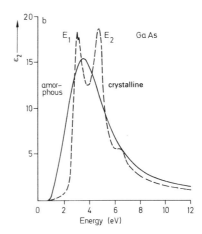

Fig. 4.8. (a) The imaginary part of the dielectric function of amorphous [4.35, 36] and crystalline germanium: sputtered onto substrates at 25 °C and at 350 °C; evaporated onto substrates at 25 °C and crystalline material

(b) The imaginary part of the dielectric function of amorphous [4.37] and crystalline gallium arsenide. The amorphous material was flash evaporated onto substrates at − 50 °C

crystalline counterparts. In most ways, these materials serve as prototypes for all the materials of this group and together they demonstrate the intricacies and problems of interpreting optical spectra.

It is appropriate to consider first the preparation independent properties and attempt to identify the relation between the crystalline and amorphous spectra. It is seen that the peak of $\varepsilon_2(\omega)$ in the amorphism lies close to the crystalline E_1 peak. Moreover, the crystalline E_2 peak is absent entirely from the amorphous spectrum. *Kramer* [4.38] argued that this selective effect of disorder on the crystalline $\varepsilon_2(\omega)$ arises from the different origins of the E_1 and E_2 peaks. On the one hand, E_1 arises from transitions in the [111] directions. Since these directions are also coincident with the directions of the bonds within the tetrahedral molecular units of the crystal, it might be expected that disorder which left the molecular units unchanged would not completely destroy E_1. On the other hand, E_2 arises from transitions in the [100] directions and consequently is sensitive to the details of the connectivity of the molecular units.

In particular, the strength of the E_2 peak in the crystal derives from a large umklapp contribution to the matrix element, which according to (4.8) is absent in the amorphous phase. This argument is confirmed in large extent by calculations of $\varepsilon_2(\omega)$ for the crystalline polytypes of germanium and silicon, in which the E_2 peak is selectively lowered with respect to the E_1 peak as the number of atoms per unit cell, and therefore in a sense the disorder, increases [4.15]. In fact, for the most disordered polytype, ST-12, $\varepsilon_2(\omega)$ has a single peak that is similar in shape to the amorphous one.

The importance of the preservation of the crystalline molecular unit in the amorphous phase is further emphasized when the data in Fig. 4.8 are analyzed in terms of the Penn model. If we ignore for the moment the small variations that occur with sample preparation and assume accordingly that $\varrho = \varrho^0$ in (4.29), the Penn gap may be estimated directly from $\tilde{\varepsilon}(\omega)$ [4.30]. The result is a reduction by at most about 5% upon transformation from the crystalline to the fully coordinated amorphous state. This presumably represents the weakening of the bond strength in the amorphous phase by bond bending, a result that is anticipated in *Pauling*'s [4.39] work. He showed that the energy of a bond is about proportional to the product of the magnitudes of bond orbitals of the two atoms in their angular dependence. Thus for hybridized s–p tetrahedral orbitals, the bond energy for a 10^0 distortion, typical of amorphous germanium and gallium arsenide, is reduced by 3% from the undistorted value. The small effect of disorder on the Penn gap of gallium arsenide is particularly significant since it indicates that chemical disorder is not large and the number of like-atom bonds is small, a result that has proved most difficult to establish by direct structural investigation [4.40].

The relation between the crystalline and amorphous phases is more apparent when use of the optical properties of the amorphous material is made to deduce the features of the electronic density of states. To this end, (4.9) may be applied to the germanium data of Fig. 4.8 with $|P_{am}(\omega)|^2$ taken to vary as ω^{-2} for $\hbar\omega > \hbar\omega_g$ and as ω for $\hbar\omega_g$ [30]. The density of valence states so obtained is shown in Fig. 4.9, with photoemission-derived results of *Ley* et al. [4.41] included for comparison. The discrepancy in the two sets of data at the top of the valence band arises from approximating the finite width of the lower edge of the conduction band by the step function $g_f(E_1)$ in (4.9), but on the whole the agreement is excellent.

Theoretical investigations of the density of states of the various forms of germanium suggest that this upper part of the valence band is derived from p states and is a broadened δ function containing two electrons per atom [4.8]. To a large extent, it is separated from the remainder of the states in the valence band. Therefore, the number of electrons per atom, $n_{eff}(\omega)$, calculated from optical transitions up to a photon energy $\hbar\omega$, should reach close to two at the same energy, evidently about 5 eV from Fig. 4.9, in both the crystalline and amorphous phases. This is demonstrated in Fig. 4.10 in which $n_{eff}(\omega)$ reaches approximately 1.85 electrons per atom at 4.5 eV in both the amorphism and the crystal [4.35].

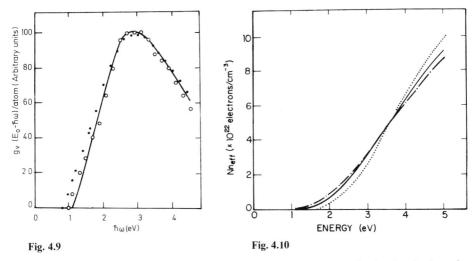

Fig. 4.9

Fig. 4.10

Fig. 4.9. The valence band density of states, measured from the conduction band edge, of amorphous germanium: sputtered onto substrates at 25 °C (···) and at 350 °C (OOO) [4.30]; evaporated material measured by photoemission (———) [4.41]

Fig. 4.10. The effective number of electrons involved in transition up to energy $\hbar\omega$ in amorphous germanium: sputtered onto substrates at 25 °C (—·—) and at 350 °C (———); crystalline material (···). [35]

An exactly similar analysis can be applied with equivalent success to amorphous gallium arsenide to obtain the upper part of the valence-band density of states [4.42]. It is found that the density of states so obtained appears to be generated by a uniform broadening of the crystalline density of states, in contrast to the density of states in germanium which is skewed to higher energy in the amorphous phase. This result has structural implications. The broadening of the δ function in the tight binding approximation is produced by the interaction with next-nearest bonds [4.43] and thus depends sensitively on the dihedral angle distribution. However, the states on the high energy side of the p-bonding peak of the crystal localize electrons more tightly in the bonds than the states on the lower energy side [4.44]. Thus the latter suffer a larger decrease in effective volume and a larger increase in kinetic energy in structures with eclipsed bonds than the states nearer the gap. The optical densities of states imply then that a relatively smaller number of eclipsed bonds occurs in amorphous gallium arsenide than in amorphous germanium. Direct structural measurements obtained this same result only with great difficulty.

The preparation dependent properties must now be considered. As shown in Fig. 4.8, variations in both the height and location of the $\varepsilon_2(\omega)$ peak and in the magnitude of $\varepsilon_1(0)$ occur in amorphous germanium. The unusual result that neither of these correlate directly with density is immediately apparent however when the data on evaporated material is compared with that on sputtered. The resolution of this paradox is found by considering the distribution of void sizes as well as the total void volume in the materials.

On the one hand, the sputtered specimens contain voids that are about 7 Å in diameter on average, and the variation of their number generates the observed density changes [4.27]. Furthermore, the careful structural study indicated that $f = 0.6$ in (4.29), and thus

$$\ln[\varepsilon_1(0) - 1] = \text{const} - 1.4 \ln \varrho \qquad (4.31)$$

is predicted and confirmed by experiment [4.30]. Similarly, the shift of the $\varepsilon_2(\omega)$ peak to higher energy follows correctly the predicted increase in the Penn gap.

On the other hand, the evaporated sample contains a significant fraction of voids that are much larger than typical atomic dimensions [4.28]. Unfortunately, the value of f is unknown, but it is unquestionably less than 0.25 for many of the larger voids. Consequently a much weaker or even opposite dependence of the $\varepsilon_2(\omega)$ peak and $\varepsilon_1(0)$ on density can be rationalized. These last trends can be put on a somewhat more quantitative basis using the effective medium theory [4.26] expressed by (4.25). Assuming a Lorentzian model for the optical dispersion of the matrix,

$$\tilde{\varepsilon}_m(\omega) = 1 - \omega_{mp}^2/(\omega_{mg}^2 - \omega^2 - i\gamma\omega), \qquad (4.32)$$

with

$$\gamma = 0.4\,\text{eV}, \hbar\omega_{mg} = 3\,\text{eV}, \quad \text{and} \quad \hbar\omega_{mp} = 16\,\text{eV}, \quad \text{and} \quad \tilde{\varepsilon}_v(\omega) = 1 \qquad (4.33)$$

allows the results in Fig. 4.11 to be obtained. With increasing density $\varepsilon_2(\omega)$ [and $\varepsilon_1(0)$] increase, but no shift is apparent in the peak location. Thus, as the density of evaporated material approaches that of the high density form of sputtered material, the optical properties tend to converge.

Nothing is yet known experimentally about the effects of voids on the optical properties of amorphous gallium arsenide above the absorption edge. Nevertheless, from studies of the edge itself discussed in the next subsection, it is clear that voids are present and probably modify the complex dielectric function to the same extent as in amorphous germanium.

b) The Absorption Edge

The absorption edges of amorphous germanium [4.31] and amorphous gallium arsenide [4.45] are shown in Fig. 4.12. The materials were prepared by sputtering onto cool substrates in an atmosphere of argon to which different quantities of hydrogen were added. The incorporation of the hydrogen into the samples has been studied in great detail, but for the purpose of this discussion, it is sufficient to recognize that a large fraction forms bonds with individual dangling bonds on the void surfaces. Thus, the optical properties of fully bonded but hydrogenated

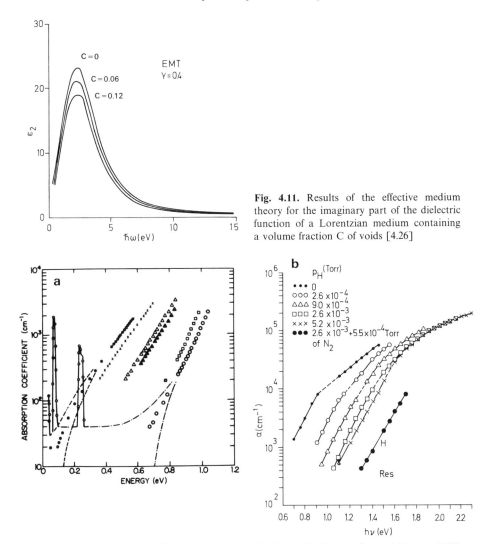

Fig. 4.11. Results of the effective medium theory for the imaginary part of the dielectric function of a Lorentzian medium containing a volume fraction C of voids [4.26]

Fig. 4.12. (a) Absorption coefficient versus photon for $Ge_{1-x}H_x$ films: $\cdots$ 0, xxx 0.01, $\triangle\triangle\triangle$ 0.028, ▲▲▲ 0.03, ☐☐☐ 0.051, and ◯◯◯ 0.08. Estimates of the error in α are shown for two of the films [4.31]. **(b)** Absorption coefficient versus photon energy for hydrogenated GaAs films [4.45]

material, which near the gap will resemble those of fully coordinated material, can in principle be established. One feature is immediately clear. The absorption edge position and, to a lesser extent, shape are dominated by the effects of voids and the optical properties of fully coordinated material cannot be straightforwardly deduced.

Another complication exists and is exhibited in Fig. 4.13. The hydrogenated germanium specimens of Fig. 4.12 were annealed for one hour at 250 °C and their

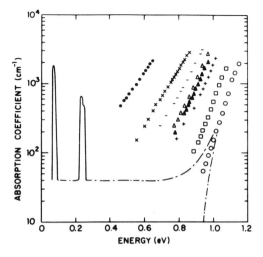

Fig. 4.13. Absorption coefficient versus photon energy for $Ge_{1-x}H_x$ films after at 250 °C anneal for one hour: $\cdots$ 0, xxx 0.01, △△△ 0.028, ▲▲▲ 0.03, □□□ 0.051, and ○○○ 0.08. Data for films produced at $T_s = 250$ °C are also shown: $---$ 0.01 and $+++$ 0.042. Estimates of the error in α are shown for one of the films [4.31]

absorption edges remeasured. Although the changes in absorption were largest in samples with least hydrogen, changes were observed even in those which had complete hydrogen coverage of the void surfaces. It must be assumed then that annealing produces absorption edge shifts both by dangling bond removal and by network reorganization, and even in fully coordinated material, the absorption edge might not have a unique form.

A more quantitative examination of these data for amorphous germanium and gallium arsenide indicates that for any particular thermal history the absorption edge, and $\varepsilon_1(0)$ also, exhibit asymptotic behavior with hydrogen coverage. In this case, the absorption below 10^4 cm^{-1} depends exponentially on photon energy and is represented by (4.14), whereas above 10^4 cm^{-1}, it is represented by the power law of (4.13). However, the exponential edge is steeper and the power law edge begins at higher energy than in unhydrogenated material. For example, in annealed germanium specimens, β changes from 8.3 to 16.7 eV^{-1} and E_0 increases from 0.8 to 1.1 eV with hydrogenation. In the remainder of this section, specific models for these phenomena will be assessed.

Exponential Absorption Region

The exponential absorption in unhydrogenated material is consistent with a model in which each band edge has an exponential tail. If it is further assumed that the valence and conduction bands are symmetric and the Fermi level is pinned at the center of the gap, it is possible to make a rough estimate of the density of states in the gap from the absorption measurements directly, using (4.14, 15). The not unreasonable result is a density of states at the Fermi level of 3×10^{19} cm^{-3} [4.30]. However, the uniqueness of this estimate will be taken up again later.

An identical analysis of annealed or hydrogenated material would obviously demonstrate a decrease in the density of gap states and it seems appropriate therefore to associate the exponential absorption in partially coordinated material with transitions involving the electronic states of the void surfaces. A quantitative relation of this kind was first suggested for evaporated amorphous silicon by *Brodsky* and *Kaplan* [4.46]. They argued that at an energy E the density of states $\varrho(E)$ could be approximately written as

$$\varrho(E) \approx \varrho_0(E) + \varrho_s(E)\, n_s, \tag{4.34}$$

where $\varrho_0(E)$ is the bulk density of states, $\varrho_s(E)$ is proportional to the surface density of states per dangling bond, and n_s, the electron spin density, is assumed to be a direct measure of the number density of dangling bonds. Then using (4.6), the absorption coefficient becomes

$$\alpha(\omega) \approx a_0(\omega) + a_1(\omega)\, n_s + a_2(\omega)\, n_s^2, \tag{4.35}$$

where a_0, a_1, and a_2 are constant coefficients at fixed photon energy. Thus the observed dependence of the absorption edge on the spin concentration not only permits the derivation of the absorption edge in fully coordinated material, as will be described later, but also emphasizes the extent to which void-related states create the exponential absorption. In the case of unhydrogenated sputtered amorphous germanium and evaporated amorphous silicon, the major role of voids is incontrovertible. Nevertheless, the extent to which potential fluctuations associated with these voids generate the exponential edge by establishing an exponentially varying matrix element is not known, and consequently the estimate of the number of gap states based on an exponential distribution of them must be treated with some caution.

From the dependence on hydrogenation, it also appears that the major contribution to the exponential absorption in unhydrogenated amorphous gallium arsenide is from transitions involving surface states on voids rather than from those involving like-bonds. One possibility is shown in Fig. 4.4. Because of surface relaxation, the onset of optical absorption arises from transitions between the valence band edge and empty p states on three coordinated gallium atoms [4.32]. This scheme thus rationalizes not only the effects of hydrogenation but also the puzzling result, found in early investigations of amorphous III–IV compounds, that the location of the exponential edge depended predominantly on the group III atom and to a much lesser extent on the group V atom.

The possibility that potential fluctuations associated with hydrogen-covered void surfaces give rise to the steeper exponential absorption in fully hydrogenated amorphous germanium must be recognized a priori. Nevertheless, *Street* [4.47] points out another contribution to the absorption, at least in glow-discharge-deposited amorphous silicon, that arises from transitions between the states of the fully coordinated material. In Fig. 4.14, the luminescence line at 1.3 eV is shown in relation to its excitation spectrum. The relative position of them indicates a Stokes shift of about 0.5 eV and therefore a distortion energy W_p of about 0.25 eV. A phonon-broadened edge, given by (4.19), must therefore

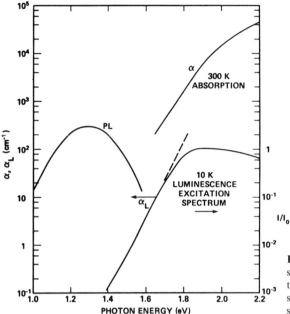

Fig. 4.14. Photoluminescence spectrum (PL), excitation spectrum, and room-temperature absorption in glow-discharge-deposited amorphous silicon [4.47]

occur. By applying detailed balance, the absorption strength associated with the luminescence band may be calculated from its radiative lifetime and the result obtained is also shown in Fig. 4.14. Clearly, a phonon-broadened component of absorption makes significant contributions to values of 10^2–10^3 cm^{-1}, and its magnitude and energy dependence is not greatly different from the measured absorption.

Power Law Absorption Region

If the densities of states of the valence and conduction band edges are assumed to have the same parabolic dependence, $g(E)$, they may be found by applying (4.13) directly to the absorption spectrum. In the particular examples shown in Fig. 4.15, the separation between the bands, E_0, and $g(E)$ of sputtered amorphous germanium increase as the deposition temperature is increased from 25 to 350 °C [4.30]. It is particularly interesting to note however that in both cases $g(E)$ is very close to the free electron value, $g_f(E)$, given by

$$g_f(E) = 6.7 \times 10^{21} (\Delta E)^{1/2}, \tag{4.36}$$

where ΔE is the departure from the band edge. Qualitatively similar results are also obtained for the effects of hydrogenation although a larger variation of E_0 can be induced. In fact, at full coverage in annealed specimens, E_0 reaches about 1.1 eV.

In the early literature, the relation of E_0 to the band gap of the crystal was much considered, but the effects of voids often produced misleading or evidently

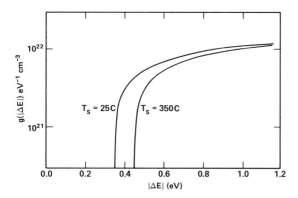

Fig. 4.15. The valence and conduction band density of states, measured from the center of the gap, for sputtered amorphous germanium films deposited at 25 and 350 °C [4.30]

conflicting experimental results. From the above discussion, however, it would seem that E_0 is larger, though not directly related to the crystalline gap. The same conclusion is also reached in amorphous silicon for which E_0 in the void-free materials is estimated, either from (4.35) [4.46] or from hydrogenated material [4.47], to be about 1.8 eV. Inasmuch as the band gap in theoretical calculations is extremely sensitive to the topology of the network, such a result is not surprising. In fact, it may represent the effect of five-membered atomic rings in the amorphous phase, masked by the opposing effect of quantitative disorder.

As yet, little work has been done on establishing the value of E_0 in fully coordinated amorphous III–V compounds. From the hydrogenation studies on amorphous gallium arsenide in Fig. 4.12, it appears that E_0 reaches at least 1.4 eV. To what extent this value is then influenced by the existence of like gallium bonds as indicated in Fig. 4.7, or can be modified by network relaxation, must still be determined. Nevertheless, a value comparable though somewhat larger than the crystalline value might again be expected.

4.2.2 Chalcogenide Semiconductors

Many of the chalcogenide semiconductors can be made as glasses by cooling from the melt. Prepared in this way, they contain defects that exist in the liquid at the glass transition temperature. Thus, coordination variation and like bonds must be expected but voids are unlikely. Moreover, relatively fewer defects exist than in materials of the same chemical composition but deposited from the vapor. The effects of large numbers of defects in vapor deposited systems can therefore be clearly established experimentally, a situation that contrasts dramatically with that in tetrahedral semiconductors.

a) Excitation Above the Absorption Edge

Only the preparation independent properties have been ascertained in this part of the optical spectrum as yet. In Figs. 16, 17, the imaginary parts of the complex dielectric functions of vitreous and trigonal selenium [4.48] and vitreous As_2Se_3

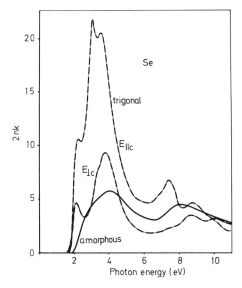

Fig. 4.16. The imaginary part of the dielectric function of amorphous and trigonal selenium [4.48]

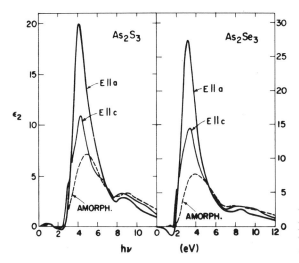

Fig. 4.17. The imaginary part of the dielectric function of amorphous As₂Se₃ and orpiment. Analogous data for As₂S₃ are also shown [4.49]

and orpiment [4.49] are shown as examples. The sharp features between 2 and 6 eV in each material that occur in the crystal as a result of singularities associated with symmetry points in the Brillouin zone do not occur for the amorphism but rather are replaced by a broad featureless band of the same overall shape. In contrast, the trough near 6 eV survives the transition to the disordered state and represents the onset of a second absorption process. Therefore, at first sight, the optical properties appear to suggest the partition of the valence band into two separated components.

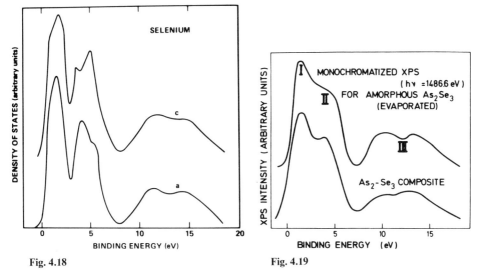

Fig. 4.18   **Fig. 4.19**

Fig. 4.18. Density of valence states of amorphous and crystalline selenium [4.50]

Fig. 4.19. Density of valence states of amorphous As_2Se_3 compared with a composite density of states constructed from those of the constituent elements [4.51]

The densities of valence states determined by photoemission [4.50, 51] are shown in Fig. 4.18, 19. In selenium, s–p hybridization is weak, because the atomic s and p levels are separated by about 10 eV. Thus a band of s states in both the amorphous and crystalline forms is expected to lie well below the p states. The p states themselves are expected to lie in two bands. One p state cannot be involved in bonding and therefore forms a nonbonding lone-pair band that lies above a bonding p band by approximately one-half the bonding–antibonding splitting. The lone-pair band is narrower than the others, reflecting the smaller overlap of the lone-pair electron distributions relative to that of the bonding electron distributions. These features are readily apparent in Fig. 4.18.

The origin of the density of valence states of amorphous As_2Se_3 can be argued similarly. In this case, peak I, so labeled in Fig. 4.19, is caused by the lone-pair electrons of selenium, peak II arises from p-bonding states of arsenic and selenium, and peak III is derived from the s states of arsenic and selenium. It is clear then that in both Se and As_2Se_3 the expectation from $\varepsilon_2(\omega)$ is born out, in that the s-derived bands are well separated from the p-derived bands.

No attempt has been made to analyse the optical properties of these materials according to (4.9) as yet, although attempts to reconstruct the optical properties from the photoemission density of states using (4.6) have been made with some success. In the case of selenium, the average crystalline matrix elements have been calculated by *Kramer* et al. [4.52]. It appears that the large difference in magnitude between 4 eV peaks in $\varepsilon_2(\omega)$ of the crystal and amorphism arises through umklapp-enhancement of the crystalline matrix element and its removal

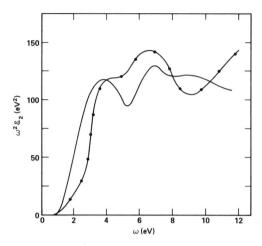

Fig. 4.20. Comparison of experimental (———) and calculated (●●●) optical transition strength of amorphous As_2Se_3 [4.53]

when applying (4.6) leads to quantitative agreement between the measured and calculated $\varepsilon_2(\omega)$ of the amorphism. A less detailed analysis has been made of As_2Se_3, but the result obtained from (4.6), when an energy-independent matrix element is used [4.53], is shown in Fig. 4.20. The positions of the peaks near 4 and 7 eV in $\omega^2\varepsilon_2(\omega)$ are reproduced approximately, suggesting that only the inclusion of an energy-dependent matrix element is required for a more quantitative description.

The similarity between the densities of valence states in the crystalline and amorphous forms has some important structural implications. In selenium, the most significant difference between them is the reversal in the relative intensities of the two peaks in the bonding p band. It appears that this is not a topological effect but rather occurs because of changes in the interchain interaction [4.54] and average dihedral angle distribution [4.55]. Thus these results suggest that atomic chains predominate in amorphous selenium, since in this way the bonding topology of the trigonal form is preserved to a large extent. Similarly, it would appear that the amorphous form of As_2Se_3 is composed of weakly interacting polymer networks similar to the molecular layers of orpiment. Of course, the chains in one case and the layers in the other are convoluted and do not exhibit the long-range order of their crystalline counterparts.

b) The Absorption Edge

It is fruitful to consider first the absorption edge in glasses containing relatively few defects, and in Fig. 4.21, the absorption edge of vitreous selenium is shown as a prototype. It has the exponential form of (4.14) over about four orders of magnitude with β equal to $17.2 \, eV^{-1}$. A clue to its origin was found by *Hartke* and *Regensburger* [4.56]. Noting that the quantum efficiency of photoconductivity does not rise to unity until about 2.6 eV, they separated the absorption edge into photoconductive and nonphotoconductive components and found

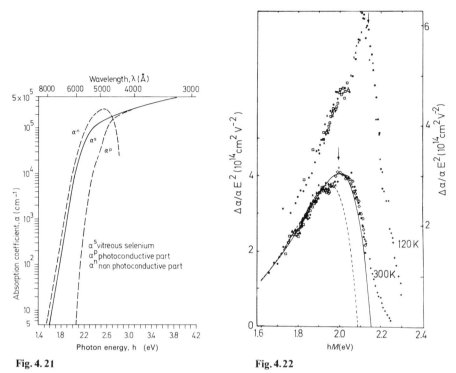

Fig. 4.21

Fig. 4.22

Fig. 4.21. Optical absorption (α^s) in amorphous selenium resolves into photoconductive (α^p) and nonphotoconductive (α^n) components on the basis of quantum efficiency [4.56]

Fig. 4.22. Electroabsorption spectra of amorphous selenium at 120 and 300 K. The arrows mark the upper limit of the exponential tail region at each temperature [4.57]

conclusively that absorption in the exponential region does not generate free carriers. The edge therefore is not generated by transitions between localized states and states beyond the mobility edges.

Electroabsorption measurements in the exponential region of the selenium edge [4.57], as shown in Fig. 4.22, give

$$\Delta\alpha = \text{const} \cdot \exp[-\beta'(E_e - \hbar\omega)], \qquad (4.37)$$

where β' is a constant that is larger than β. Thus, the data indicate that there is a component of the absorption that is sensitive to electric field but agreement with the predictions of the electric-field-broadened exciton model of (4.18) can only be achieved by assuming that there is also a field-insensitive component of absorption that dominates the absorption below about 10^2–10^3 cm^{-1}. *Sussman* et al. [4.57] suggest that this might represent the maintenance of structural anisotropy in microscopic regions of the amorphism. In this case, the dependence of the absorption on the electric field in a microregion would be largest for a

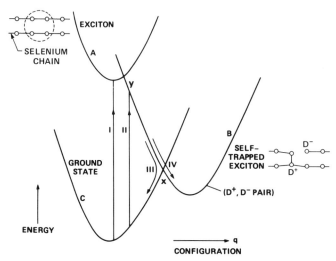

Fig. 4.23. Configuration coordinate diagram illustrating the self-trapping of excitons in chalcogenide glasses. Nonradiative recombination can occur directly to the ground state by path III or to the metastable self-trapped exciton state via path IV [4.24]

particular direction of the light vector in relation to a local chain segment and as a result, the bulk absorption would consist of two components with different sensitivities to an external electric field.

An alternative explanation for the exponential absorption is broadening by a strong electron–phonon interaction. This has been argued in detail for the analogous case of amorphous As_2Se_3 [4.24], and the situation is envisaged in Fig. 4.23. Optical excitation can occur by two distinct mechanisms. One is a transition between the ground state and the uncoupled exciton. This gives an exciton line and a continuum at higher energies. Of course, the exciton will be broadened by the internal electric fields that accompany the disorder and defect states and this component is expected to be sensitive to externally applied electric fields. Furthermore, it should lead to the observed dependence of the quantum efficiency. The second is a transition from the ground state to the self-trapped exciton. The initial and final configurations of this transition are sketched in Fig. 4.23. The uncoupled exciton is envisaged as spanning selenium atoms on different chains in selenium or plains in As_2Se_3. When the separation between electron and hole is small, self-trapping can occur by the bond breaking and reforming mechanism indicated to generate a close D^+–D^- pair, which according to Fig. 4.23 can be metastable when the electron–phonon interaction is strong enough. Thus, the transition in question gives the absorption of (4.19) when linear electron–phonon coupling exists [4.24] and in Fig. 4.24, the absorption edge so calculated, using $\omega_p = 2\,eV$ and $\hbar\omega_0 = 0.03\,eV$, is compared with that of As_2Se_3 for which the parameters are appropriate. Although the shapes of the experimental and predicted curves are different in detail, it is clear that the broadening created by strong electron–phonon coupling is significant and could very well represent the component that is insensitive to external electric fields in the electroabsorption experiments.

Strong electron–phonon coupling also appears to provide an explanation of photoinduced structural changes in chalcogenide glasses. These occur when the materials are irradiated with a sufficient flux of high-energy photons [4.58] and may be observed as changes in thickness, X-ray diffraction, chemical reactivity, refractive index, and optical band gap. In Fig. 4.25, the effect is demonstrated in the absorption edge of vitreous As_2S_3 [4.59]. The size of the shift increases as the temperature under which illumination occurs decreases, and reaches -0.2 eV for illumination at 14 K. Figure 4.26 shows that the excitation spectrum for this photodarkening is independent of photon energy above absorption coefficients of about 10^3 cm^{-1}, but at lower values the effect decreases rapidly [4.60]. It should also be noted that the efficiency curve is almost exactly complementary to the excitation spectrum of luminescence from free excitons trapped on D$^-$ centers [4.61]. The photodarkening can be reversed by annealing and an activation energy of about 0.45 eV has been obtained for the process in both As_2S_3 [4.59] and As_2Se_3 [4.62].

The structural changes accompanying the optical edge shift are shown in Fig. 4.27 by the effect of illumination at room temperature and 77 K on the first X-ray diffraction peak of As_2S_3 [4.63]. Unfortunately, the data are not sufficiently detailed to allow the differential radial distribution function to be obtained. Nevertheless, the effect is accompanied by increased chemical reactivity, manifested for example by an increased oxidation rate and enhanced diffusion of many metals, and bond breaking is therefore likely to be involved.

A specific explanation [4.24] of these phenomena is available in Fig. 4.23. Since close D$^+$–D$^-$ pairs or equivalently self-trapped excitons are metastable, particularly at low temperatures, optical excitation to free exciton states can create them with some small efficiency. It is then argued that both the enhanced chemical reactivity and the modified optical properties result from the new local order of the network in the metastable state and the activation energy for the annealing process represents the barrier between the metastable self-trapped exciton state and the deformed ground state. Of course, the creation of close D$^+$–D$^-$ pairs competes with luminescence from free excitons trapped on D$^-$ centers, so that the excitation spectra of luminescence and photostructural change are complementary.

In Fig. 4.28, an additional effect of prolonged illumination at low temperatures with high-energy photons is demonstrated for As_2S_3 [4.64] and As_2Se_3 and must be differentiated from the effects of photostructural changes. In this case, an absorption band, extending down to approximately midgap, is induced by illumination. The effect can be reversed either by thermal annealing or by illumination at photon energies within the absorption band, behavior that is also found for the fatiguing of luminescence and for photoinduced electron spin resonance. The explanation given for these results [4.65] involves the presence of D$^-$ centers within the specimens prior to illumination. Thus, a free exciton may be captured by a D$^-$ center and, since the electron is then no longer bound by

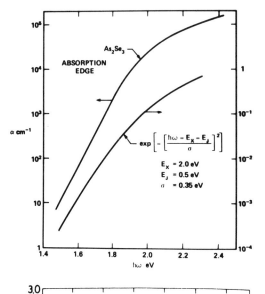

Fig. 4.24. Comparison of the observed absorption edge of amorphous As$_2$Se$_3$ and the predicted absorption for transitions to the self-trapped exiton [4.24]

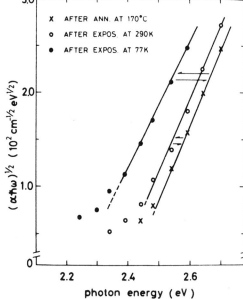

Fig. 4.25. The reversible changes in the optical absorption edge produced by repeated cycles of illumination and annemaling in As$_2$S$_3$ films [4.59]

Coulombic attraction to the metastable D^0 center so formed, it may hop away. Prolonged excitation can therefore induce a large fraction of the D$^-$ centers to be converted to D^0 states at low temperatures and an induced absorption tail arises from transitions between these centers and the band edges.

Also shown in Fig. 4.28 is the effect of the coordination defects themselves on the absorption edge. The small absorption tail in unilluminated material at

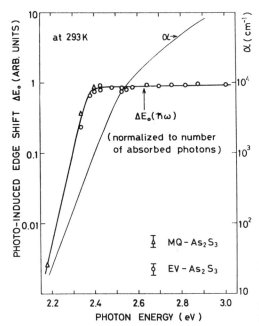

Fig. 4.26. Spectral dependence of the efficiency of exciting the photostructural transformation in As_2S_3. The absorption edge is also shown before illumination for comparison [4.60]

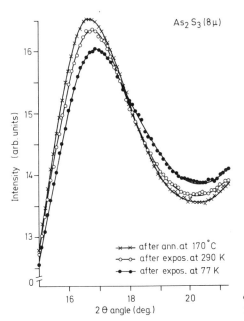

Fig. 4.27. The reversible changes in the X-ray diffraction curves of As_2S_3 produced by illumination and annealing [4.63]

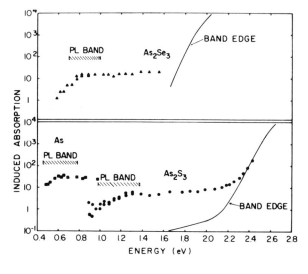

Fig. 4.28. Low-temperature optically induced absorption optically induced absorption for glassy As₂Se₃, As₂S₃, and arsenic. Pre-illumination absorption data show the broad low energy tail in As₂S₃. The approximate position of the luminescence band is also shown [4.64]

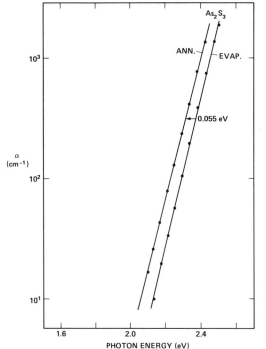

Fig. 4.29. The absorption edges of virgin and annealed evaporated amorphous As₂S₃ [4.66]

absorption coefficients below $1 \, \text{cm}^{-1}$ arises from the optical excitation of D^- and D^+ centers and is the Stokes shifted band of the luminescence, as indicated in Fig. 4.6.

It is now appropriate to examine briefly the optical properties of vapor-deposited chalcogenide films. In Fig. 4.29, the absorption edges of virgin and annealed As_2S_3 are shown, and a shift of 0.055 eV to lower energies with annealing is evident [4.66]. Interestingly, the absorption edge of the annealed material is very similar to that of the glass.

An explanation of these effects is suggested by X-ray, Raman, and EXAFS studies [4.66]. These measurements indicate that arsenic and selenium like-bonds, present in the virgin films at the 5 atomic % level, are removed by annealing. Whether this occurs by polymerization of As_4S_4 molecules or by removal of like-bonds from the network is not entirely clear, but that it creates the optical effects is well established. The possibility has been suggested that the optical properties of the virgin material can be related to those of off-stoichiometric glasses, which of necessity contain like-bonds and indeed the sign and semi-quantitative prediction for the edge shifts with annealing of a number of chalcogenide materials can be predicted from structural measurements.

4.3 Conclusion

A number of areas for concern should have become apparent in the preceding description of the optical properties of amorphous semiconductors. One example is that theoretical descriptions of optical excitation are semiquantitative at best and qualitative at worst. This arises as much from the complexities of calculating the electronic density of states in the absence of long-range order as from the assumptions about the matrix elements for optical transitions between filled and empty states. It is unlikely that this situation will improve significantly in the near future. Another example is that the experimental investigations have been unable in most cases to establish the detailed form of the absorption edge in fully coordinated material. This arises, in some cases, from the difficulty of preparing defect-free material, and in other cases, from an inability to measure precisely the role and magnitude of the electron–phonon interaction. Both of these problems currently plague the work on doped amorphous silicon, in which there is enormous interest because of its potential as a large-area solar cell material [4.67]. It might be hoped however that definitive experiments in this area will be performed in the not-too-distant future. Despite these problems, the measurement of the optical properties has played and continues to play a significant role in developing an overall understanding of the structural and electronic properties of amorphous semiconductors.

References

4.1   M. Lax: Phys. Rev. **109**, 1921 (1958)
4.2   M. Kastner, D. Adler, H. Fritzsche: Phys. Rev. Lett. **37**, 1504 (1976)
4.3   M. Gorman, S. A. Solin: Solid State Commun. **18**, 1401 (1976)
4.4   H. Krebs, R. Steffen: Z. Anorg. Allg. Chem. **327**, 224 (1964)

4.5   M.H.Brodsky, R.S.Title: Phys. Rev. Lett. **23**, 581 (1969)
4.6   N.J.Shevchik, W.Paul: J. Non-Cryst. Solids **13**, 1 (1973–74)
4.7   A.F.Ioffe, A.R.Regel: Prog. Semicond. **4**, 237 (1960)
4.8   D.Weaire, M.F.Thorpe: Phys. Rev. B**4**, 2508, 3518 (1971)
4.9   P.W.Anderson: Phys. Rev. **109**, 1492 (1958)
4.10 N.F.Mott: Adv. Phys. **16**, 49 (1967)
4.11 F.L.Galeener: Phys. Rev. Lett. **27**, 1716 (1971)
4.12 J.Tauc, R.Grigorovici, A.Vancu: Phys. Status Solidi **15**, 627 (1966)
4.13 N.K.Hindley: J. Non-Cryst. Solids **5**, 17 (1970)
4.14 K.Maschke, P.Thomas: Phys. Status Solidi (b) **41**, 743 (1970)
4.15 J.D.Joannopoulos, M.L.Cohen: Phys. Rev. B**8**, 2733 (1973)
4.16 D.Penn: Phys. Rev. **128**, 2093 (1962)
4.17 M.Cardona: In Proc. Int. School Phys. "Enrico Fermi", Varenna, Course 52, ed. by J.Tauc (Academic Press, New York 1972) p. 514
4.18 E.A.Davis, N.F.Mott: Philos. Mag. **22**, 903 (1970)
4.19 F.Urbach: Phys. Rev. **92**, 1324 (1953)
4.20 A.A.Andreev, B.T.Kolomiets, T.F.Mazets, A.L.Manukyan, S.K.Pavlov: Sov. Phys. Solid State **18**, 29 (1976)
4.21 J.D.Dow, J.J.Hopfield: J. Non-Cryst. Solids **8–10**, 664 (1972)
4.22 J.Tauc: Mat. Res. Bull. **5**, 721 (1970)
4.23 J.D.Dow, D.Redfield: Phys. Rev. B**5**, 594 (1972)
4.24 R.A.Street: Solid State Commun. **24**, 363 (1977)
4.25 T.H.Keil: Phys. Rev. **140**, A601 (1965)
4.26 I.Webman, J.Jortner, M.H.Cohen: Phys. Rev. B**15**, 5712 (1977)
4.27 R.J.Temkin, W.Paul, G.A.N.Connell: Adv. Phys. **22**, 581 (1973)
4.28 N.J.Shevchik, W.Paul: J. Non-Cryst. Solids **16**, 55 (1974)
4.29 J.C.Phillips: Phys. Status Solidi (b) **44**, K1 (1971)
4.30 W.Paul, G.A.N.Connell, R.J.Temkin: Adv. Phys. **22**, 529 (1973)
4.31 G.A.N.Connell, J.R.Pawlik: Phys. Rev. B**13**, 787 (1976)
4.32 G.A.N.Connell, W.Paul: Bull. Am. Phys. Soc. **22**, 405 (1977)
4.33 R.A.Street, N.F.Mott: Phys. Rev. Lett. **35**, 1293 (1975)
4.34 M.Kastner: Phys. Rev. Lett. **28**, 355 (1972)
4.35 G.A.N.Connell, R.J.Temkin, W.Paul: Adv. Phys. **22**, 643 (1973)
4.36 T.M.Donovan, W.E.Spicer, J.M.Bennett, E.J.Ashley: Phys. Rev. B**2**, 397 (1970)
4.37 J.Stuke, G.Zimmerer: Phys. Status Solidi (b) **49**, 513 (1972)
4.38 B.Kramer: Phys. Status Solidi (b) **47**, 501 (1971)
4.39 L.Pauling: *Nature of the Chemical Bond* (Cornell University Press, Ithaca, NY 1967) pp. 108–120
4.40 R.J.Temkin: Solid State Commun. **15**, 1325 (1974)
4.41 L.Ley, S.Kowalczyk, R.Pollak, D.A.Shirley: Phys. Rev. Lett. **29**, 1088 (1972)
4.42 G.A.N.Connell: Solid State Commun. **14**, 377 (1974)
4.43 D.Henderson, I.B.Ortenburger: *Computational Methods for Large Molecules and Localized States in Solids* (Plenum Press, New York 1972)
4.44 J.P.Walter, M.L.Cohen: Phys. Rev. B**4**, 1877 (1971)
4.45 W.Paul, T.D.Moustakas, D.A.Anderson, E.Freeman: In *7th Int. Conf. on Liquid and Amorphous Semiconductors*, ed. by W.E.Spicer (CICL Univ., Edinburgh 1977) p. 467
4.46 M.H.Brodsky, D.M.Kaplan, J.F.Ziegler: In *Proc. 11th Int. Conf. on Phys. Semicond.* (PWN – Polish Scientific Publ., Warsaw 1972) p. 529
4.47 R.A.Street: Philos. Mag. B **37**, 35 (1978)
4.48 J.Stuke: J. Non-Cryst. Solid **4**, 1 (1970)
4.49 R.E.Drews, R.L.Emerald, M.L.Slade, R.Zallen: Solid State Commun. **10**, 293 (1972)
4.50 N.J.Shevchik, M.Cardona, J.Tejeda: Phys. Rev. B**8**, 2833 (1973)
4.51 S.G.Bishop, N.J.Shevchik: Phys. Rev. B**12**, 1567 (1975)
4.52 B.Kramer, K.Maschke, P.Thomas, J.Treusch: Phys. Rev. Lett. **25**, 1020 (1970)

4.53 B.Velicky, M.Zavetova, L.Pajasova: In *Proc. 6th Int. Conf. on Liquid and Amorphous Semiconductors*, ed. by B.T.Kolomiets (Acad. Sciences, USSR, Leningrad 1976) p. 273

4.54 J.D.Joannopoulos, M.Schulter, M.L.Cohen: In *Proc. 12th Int. Conf. on Semiconductors*, ed. by M.H.Pilkuhn (B.G.Tenbaer, Stuttgart 1974) p. 1304

4.55 N.J.Shevchik: J. Phys. C**8**, 3767 (1975)

4.56 J.L.Hartke, P.J.Regensburger: Phys. Rev. **139**, A970 (1965)

4.57 R.S.Sussman, I.G.Austin, T.M.Searle: J. Phys. C**8**, L182 (1975)

4.58 J.P. de Neufville: In *Optical Properties of Solids – New Developments*, ed. by B.O.Seraphin (North-Holland, Amsterdam 1976) p. 437

4.59 K.Tanaka: AIP Conf. Proc. **31**, 148 (1976)

4.60 K.Tanaka, H.Hamanaka, S.Iizima: In *Proc. 7th Int. Conf. on Amorphous and Liquid Semiconductors*, ed. by W.E.Spicer (CICL Univ., Edinburgh 1977) p. 787

4.61 R.A.Street: Adv. Phys. **25**, 397 (1976)

4.62 Y.Asahara, T.Izumitani: J. Appl. Phys. **47**, 4882 (1976)

4.63 K.Tanaka: Appl. Phys. Lett. **26**, 243 (1975)

4.64 S.G.Bishop, U.Strom, P.C.Taylor: Phys. Rev. B**15**, 2278 (1977)

4.65 R.A.Street, T.M.Searle, I.G.Austin: Amorphous and Liquid Semiconductors, ed. by J.Stuke (W.Brenig (Taylor and Francis, London 1974) p. 953

4.66 R.J.Nemanich, G.A.N.Conell, T.M.Hayes and R.A.Street: Phys. Rev. B **18**, 6900 (1978)

4.67 D.E.Carlson, C.R.Wronski: Appl. Phys. Lett. **28**, 671 (1976)

5. Electronic Transport in Amorphous Semiconductors

P. Nagels

With 11 Figures

In recent years, exciting advances have been made in understanding the problem of how the disorder in amorphous semiconductors influences the band structure and hence the electrical properties. The aim of this chapter is to provide a survey of certain salient aspects of electronic transport in amorphous materials. The first part deals with the various models which have been suggested for the energy distribution of the density of states and is mainly concerned with the important question of the existence of localized states in tails of the valence and conduction band and of a mobility edge, separating extended from localized states. On the basis of these models the different mechanisms for conduction in amorphous semiconductors will be briefly discussed. The next section is devoted to a derivation of some basic expressions, in particular of those transport coefficients which have contributed considerably to our present knowledge about the conduction in amorphous semiconductors. In this part we shall treat in some detail dc and ac conductivity, thermoelectric power, Hall effect and photoconductivity. Special attention will be paid to variable-range hopping in localized states near the Fermi level, a type of conduction which is unique for amorphous semiconductors. The last and longest section is concerned with some experimental results obtained on chalcogenide glasses which have received a great deal of attention. Here, we shall present various pieces of experimental evidence that have been found to support the idea of localized band tail states and gap states. Although considerable progress in the knowledge of the amorphous state has been made in the last decade, many problems still remain unsettled. An important problem, which urgently asks for clarity, is the role played by lattice deformation in noncrystalline solids. If strong enough this deformation may cause small-polaron formation. In the last section we shall discuss therefore Emin's suggestion to explain the transport data of covalent amorphous semiconductors in terms of the existing small-polaron theory developed for crystalline semiconductors. It will become clear that some transport experiments fail to distinguish between mechanisms of quite different physical origin.

5.1 Introductory Remarks

The study of the physical properties of amorphous semiconductors has become an active field in solid state physics over the last decade. In particular, the nature of electronic transport in these materials has attracted much attention during recent years. The interest arose from two sides. In 1968 Ovshinsky reported the discovery of nondestructive fast electrical switching in thin films of multicomponent chalcogenide glasses. This phenomenon was observed in experiments where the voltage–current characteristics were measured as a function of increasing applied electric field. For low electric fields, the current is ohmic and the material is in an almost nonconductive state, in which the resistivity is typically in the range 10^8–10^{10} Ω cm at room temperature. When a certain critical voltage is reached, the resistance drops drastically and the material is converted in a very short time from a highly resistive into a highly conductive state. When the voltage is decreased, this state may be maintained, or the material may revert back to its original state, depending on the type of glass used. This discovery immediately initiated long discussions about the physical interpretation of the effects, the fundamental question being whether the switching is either of thermal or of purely electronic nature.

From a fundamental point of view, the noncrystalline material is of considerable intrinsic interest. For many years the crystalline solid has been studied and a high degree of understanding of its various physical properties has been attained. The crystalline state is characterized by a regular periodicity in the atom or ion positions over long distances. The salient features of crystalline semiconductors, e.g., the existence of sharp edges at the valence and conduction bands leading to a well-defined forbidden energy gap, are direct consequences of the short- and long-range order. This band structure determines the electrical transport properties that are commonly measured, such as the electrical conductivity and the Hall coefficient.

Materials which can be transformed into the semiconducting amorphous state generally contain elements of the groups III, IV, V and VI of the periodic table. These elements possess highly directional interatomic binding forces and, as a consequence, show rather strong local order in their glassy compounds. X-ray diffraction measurements have indeed revealed that the nearest-neighbor configuration in many amorphous materials is very similar to that of the crystalline state. Small deviations in the interatomic distances and in the bond angles lead, however, to a complete loss of translation order after a few coordination spheres. This immediately raised the important question concerning the applicability of special characteristics of the crystalline material, in particular its band structure model, to the amorphous state.

At present, there appears to be widespread agreement that, if the short-range order is the same in the amorphous state as in the crystalline one, some basic features of the electronic structure of the crystal are preserved. This can be understood intuitively if one considers that the tight-binding approximation theory uses the atomic wave functions of the individual atoms perturbed only by the presence of the nearest neighbors. In order to account for the translational disorder, accompanied by a possible compositional disorder in multicomponent systems, modifications have been proposed for the band structure of the amorphous solid. These are the well-known *Cohen–Fritzsche–Ovshinsky* (CFO) [5.1] and *Davis–Mott* [5.2] models, which will be briefly discussed in the next section. Introducing the basic idea of the presence of localized states at the band extremities, these models have been widely used to interpret experimental data in the field of electrical and optical properties. Nevertheless, in many respects the situation is far from clear at present, partly due to the absence of sophisticated theoretical work.

In contrast to the CFO and Davis–Mott ideas, a different approach to the understanding of the electrical properties of amorphous semiconductors has been put forward by *Emin* [5.3]. He suggested that the charge carriers in some amorphous materials might be small polarons. It is generally accepted that hopping of small polarons is the mechanism responsible for electrical transport in oxide glasses, in which the major constituent is a transition-metal oxide. When the composition of the oxide deviates from stoichiometry, the transition-metal ion can occur in two valency states and conduction is due to hopping of charge carriers between aliovalent ions. A disagreement exists, however, in the interpretation of the electrical data obtained on covalent amorphous semiconductors, such as a-Si, a-As, and the chalcogenide glasses. Here, it is not at all clear whether lattice deformation, induced by the incorporation of an extra charge at a given site, is strong enough to cause small-polaron formation.

This review of the electronic properties of amorphous semiconductors will be divided into three main parts. In Sect. 5.2 the various models describing the electronic density of states will be presented. Here, the fundamental question arises whether the features of the band structure of the crystalline material are preserved when going from the crystalline to the disordered state. On the basis of these models the different channels for conduction in amorphous semiconductors will be briefly discussed.

Section 5.3 is concerned with some basic electrical properties of amorphous semiconductors, particularly with dc and ac conductivity, thermoelectric power, Hall effect, and photoconductivity. Formulas describing various transport coefficients will be presented and will be compared with crystalline formulations where appropriate. As already mentioned, an alternative explanation of the transport data of covalent amorphous semiconductors was suggested by *Emin*. He was able to analyze experimental data of dc conductivity, thermopower, and Hall mobility obtained on some chalcogenide glasses in the framework of the existing small-polaron theory developed for crystalline semiconductors. The notion of small polaron will be introduced and the formulas of the transport coefficients used in the analysis will be briefly discussed.

Section 5.4 deals with some typical experimental results obtained on amorphous semiconductors showing covalent bonds which form one family of materials. It is customary to classify the amorphous semiconductors into two broad groups, mainly depending on the nature of the chemical bond. This classification, also used for crystalline solids, is based on an old idea of Ioffe that the fundamental electronic properties of the solid state are primarily determined by the character of the chemical bond between the nearest neighbors. In this view, two groups can be distinguished which contain, on the one hand, ionic, and, on the other hand, covalent, materials. The ionic materials that have been studied most extensively are those based on glass-forming oxides such as SiO_2, P_2O_5, and B_2O_3. Electrical conduction in these glasses may be ionic, when significant amounts of alkali metal ions are present, or electronic by electron transfer between metal ions of different valency states.

The covalent amorphous semiconductors are generally further divided into two groups. The first class contains materials with tetrahedral coordination such as the elements Si, Ge, and the III–V compounds. These materials can only be prepared in the amorphous phase by thin-film deposition. The second group contains the so-called chalcogenide glasses which are based on the elements S, Se, and Te, and to which other elements, e.g., Si, Ge, As, can be added. Section 5.4 of this review will be mainly concerned with this last class of materials. They have been selected because it is generally accepted that they are the best suited to test the band models proposed for amorphous semiconductors. Indeed, these covalent alloys may show compositional disorder in addition to translational disorder encountered in elemental and compound amorphous semiconductors. In the past a considerable amount of experimental work has been published on the transport properties of chalcogenide glasses. We shall not attempt to summarize all the published data, but we shall focus our attention on some results which have contributed some insight in the electronic structure of these materials. Therefore, the main purpose will be to find out whether the experimental results along with the transport formulas of Sect. 5.3 can yield evidence for the band structure of amorphous semiconductors described in Sect. 5.2.

5.2 Band Models

Experimental data of electrical transport properties can only be properly interpreted if a model for the electronic structure is available. Since the pioneering work of Bloch it is known that the electronic structure of the crystal shows universal characteristics. For semiconductors, the main features of the energy distribution of the density of electronic states $N(E)$ of crystalline solids are the sharp structure in the valence and conduction bands, and the abrupt terminations at the valence band maximum and the conduction band minimum. The sharp edges in the density of states produce a well-defined forbidden energy gap. Within the band the states are extended, which means that the wave functions occupy the entire volume. The specific features of the band structure are consequences of the perfect short-range and long-range order of the crystal. In an amorphous solid, the long-range order is destroyed, whereas the short-range order, i.e., the interatomic distance and the valence angle, is only slightly changed. The concept of the density of states is also applicable to noncrystalline solids.

Based on *Anderson*'s theory [5.4], *Mott* [5.5] argued that the spatial fluctuations in the potential caused by the configurational disorder in amor-

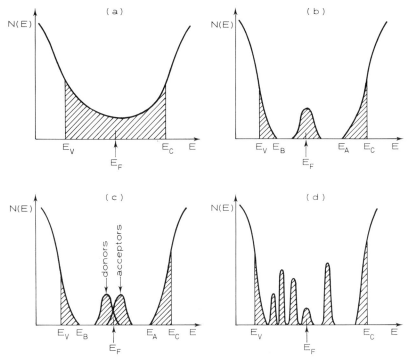

Fig. 5.1a–d. Schematic density of states diagrams for amorphous semiconductors. (a) The Cohen–Fritzsche–Ovshinsky model, (b) the Davis–Mott model showing a band of compensated levels near the middle of the gap, (c) modified Davis–Mott model, (d) a "real" glass with defect states

phous materials may lead to the formation of localized states, which do not occupy all the different energies in the band, but form a tail above and below the normal band. *Mott* postulated furthermore that there should be a sharp boundary between the energy ranges of extended and localized states. The states are called localized in the sense that an electron placed in a region will not diffuse at zero temperature to other regions with corresponding potential fluctuations.

Several models were proposed for the band structure of amorphous semiconductors, which were the same to the extent that they all used the concept of localized states in the band tails. Opinions vary, however, as to the extent of this tailing. Figure 5.1 illustrates, schematically, the main features of these various models.

5.2.1 The Cohen–Fritzsche–Ovshinsky Model

The CFO model [5.1], shown in Fig. 5.1a, assumes that the tail states extend accross the gap in a structureless distribution. This gradual decrease of the localized states destroys the sharpness of the conduction and valence band edges. The CFO model was specifically proposed for the multicomponent chalcogenide glasses used in the switching devices. More recently, a similar model [5.6] was used to calculate barrier formation and photoconductivity in amorphous Si. The authors suggested that in the chalcogenide alloys, the disorder is sufficiently great that the tails of the conduction and valence bands overlap, leading to an appreciable density of states in the middle of the gap. A consequence of the band overlapping is that there are states in the valence band, ordinarily filled, that have higher energies than states in the conduction band that are ordinarily unfilled. A redistribution of the electrons must take place, forming filled states in the conduction band tail, which are negatively charged, and empty states in the valence band, which are positively charged. This model, therefore, ensures self-compensation, and pins the Fermi level close to the middle of the gap, a feature required by the electrical properties of these materials. One of the major objections against the CFO model was the high transparency of the amorphous chalcogenides below a well-defined absorption edge. It is now almost certain from different observations that the extent of tailing in chalcogenides is rather limited. For a-Si, the model may be more relevant.

5.2.2 The Davis–Mott Model

According to *Davis* and *Mott* [5.2] the tails of localized states should be rather narrow and should extend a few tenths of an electron volt into the forbidden gap. They proposed furthermore the existence of a band of compensated levels near the middle of the gap, originating from defects in the random network, e.g., dangling bonds, vacancies, etc. Figure 5.1b sketches the Davis–Mott model; here, E_C and E_V represent the energies which separate the ranges where the states are localized and extended. The center band may be split into a donor and an acceptor band, which will also pin the Fermi level (Fig. 5.1c). *Mott* suggested that at the transition from extended to localized states the mobility drops by several orders of magnitude producing a mobility edge. Indeed, the concept of localized states implies that the mobility is zero at $T = 0 \, K$. The interval between the energies E_C and E_V acts as a pseudogap and is defined as the mobility gap. *Cohen* [5.7] proposed a slightly different picture for the energy dependence of the mobility. He suggested that there should not be an abrupt but rather a continuous drop of the mobility occurring in the extended states just inside the mobility edge. In this intermediate range the mean free path of the carriers becomes of the order of the interatomic spacing, so that the ordinary transport theory based on the Boltzmann equation cannot be used. Cohen described the transport as a Brownian motion in which the carriers are under the influence of a continuous scattering.

In recent years experimental evidence, mainly coming from luminescence, photoconductivity and drift mobility measurements, has been found for the existence of various localized gap states, which are split off from the tail states and are located at well-defined energies in the gap. These states are associated with defect centers, the nature of which is not always known. The presence of appreciable densities of different types of structural defects is not surprising if one realizes that most of the amorphous semiconductors are prepared by fast deposition from the vapour phase or by freezing-in the liquid state by quenching. At the present time special effort is made in order to unravel the nature of the defect centers, a problem which can also be attacked by studying the slightly disordered crystalline state. A first proposal of a model showing bands of donors and acceptors in the upper and lower halves of the mobility gap was already introduced by *Marshall* and *Owen* in 1971 [5.8], and is therefore sometimes called the Marshall–Owen model. It is clear now that the density of states of a "real" amorphous semiconductor does not decrease monotonically into the gap but shows many peaks which can be well separated from each other, as is the case in some chalcogenide glasses. The position of the Fermi level is largely determined by the charge distribution in the gap states. Figure 5.1d illustrates such a picture for the density of states distribution for glassy As_2Se_3 [5.9]. More detailed information about this problem will be presented in the last part of this review.

The interpretation of electrical transport data is closely interwoven with the energy distribution of the density of states. On the basis of the Davis–Mott model, there can be three processes leading to conduction in amorphous semiconductors. Their relative contribution to the total conductivity will predominate in different temperature regions. At very low temperatures conduction can occur by thermally assisted tunneling between states at the Fermi level. At higher temperatures charge carriers are excited into the localized states of the band tails; carriers in these localized states can take part in the electric charge transport only by hopping. At still higher temperatures carriers are excited across the mobility edge into the extended states. The mobility in the extended states is much higher than in the localized states. From this it follows that electrical conductivity measurements over a wide temperature range are needed to study the electronic structure of amorphous semiconductors. The formulas of electrical conductivity, thermoelectric power, and Hall coefficient dealing with these three types of conduction processes will be discussed in the next section.

5.2.3 Small-Polaron Model

The role of lattice distortion in the presence of an extra charge carrier in an amorphous solid has been discussed in detail by *Emin* [5.3]. He suggested that the charge carriers in some amorphous semiconductors may enter a self-trapped (small-polaron) state as a result of the polarization of the surrounding atomic lattice. In support of this hypothesis, *Emin* argued that the presence of disorder in a noncrystalline solid tends to slow down a carrier. This slowing down may lead to a localization of the carrier and, if the carrier stays at an atomic site sufficiently long enough for atomic rearrangements to take place, it may induce displacements of the atoms in its immediate vicinity, causing small-polaron formation. Since the small polaron is local in nature, the absence of long-range order in noncrystalline solids may be expected to have no significant influence on its motion. In support of the validity of his ideas, *Emin* was able to analyze experimental data of dc conductivity, thermopower, and Hall mobility obtained on some chalcogenide glasses in the framework of the existing small-polaron theories developed for crystalline solids. Because of the importance of this view we shall present in the next section in an elementary way some essential features of the theory of small-polaron motion and summarize very briefly the formulas governing the main transport properties.

5.3 Electrical Properties

In the preceding section it was pointed out that two different starting points have been adopted to describe the transport properties of amorphous semiconductors. One way is based on the acceptance of the Davis–Mott model. A meaningful interpretation to this model can only be made if one possesses adequate formulas for the transport coefficients in the three possible conduction channels. A major difficulty in deriving the expressions is the fact that from theoretical side the actual knowledge about the details of the distribution of the density of states is still relatively poor. In the next part we shall treat some of the most commonly used transport properties, which may in principle provide some information about the band structure. In the absence of well-settled details, we shall make use of very simplified models for the energy distribution of the electron states. The formulas derived in this way will be compared in the last section with some experimental results obtained on chalcogenide glasses.

5.3.1 dc Electrical Conductivity

The essential features of the Davis and Mott model for the band structure of amorphous semiconductors are the existence of narrow tails of localized states at the extremities of the valence and conduction bands and furthermore of a band of localized levels near the middle of the gap. This leads to three basically different channels for conduction.

a) Extended State Conduction

The conductivity for any semiconductor can be expressed in the form

$$\sigma = -e \int N(E)\mu(E)kT\frac{\partial f(E)}{\partial E}\,dE,$$ (5.1)

where $f(E)$ is the Fermi–Dirac distribution function,

$$f(E) = \frac{1}{1 + \exp[(E - E_F)/kT]}.$$

Using the relationship

$$\frac{\partial f(E)}{\partial E} = -f(E)[1 - f(E)]/kT,$$

σ can be written as

$$\sigma = e \int N(E)\mu(E)f(E)[1 - f(E)]\,dE. \tag{5.2}$$

In the Davis–Mott model, the Fermi level E_F is situated near the middle of the gap and thus sufficiently far from E_C, the energy which separates the extended from the localized states, so that Boltzmann statistics can be used to describe the occupancy of states,

$$f(E) = \exp[-(E - E_F)/kT].$$

According to *Mott*'s view, the mobility drops sharply at the critical energy E_C (or E_V) but at present it is not exactly known how the mobility depends on the energy in both conduction regimes.

In the nondegenerate case and under the assumption of a constant density of states and constant mobility, the conductivity due to electrons excited beyond the mobility edge into the extended states is given by

$$\sigma = e N(E_C)kT\,\mu_C \exp[-(E_C - E_F)/kT], \tag{5.3}$$

where μ_C is the average mobility. The number of electrons is given by

$$n = \int_{E_C}^{\infty} N(E_C)\exp[-(E - E_F)/kT]\,dE$$

$$= N(E_C)kT \exp[-(E_C - E_F)/kT]. \tag{5.4}$$

In order to get an idea of the order of magnitude of μ_C we shall follow *Mott*'s treatment [5.10]. We define $\sigma(E_C) = e N(E_C)\mu_C kT$. If $N(E_C) \simeq \langle N(E)\rangle/3$, where $\langle N(E)\rangle$ is the average density of states over the band, then

$$\sigma(E_C) = e\langle N(E)\rangle\mu_C kT/3.$$

Mott calculated the lowest value of the electrical conductivity before the start of an activated process, i.e., just at E_C. This quantity he called the "minimum metallic conductivity." He derived the expression

$$\sigma_{min} = const \cdot e^2/\hbar a, \tag{5.5}$$

where the constant lies in the range between 0.026 and 0.1 ; σ_{min} is usually of the order 200–300 Ω^{-1} cm^{-1}. Taking const $=0.026$, one finds for the mobility

$$\mu_C = 0.078(e/\hbar a)/\langle N(E)\rangle kT.$$

In the nearly free electron model, $N(E)$ is given by

$$N(E) = km/2\pi^2\hbar^2 .$$

On the other hand, the maximum energy E_{max} of the band, which also yields the width B of the band, is expressed by

$$E_{max} = \frac{\hbar^2\pi^2}{2ma^2} \quad and \quad k_{max} = \frac{\pi}{a}.$$

This yields $N(E) \simeq 1/a^3 B$.

Introducing this result in the expression for μ_C one gets

$$\mu_C = 0.078\, ea^2 B/\hbar kT. \tag{5.6}$$

Taking $a=2$ Å, $B=5$ eV one finds at room temperature that $\mu_C \simeq 10$ cm^2 V^{-1}s^{-1}. This value corresponds to a mean free path comparable or less than the interatomic distance. *Cohen* [5.7] suggested that conduction in this case would be more properly described as a diffusive or Brownian-type motion. In this regime the mobility can be obtained with the help of Einstein's relation,

$$\mu = eD/kT.$$

The diffusion coefficient D may be written as

$$D = (1/6)va^2 ,$$

where v is the jump frequency and a the interatomic separation. The mobility in the Brownian-motion regime is then given by

$$\mu_C = \tfrac{1}{6}\frac{ea^2}{kT}v. \tag{5.7}$$

This expression yields the same temperature dependence as (5.6) derived by *Mott*. Since $\mu_C \propto 1/kT$, one expects that the expression for the conductivity is of the form

$$\sigma = \sigma_0 \exp[-(E_C - E_F)/kT]. \tag{5.8}$$

Optical absorption measurements made on amorphous semiconductors have shown that the band gap decreases with increasing temperature. The energy distance $E_C - E_F$ therefore will show a similar behavior, and, under the assumption of a linear temperature dependence,

$$E_C - E_F = E(0) - \gamma T, \tag{5.9}$$

the expression for the conductivity becomes

$$\sigma = \sigma_0 \exp(\gamma/k) \exp[-E(0)/kT]. \tag{5.10}$$

Here $E(0)$ is the energy distance at $T = 0$ K.
    We can write this formula in the form

$$\sigma = C_0 \exp[-E(0)/kT], \tag{5.11}$$

where

$$C_0 = e N(E_C) kT \mu_C \exp(\gamma/k). \tag{5.12}$$

As seen before μ_C is proportional to $1/T$, so that the preexponential factor C_0 is temperature independent. *Mott* [5.5] has made an estimate of the pre-exponential σ_0. In general σ_0 may lie between 10 and $10^3\,\Omega^{-1}\,\mathrm{cm}^{-1}$ in most amorphous semiconductors. An estimate of γ can be obtained from the temperature dependence of the optical gap. In chalcogenide glasses the temperature coefficient of the optical gap generally lies between 4×10^{-4} and 8×10^{-4} eV deg^{-1}. As the Fermi level is situated near the middle of the gap, values of γ of approximately half this magnitude are expected and hence values of $\exp(\gamma/k)$ in the range 10–100 seem to be most probable.
    A different approach based on a so-called random phase model was used by *Hindley* [5.11] and *Friedman* [5.12] to calculate the mobility in the extended states near E_C or E_V. The basic feature of this model, used to describe an idealized situation of the amorphous state, was first put forward by *Mott* [5.5] and *Cohen* [5.7] and is related to the character of the extended state wave function. *Hindley* and *Friedman* made the ad hoc assumption that the phase of the probability amplitude for finding an electron on a particular atomic site varies randomly from site to site. Assuming a constant density of states $N(E_C)$ near the mobility edge, they derived the following expression for the dc conductivity in the extended states:

$$\sigma = \frac{2\pi e^2}{3\hbar a} \{z a^6 J^2 [N(E_C)]^2\} \exp[-(E_C - E_F)/kT]. \tag{5.13}$$

Here a is the interatomic spacing, z the coordination number, and J the electronic transfer integral. The dc conductivity within the random phase model follows a simple exponential law:

$$\sigma = \sigma_0 \exp[-(E_C - E_F)/kT]. \tag{5.14}$$

The conductivity mobility can be derived from (5.13) by dividing by ne, where the electron density is given by (5.4). This yields

$$\mu_C = \frac{2\pi e a^2}{3\hbar} z \left[\frac{J}{kT} a^3 J N(E_C)\right]. \tag{5.15}$$

The mobility decreases with increasing temperature as T^{-1}. Inserting typical numbers $a = 2\,\text{Å}$, $z = 4$, $J = 1\,\text{eV}$, $N(E_C) = 10^{21}\,\text{cm}^{-3}\,\text{eV}^{-1}$, $T = 300\,\text{K}$, one finds a mobility of $2\,\text{cm}^2\,\text{V}^{-1}\,\text{s}^{-1}$.

b) Conduction in Band Tails

If the wave functions are localized, so that $\sigma(E) = 0$, conduction can only occur by thermally activated hopping. Every time an electron moves from one localized state to another it will exchange energy with a phonon. It may be expected that the mobility will have a thermally activated nature,

$$\mu_{\text{hop}} = \mu_0 \exp[-W(E)/kT]. \tag{5.16}$$

The preexponential μ_0 has the form

$$\mu_0 = (1/6)v_{\text{ph}}eR^2/kT, \tag{5.17}$$

where v_{ph} is the phonon frequency and R the distance covered in one hop.

For a typical phonon frequency $v_{\text{ph}} = 10^{13}\,\text{s}^{-1}$ and $W \simeq kT$, (5.16) yields a mobility of the order $10^{-2}\,\text{cm}^2\,\text{V}^{-1}\,\text{s}^{-1}$ at room temperature. Comparison of this value with the one calculated for conduction in the extended states suggests, as postulated by *Mott*, that the mobility may drop by a factor of at least 100 at the energy which separates the localized and nonlocalized states.

The conductivity, being an integral over all available energy states, will depend on the energy distribution of the density of localized states. If one assumes that the density of states $N(E)$ behaves as some power s of E,

$$N(E) = \frac{N(E_C)}{(\Delta E)^s}(E - E_A)^s \tag{5.18}$$

with $\Delta E = E_C - E_A$, then the conductivity σ_{hop} due to electrons can be easily calculated starting from (5.2):

$$\sigma_{\text{hop}} = \sigma_{0\,\text{hop}}\left(\frac{kT}{\Delta E}\right)^s C \exp[-(E_A - E_F + W)/kT], \tag{5.19}$$

where

$$\sigma_{0\,\text{hop}} = (1/6) v_{\text{ph}} e^2 R^2 N(E_C),\tag{5.20}$$

and

$$C = s! - \left(\frac{\Delta E}{kT}\right)^s \exp\left(-\frac{\Delta E}{kT}\right)\left[1 + s\left(\frac{kT}{\Delta E}\right) + s(s-1)\left(\frac{kT}{\Delta E}\right)^2 + \dots\right].\tag{5.21}$$

For the specific case of $s = 1$ (linear variation), the conductivity is given by

$$\sigma_{\text{hop}} = \sigma_{0\,\text{hop}} \frac{kT}{\Delta E} C_1 \exp\left[-(E_A - E_F + W)/kT\right],\tag{5.22}$$

with

$$C_1 = 1 - \exp\left(-\frac{\Delta E}{kT}\right)[1 + (\Delta E/kT)].\tag{5.23}$$

c) Conduction in Localized States at the Fermi Energy

If the Fermi energy lies in a band of localized states, as predicted by the Davis–Mott model, the carriers can move between the states via a phonon-assisted tunneling process. This is the transport analogous to impurity conduction observed in heavily doped and highly compensated semiconductors at low temperatures. An estimate for the temperature dependence of the hopping conductivity at E_F has been given by *Mott* [5.13]. We shall follow here his original derivation. Let us consider an electron that is scattered by phonons from one localized state to another. The energy difference between the states is denoted by W. The probability p that an electron will jump from one state to another is determined by three factors, which are the following:

I) The probability of finding a phonon with an excitation energy equal to W, given by a Boltzmann expression $\exp(-W/kT)$.

II) An attempt frequency v_{ph}, which cannot be greater than the maximum phonon frequency (in the range $10^{12} - 10^{13}\,\text{s}^{-1}$).

III) The probability of electron transfer from one state to another. This factor depends on the overlapping of the wave functions and should be given by $\exp(-2\alpha R)$. Here R is the jumping distance, which at high temperatures equals the interatomic spacing, and α is a quantity which is representative for the rate of fall-off of the wave function at a site. If overlapping of the wave functions is important, the factor $\exp(-2\alpha R)$ becomes of the order of one.

The probability p that an electron jumps will then be expressed by

$$p = v_{\text{ph}} \exp(-2\alpha R - W/kT).\tag{5.24}$$

By making use of Einstein's relation

$$\mu = \frac{eD}{kT}$$

with $D = (1/6)pR^2$, the conductivity can be written as

$$\sigma = (1/6)e^2 pR^2 N(E_F).$$

Here $N(E_F)$ is the density of states at the Fermi level and $N(E_F)\,kT$ is the number of electrons that contribute to the conductivity.

Using (3.24) the conductivity is represented by,

$$\sigma = (1/6)e^2 R^2 v_{ph} N(E_F) \exp(-2\alpha R) \exp(-W/kT). \tag{5.25}$$

As the temperature is lowered the number and energy of phonons decrease, and the more energetic phonon-assisted hops will progressively become less favorable. Carriers will tend to hop to larger distances in order to find sites which lie energetically closer than the nearest neighbors. This mechanism is the so-called variable range hopping. The factor $\exp(-2\alpha R - W/kT)$ will not have its maximum value for the nearest neighbors. In order to find the most probable hopping distance, *Mott* used an optimization procedure. This is as follows. If $N(W)$ is the density of states per unit volume and per unit energy, then the number of states with energy difference W within a distance R from a particular atom is given by

$$\frac{4\pi}{3} R^3 N(W) W.$$

The electron can leave its site only if the number of accessible sites is at least one. Taking this into account one gets for the average energy spacing between states near the Fermi level

$$W = \frac{3}{4\pi R^3 N(E_F)}, \tag{5.26}$$

and for the jump probability

$$p = v_{ph} \exp\{-2\alpha R - [(4\pi/3)N(E_F)R^3 kT]^{-1}. \tag{5.27}$$

The most probable jump distance is found by minimizing the exponent of the above expression as a function of R:

$$R = \left[\frac{9}{8\pi\alpha N(E_F)kT}\right]^{1/4}. \tag{5.28}$$

This gives a jump frequency of the form

$$p = v_{ph} \exp\left(-\frac{A}{T^{1/4}}\right),$$ (5.29)

where

$$A = 2.1[\alpha^3/k N(E_F)]^{1/4}.$$ (5.30)

Mott's treatment of variable range hopping leads to a temperature dependence for the conductivity of the form

$$\sigma = (1/6)e^2 R^2 v_{ph} N(E_F) \exp\left(-\frac{A}{T^{1/4}}\right),$$ (5.31)

or

$$\sigma = \sigma_0(T) \exp\left(-\frac{A}{T^{1/4}}\right).$$ (5.32)

There have been several more derivations of the conductivity formula for variable range hopping. In general, the $T^{-1/4}$ relationship remains unchanged, but slightly different values were found for the numerical factor in A. In *Mott's* derivation the prefactor $\sigma_0(T)$ is given by

$$\sigma_0(T) = \frac{e^2 v_{ph} N(E_F)}{6} \left[\frac{9}{8\pi\alpha N(E_F)kT}\right]^{1/2} = \frac{e^2}{2(8\pi)^{1/2}} v_{ph} \left[\frac{N(E_F)}{\alpha kT}\right]^{1/2}.$$ (5.33)

In principle, the two parameters α and $N(E_F)$ can be evaluated from the slope of a plot of $\ln[\sigma(T)T^{1/2}]$ vs $T^{-1/4}$ and from the intercept at $T^{-1/4} = 0$, if one makes a reasonable assumption for v_{ph}. Alternatively, one can get an idea of the most probable hopping distance R at a given temperature by inserting values for α and $N(E_F)$ in (5.28). Assuming $\alpha^{-1} = 10$ Å and $N(E_F) = 10^{19}$ cm^{-3} eV^{-1}, one finds $R = 250\, T^{-1/4}$ Å which yields 80 Å at 100 K.

In the recent literature many experimental studies, especially on the tetrahedral bonded materials, dealt with variable range hopping. In many cases, unreasonably high values were calculated for $N(E_F)$, very often by use of the preexponential factor $\sigma_0(T)$. Although the $T^{-1/4}$ dependence of $\ln \sigma$ is widely observed, these findings seem to indicate that the above expressions do not describe in an exact quantitative way the conductivity. It must be mentioned here that *Mott's* derivation of variable range hopping implies a large number of simplifying assumptions of which the most important are: energy independence of the density of states at E_F, neglection of correlation effects in the tunneling process, omission of multiphonon processes, and neglection of the electron–phonon interaction. Several authors have investigated the effect of departure from a constant density of states and have clearly demonstrated that the energy distribution of the density of states is of major importance in the theory of variable range hopping. A review article on hopping conductivity in disordered solids has recently been published by *Overhof* [5.14].

5.3.2 Thermopower

Important information about the transport mechanism in amorphous semiconductors has been obtained from thermopower data. In this section we shall derive the formulas for the thermopower associated with the three possible processes of conduction.

Fritzsche [5.15] has given a general expression for the thermopower S,

$$S = -\frac{k}{e} \int \frac{E - E_F}{kT} \frac{\sigma(E)}{\sigma} dE. \tag{5.34}$$

If we combine this with (5.1) for the conductivity, then we have,

$$S = -\frac{k}{e} \frac{\int \mu(E)N(E)kT[(E-E_F)/kT]\frac{\partial f}{\partial E}dE}{\int \mu(E)N(E)kT\frac{\partial f}{\partial E}dE}, \tag{5.35}$$

where f is the Fermi–Dirac distribution function.

When we use the relationship

$$\frac{\partial f}{\partial E} = -f(1-f)/kT, \tag{5.36}$$

then S can be written as

$$S = -\frac{k}{e} \frac{\int \mu(E)N(E)[(E-E_F)/kT]f(1-f)dE}{\int \mu(E)N(E)f(1-f)dE}. \tag{5.37}$$

For a nondegenerate semiconductor classical Boltzmann statistics are appropriate. In this case the factor $f(1-f)$ in (5.37) reduces to a Boltzmann factor $f = \exp[-(E-E_F)/kT]$.

a) Conduction in Extended States

Under the assumption of a constant density of states and an energy-independent mobility the thermopower can be readily found by integrating

(5.37). This yields an expression of the familiar form for band conduction of electrons

$$S = -\frac{k}{e}\left(\frac{E_C - E_F}{kT} + A\right),\tag{5.38}$$

with $A = 1$.

In crystalline semiconductors it is well known that the kinetic term A depends on the scattering mechanism.

Comparison of (5.8) and (5.38) shows that a plot of $\ln\sigma$ and of S vs $1/T$ should have the same slope if conduction takes place in the extended states. *Hindley* [5.11] has found a similar expression for S using the random phase model.

If both electrons and holes contribute to the conductivity then the thermopower is the algebraic sum of the individual contributions S_e and S_h but each weighed according to the ratio of its conductivity to the total conductivity. Thus we have

$$S = \frac{S_e\sigma_e + S_h\sigma_h}{\sigma_e + \sigma_h}.\tag{5.39}$$

b) Conduction in Localized States Near the Mobility Edge

The thermopower built up by the carriers conducting in the localized states of the band tails will be given by

$$S = -\frac{k}{e}\frac{\int[(E - E_F)/kT]\exp[-(E - E_F)/kT]N(E)dE}{n}.\tag{5.40}$$

If it is again assumed that the density of localized states behaves like some power s of the energy E,

$$N(E) = \frac{N(E_C)}{(E_C - E_A)^s}(E - E_A)^s,$$

then one obtains

$$S = -\frac{k}{e}\left(\frac{E_A - E_F}{kT} + \frac{C^*}{C}\right),\tag{5.41}$$

where C is defined by (5.21), and

$$C^* = \int_0^{\frac{\Delta E}{kT}} e^{-x}x^{s+1}dx.$$

Here $\Delta E = E_C - E_A$ is the width of the tail and $x \equiv (E - E_A)/kT$.

The value of the term C^*/C depends on the energy distribution of the density of localized states. For a linear variation of $N(E)$, S can be represented by

$$S = -\frac{k}{e}\left(\frac{E_A - E_F}{kT} + \frac{C_1^*}{C_1}\right),$$

(5.42)

with

$$C_1^* = 2 - \{\exp(-\Delta E/kT)[2 + 2(\Delta E/kT) + (\Delta E/kT)^2]\},$$

(5.43)

and

$$C_1 = 1 - \{\exp(-\Delta E/kT)[1 + (\Delta E/kT)]\}.$$

If the carriers move by hopping in the localized states, the conductivity varies nearly exponentially with temperature [apart from a small temperature-dependent preexponential term, as shown by (5.19)], and the measured activation energy is the sum of the activation energies for carrier creation and for hopping. The activation energy for hopping does not appear in the expression for S and, therefore, one expects a difference in slope between the conductivity and thermopower curve.

c) Conduction in Localized States at the Fermi Energy

At low temperature, charge transport can arise from electrons tunneling between states at E_F. *Cutler* and *Mott* [5.16] suggested that the thermopower in this regime should be identical to the equation used for metallic conduction,

$$S = \frac{\pi^2 k^2 T}{3e}\left[\frac{d \ln \sigma(E)}{dE}\right]_{E_F},$$

(5.44)

since the Fermi level lies in a region where the density of states is finite. The thermopower is expected to be small. Its sign may be positive or negative, depending on whether the major contribution to the current will lie below or above the Fermi energy. Several authors [5.17–20] have calculated the thermopower in the variable range hopping regime. Their results disagree, however, in the estimation of the temperature dependence of S.

5.3.3 Hall Effect

The basic transport properties usually measured in crystalline semiconductors are the conductivity σ and the Hall coefficient R_H. For unipolar conduction, measurements of R_H provide a reliable guide to the charge carrier concentration. For n-type semiconductors the Hall coefficient is negative and is given by the general formula,

$$R_H = -\frac{r}{ne}.$$

Here r is the scattering factor. It is usually not much greater than one. In p-type semiconductors, the Hall coefficient is positive. From σ and R_H we may determine the Hall mobility,

$$\mu_H = |R_H|\sigma = r\mu.$$

Thus the Hall mobility is greater than the conductivity mobility μ by the scattering factor r.

Interpretation of the Hall coefficient on this basis is valid for materials in which the mean free path is long compared with the interatomic spacing. In amorphous semiconductors the mobilities are found to be very low so that the carriers will move with a mean free path comparable to the interatomic distance. As a consequence the ordinary transport theory based on the Boltzmann equation cannot be used anymore. So far measurements of the Hall coefficient have been made on a limited number of amorphous semiconductors due to the high resistivities and the low carrier mobilities. In most chalcogenide glasses the Hall coefficient was found to be negative, in contrast with the thermopower which showed a positive sign.

Using the random phase model, describing the crystal wave function as a linear superposition of atomic wave functions with coefficients which have no phase correlation from site to site, *Friedman* [5.12] was able to find an expression for the Hall mobility. This approach, which was used by *Hindley* [5.11] to calculate the conductivity, may be applicable to the conduction regime in the extended states near the mobility edge. *Friedman*'s formula is

$$\mu_H = 4\pi \left(\frac{ea^2}{\hbar}\right) [a^2 J N(E_C)] \eta \left(\frac{\bar{z}}{z}\right). \tag{5.45}$$

Here J is the overlap energy integral between neighboring sites; a is the interatomic spacing; z is the coordination number and $\bar{z}$ is the number of interacting sites (a minimum of three sites is necessary in order to obtain a Hall effect); η is a parameter usually of the order of $1/3$.

From (5.45) it follows that the Hall mobility within the random phase model is temperature independent. *Friedman* made an estimate of the magnitude of μ_H and obtained $\mu_H \simeq 10^{-1}\,\mathrm{cm^2V^{-1}s^{-1}}$. The Hall mobility is related to the conductivity mobility (5.15) by a simple relation,

$$\frac{\mu_H}{\mu_C} = \frac{kT}{J}. \tag{5.46}$$

J is a quantity of the order of 1 eV ($2zJ$ is the band width obtained from a tight-binding calculation), so that $J \gg kT$. The Hall mobility will be considerably smaller than the conductivity mobility, the ratio being at least 1/10. An important feature resulting from *Friedman's* treatment is connected with the sign of the Hall coefficient. For a configuration of three interacting sites the Hall coefficient is negative, even if holes are responsible for the conductivity. This result is consistent with the *p–n* anomaly between the thermopower and the Hall effect commonly encountered in amorphous semiconductors. As pointed out by *Holstein* [5.21], the basic elementary process responsible for the Hall effect is in *Friedman's* treatment in essence the same as the one used for small-polaron hopping, where it was concluded that the sign of the Hall effect is the same for holes as for electrons when the elementary jump process involves a three-site configuration. In his paper *Holstein* presented a prescription for the sign of the Hall coefficient for holes relative to that for electrons in hopping-type charge transport. The relative sign for holes is related to n, the minimum number of sites in a closed path, by $-(-1)^n$. Thus, for triangular closed paths ($n=3$), a hole yields the same sign for the Hall effect as that of an electron, namely negative. In a four-site configuration ($n=4$) (typical for a cubic lattice) the sign of the Hall effect for holes is opposite to that for electrons.

The sign of the Hall effect remains one of the most outstanding problems at the moment. Differences in sign of the carrier as determined by thermopower and Hall effect measurements have been reported for many amorphous semiconductors. *Le Comber* et al. [5.22] have observed a double sign reversal in amorphous silicon films: a *n*-type Hall effect in conjunction with a *p*-type thermopower, and a *p*-type Hall effect coexisting with a *n*-type thermopower. *Mell* [5.23] found the same anomalies in amorphous III–V compounds. Similarly, in CdGe$_x$As$_2$ glasses, a double anomaly has been measured by *Callaerts* et al. [5.24]. Moreover, in amorphous arsenic, a *p*-type Hall effect and a *n*-type thermopower has been found by *Mytilineou* and *Davis* [5.25]. In a recent paper *Emin* [5.26] demonstrated that the sign of the Hall effect in hopping conduction depends not only on the local geometry but also on the nature (*s*-like, *p*-like, bonding or antibonding) and the relative orientations of the local orbitals between which the carrier moves. He was able to show that arrangements of specific orbitals may account for the sign anomalies of the Hall effect encountered in noncrystalline solids.

5.3.4 ac Conduction

Information about the nature of the conduction mechanism in a material can be obtained from ac conductivity. In these experiments the electrical con-ductivity is measured as a function of the frequency ω of an alternating electric field. The frequency range usually covered in the measurements extends from 10^2 to 10^8 Hz. A difference in behavior of $\sigma(\omega)$ is expected when conduction occurs by motion of charge carriers in the extended states or by hopping in localized states. The two types of conduction processes will now be examined.

When carriers are excited across the mobility edges into the range of extended states, the conductivity does not depend on frequency, at least in the frequency range up to 10^8 Hz. On the other hand, when conduction occurs by phonon-assisted hopping between localized states one expects the conductivity to increase with frequency. *Austin* and *Mott* [5.27] have derived a formula when hopping conduction takes place near E_F:

$$\sigma(\omega) = \frac{\pi}{3} e^2 kT [N(E_F)]^2 \alpha^{-5} \omega [\ln (v_{ph}/\omega)]^4, \tag{5.47}$$

where v_{ph} is a phonon frequency (typically of the order of 10^{13} s^{-1}) and α describes the decay of the localized state wave function ($1/\alpha$ is the effective range of the wave function).

Two important features can be drawn from this equation: (I) $\sigma(\omega)$ varies linearly with temperature; (II) because of the presence of the logarithmic term $[\ln (v_{ph}/\omega)]^4$, the slope of a plot $\ln \sigma(\omega)$ versus $\ln \omega$ is not constant, but decreases slightly with increasing frequency of the applied field.

The above equation can be approximated by an expression of the form

$$\sigma(\omega) \simeq \text{const} \cdot \omega^s, \tag{5.48}$$

where the exponent s is defined as $d(\ln \sigma)/d(\ln \omega)$ and is given by

$$s = 1 - \frac{4}{\ln (v_{ph}/\omega)}. \tag{5.49}$$

Taking $v_{ph} = 10^{13}$ s^{-1} the exponent decreases from a value $s = 0.84$ at $\omega = 10^2$ s to $s = 0.65$ at $\omega = 10^8$ s^{-1}. One often writes the frequency dependence of the ac conductivity in the form

$$\sigma(\omega) = \text{const} \cdot \omega^{0.8},$$

which means that the logarithmic term $[\ln (v_{ph}/\omega)]^4$ varies as $\omega^{-0.2}$, in principle only strictly valid for frequency values in the neighborhood of 10^4 Hz.

Experimentally one has found that the ac conductivity of many chalcogenide glasses follows over a large frequency range the relationship given by (5.48) with s lying between 0.7 and 1.0. Values close to one are difficult to understand on the basis of this formula, since they only can be obtained for unreasonably high values of the phonon frequency v_{ph}.

If one makes a proper choice of the parameter α^{-1}, (5.47) offers a straightforward way for evaluating the density of states at the Fermi level. For a number of chalcogenide glasses estimates in the range 10^{18}–10^{20} eV^{-1}cm^{-3} were obtained. These values are considered to be much too high and are difficult to reconcile with those deduced from other experiments such as optical absorption or photoinduced ESR, which give estimates lower by several orders of magnitude. As seen in Sect. 5.3.2, another approach to the evaluation of $N(E_F)$ is based on the temperature dependence of the dc conductivity for variable range hopping at the Fermi level. By utilizing (5.31), an estimate of $N(E_F)$ should be obtainable from the slope of a ln σ versus $T^{-1/4}$ plot. This procedure has been followed for amorphous Si and Ge but cannot be applied to most of the chalcogenide glasses because of the absence of any observable variable range hopping even at the lowest temperature of measurement. It must be remarked here that the use of the Austin-Mott formula to describe to observed $\sigma(\omega) \propto \omega^s$ behavior has been strongly criticized, essentially on the grounds that a similar frequency dependence has been observed in a number of insulating noncrystalline solids, such as polymers, molten quartz, SiO_2 glasses, etc. Therefore, it is now generally accepted that there may exist many other sources of dielectric loss mechanisms, such as surface barriers, ionic dipoles, dipole layers, inhomogeneities, which could give rise to a ω^s behavior. These observations certainly indicate that the interpretation of the results on the basis of the electronic model described by (5.31) and which implies phonon-assisted hopping in localized states at E_F is not necessarily the unique one.

According to the Davis–Mott model of the band structure a third channel for conduction is available when charge carriers are excited into the localized states of the band tails. Because of the thermally activated nature of this transport process, one expects that $\sigma(\omega)$ will also increase as $\omega[\ln(v_{ph}/\omega)]^4$, i.e., approximately as $\omega^{0.8}$. A distinction between hopping near the mobility edges and nearby the Fermi energy can be found by looking at the temperature dependence of $\sigma(\omega)$. The creation of charge carriers into the band tails requires an activation energy $E_A - E_F$ for electrons ($E_F - E_B$ for holes), yielding an exponentially increasing number of charge carriers with raising temperature. This term must be incorporated in the expression for $\sigma(\omega)$, which will therefore have a temperature dependence of the form $\sigma(\omega, T) \propto kT \exp(-\Delta E/kT)$. Since the exponential term usually plays by far the predominant role, the ac conductivity should show in the case of hopping in the tails the same temperature dependence as the dc conductivity curve. For transport at E_F the number of carriers remains constant and $\sigma(\omega)$ varies linearly with temperature as can be seen from (5.47). It is clear that the mechanism yielding the highest contribution to the conductivity at a given temperature will be observed experimentally. When changing the temperature one should be able to see the transition from one mechanism to another. Experimental data of ac conductivity for some amorphous chalcogenide glasses will be presented in Sect. 5.4.

5.3.5 Transit Time

Transit time characteristics are obtained from time-of-flight experiments. In such an experiment, the sample, which is covered with two contacts, one of which is semitransparent, is illuminated for a short time with a strongly absorbed light pulse. This illumination causes a current $I(t)$ of unipolar carriers

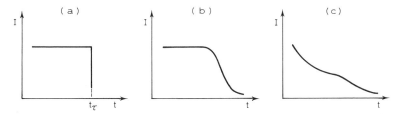

Fig. 5.2a–c. Typical carrier transit pulses, (a) ideal, (b) broadening due to Gaussian distribution of drift mobilities, (c) strong dispersive broadening

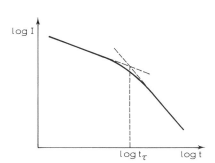

Fig. 5.3. Double logarithmic plot of transient current I versus time t, in case of strong dispersion

to flow through the sample, across which a constant voltage is applied. In the ideal case the injected carrier sheet reaches the back electrode after a definite transit time t_τ (see Fig. 5.2a), from which a drift mobility can be derived. In reality the drift velocities of the different carriers obey Gaussian statistics. As a consequence the charge sheet, which propagates with a constant mean drift velocity through the sample, is broadened. The net result is a rounding of the $I(t)$ curve (Fig. 5.2b).

Recent measurements on some amorphous materials such as a-Se at low temperatures, a-As$_2$Se$_3$ and polymeric substances doped with molecules of different ionization energies [5.28–30] show a departure from the Gaussian statistics. The $I(t)$ versus time t plot is rather featureless and no well-defined transit time (and drift mobility consequently) can be derived from such a measurement (Fig. 5.2c). A plot of log $I(t)$ versus log t usually yields two straight lines with different slopes (see Fig. 5.3). The intersection of these lines is then used to define a "transit time" t_τ. In general, this transit time and the deduced drift mobility depend on the field strength and the sample thickness and show a thermally activated behavior. This temperature dependence may be different from that of the conductivity mobility.

In order to explain the decrease with time of the mean velocity as the charge packet moves through the sample various models have been proposed. A common characteristic of these descriptions is the presence of some stochastic variable in the system, which is responsible for the observed transit time dispersion.

The possible transport mechanisms which have been considered are the following:

(I) *Multiple Trapping* [5.8]

Here, a distribution of the energies of the trapping levels has to account for the non-Gaussian behavior. The injected carrier moves in a band of extended states, but occasionally will be trapped for some time. Since the time a carrier stays at a trapping site is closely related to the energy distance between this level and the band and furthermore to the temperature, a distribution in this distance can be responsible for transit time dispersion to occur. The dispersion will be temperature dependent, i.e., the shape of the $I(t)$ curve will change with temperature.

(II) *Hopping between Localized States* [5.31]

This accounts for the temperature independent dispersion of the transit time observed in some amorphous semiconductors. In this model the temperature-independent hopping distance R is assumed to vary stochastically in the system. The reason that such a fluctuation can cause strong dispersion is directly seen from the expression which related this hopping distance to the hopping probability p [see (5.24)],

$$p = v_{ph} \exp(-2\alpha R) \exp(-W/kT).$$

Since R appears as an exponent in the tunneling factor, a small change in R may cause an amplified change of p and, hence, of the release time. Physically this means that an injected carrier moves towards the back electrode by hopping with a constant hopping energy W, but this carrier occasionally will meet a hopping site in which it stays for a rather long time. Therefore, only a small fraction of the injected carriers, i.e., the fastest ones, will reach the electrode without passing a "difficult" hopping site, while most of the other carriers are delayed by a different number of such hops. At this stage it may be remarked that inclusion of the hopping energy W as a stochastic variable will generate a temperature-dependent dispersion, which may be described in terms of the stochastic hopping model. Therefore, it might be difficult to find a distinction between the multiple trapping model and the hopping model in case of temperature-dependent dispersion. On the other hand, if one observes a temperature-independent dispersion, the hopping model may give a more consistent explanation than the trapping model.

III) *Trap-Controlled Stochastic Hopping* [5.32]

In some experiments a rather high activation energy of the drift mobility has been observed, which is difficult to associate with the hopping energy. In order to explain this observation, *Pfister* and *Scher* [5.32] presented a modification of the hopping model in the way that the presence of traps is incorporated in the system. This new model includes the features of the stochastic hopping model such as the temperature independence but also offers the possibility of carrier capture by isolated trapping sites. Therefore, the dispersion still originates from the hopping-distance fluctuations while the mobility activation energy is related to trap depth.

Although these models cannot be directly applied to explain the dark dc conductivity, the thermopower, etc. (the observation time in these cases exceeds all event times), they may offer a more detailed insight in the microscopic transport mechanism.

5.3.6 Photoconductivity

Photoconductivity is the change in the electrical conductivity when a material is exposed to electromagnetic radiation. An excess conductivity $\Delta\sigma$ appears if, under the action of absorbed light, the densities of the charge carriers n and p increase compared with their values at thermal equilibrium:

$$\Delta\sigma = e(\Delta n \mu_n + \Delta p \mu_p). \tag{5.50}$$

At low temperatures the values Δn and Δp may be considerably higher than the corresponding equilibrium densities n_0 and p_0. Under steady-state conditions, the excess densities are equal to their rate of generation g (i.e., the number of carriers generated per unit time in unit volume) multiplied by their average lifetime τ:

$$\Delta n = g\tau_n \quad \text{and} \quad \Delta p = g\tau_p. \tag{5.51}$$

The generation rate is governed by the quantum yield η, which is the number of electron-hole pairs generated by the absorption of a photon.

The nonequilibrium charge carriers exist until they disappear by recombination which may occur, in general, through three processes: direct recombination of a free electron with a free hole, capture of an electron by a center in which a hole is localized, and capture of a hole by a center in which there is a bound electron. In steady state, the generation rate of carriers is equal to the recombination velocity.

In crystalline semiconductors, steady-state photoconductivity has been extremely successful in determining the recombination center parameters, in particular, the location of the level in the forbidden gap. If one type of dominant recombination center is present, the nonequilibrium lifetime is governed by the processes of electron capture and subsequent hole capture by the local levels of the dominant center, and for this simple case the rate analysis is straightforward. However, when different defect centers act as trapping and (or) recombination levels, the analysis becomes much more complicated since one has to consider a whole set of possible electron and hole transitions and write down corresponding equations describing the rate of change of occupation of the levels. In Sect. 2.2 we saw that the band structure of a real amorphous semiconductor shows such discrete energy levels associated with defect states, and this directly indicates the complexity of the analysis of

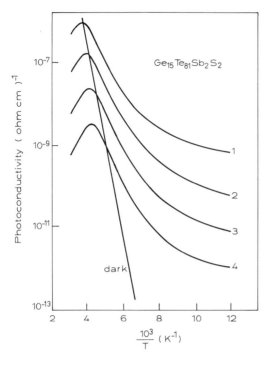

Fig. 5.4. Dark conductivity and photoconductivity as a function of temperature for $Ge_{15}Te_{81}Sb_2S_2$ for different excitation intensities [5.34]

experimental photoconductivity data. Measurements of this type have been reported in various amorphous materials and interpreted by the authors in different ways. The treatments differ in the model used for the density of gap states and the transitions involved. In general, two different approaches have been adopted. The first one considers a slowly varying trap distribution, consistent with the Cohen–Fritzsche–Ovshinsky model of the band structure. Several variants of the original model have been introduced to interpret the photoconductivity data in chalcogenide glasses [5.33–36]. A contrasting approach was put forward by *Main* and *Owen* [5.37], who start from a band model showing more or less discrete sets of localized states. This picture clearly ressembles the one which is commonly used in the crystalline state.

When measured as a function of temperature, the photoconductivity of most amorphous semiconductors shows a typical behavior illustrated in Fig. 5.4. Here the photoconductivity is plotted versus the reciprocal temperature for a $Ge_{15}Te_{81}Sb_2S_2$ sample [5.34]. In the high-temperature range (regime I), the photoconductivity increases exponentially with $1/T$, showing a well-defined activation energy. One observes furthermore a linear variation of the photoconductivity with the exciting light intensity. At lower temperatures (regime II), the photoconductivity decreases with $1/T$ and is proportional to the square root of the light intensity. At still lower temperatures, the curves seem to level off to a constant value. The maximum in the temperature dependence generally occurs near the temperature, where the dark current exceeds the photocurrent. In order to explain these photoconductivity characteristics, *Weiser* et al. [5.33] proposed a recombination model, which assumes that the electrical transport on either side of the maximum is of the same nature but that the recombination kinetics change in character. In the high-temperature range, the density of the thermally activated carriers is much larger than the photoexcited carriers, which yields a monomolecular recombination process. In regime II, however, the photocurrent is much larger than the dark current and the recombination obeys second-order kinetics. The dependence of the photo-conductivity on the excitation intensity is a well-known fact in crystalline solids. Let us, e.g., consider a simple model of a semiconductor with only one type of recombination center. If an excess electron density Δn is established by light excitation, then, assuming the material to be intrinsic ($n_0 = p_0$) and that charge neutrality exists ($\Delta n = \Delta p$), one has

$$\frac{d(\Delta n)}{dt} = g - [C_n N_r (n_0 + \Delta n) - C_n n_0^2], \tag{5.52}$$

where n_0 is the density of thermal carriers available to recombine with the excess carriers, C_n is the capture coefficient (i.e., the product of the cross section for the capture of free electrons by recombination and the average thermal velocity of the carriers), and N_r is the concentration of the recombination centers. In the steady state $d(\Delta n)/dt = 0$, and under the condition mentioned above $N_r = n_0 + \Delta n$, so one gets

$$g = C_n(\Delta n^2 + 2n_0 \Delta n). \tag{5.53}$$

This relation gives a direct insight in the dependence of Δn on the generation rate, and consequently on the light intensity, which is proportional to g. In the regime $n_0 \gg \Delta n$,

$$\Delta n = \frac{g}{2C_n n_0} \quad \text{(monomolecular)}, \tag{5.54}$$

and the photocurrent varies linearly with the excitation intensity. When $n_0 \ll \Delta n$,

$$\Delta n = \left(\frac{g}{C_n}\right)^{1/2} \quad \text{(bimolecular)}, \tag{5.55}$$

and the photocurrent is proportional to the square root of the light intensity. A rate equation of the form (5.52) has been used by *Weiser* et al. in the analysis of their photoconductivity data.

In the last section an attempt will be made to discuss some typical results obtained on chalcogenide glasses but it must be stressed that photoconductivity measurements alone cannot yield unambiguous information about gap states.

5.3.7 Small-Polaron Motion

The likehood of small-polaron formation in noncrystalline materials has been discussed by *Emin* [5.3]. He argued that the probability of small-polaron formation in the disordered state is largely increased as compared to the crystalline state. In support to this view he was able to show that several aspects of the transport data can be consistently explained by using the equations of the transport coefficients originally developed for the motion of small polarons in a crystal. If *Emin*'s ideas are found to be correct, one must accept that small polarons will dominate the conduction processes in amorphous semiconductors, throwing in this way serious doubts about the applicability of the Davis–Mott model for the distribution of the density of states. It may be remembered that this model postulates the existence of mobility edges at the energies which separate the localized tail states from the extended states. The existence of such tail states does not appear in the small-polaron picture where it is assumed that the electronic states form a small-polaron band. In the review on the chalcogenide glasses we shall summarize the main arguments used by *Emin* in order to prove on experimental ground the occurrence of small polarons in disordered systems. For the sake of clarity it seems necessary to include here a short survey of the essential features of the transport properties of small polarons.

If the charge carrier remains in the vicinity of a particular atomic site over a time interval long enough for displacements of the surrounding atoms to take place, a potential well may be created which can lead to trapping of the carrier. The unit built up by the trapped carrier and its induced lattice deformation is called a polaron. In its bound state, the carrier cannot move without changing the positions of the neighboring atoms. The polaron has a lower energy than the free electron but has a larger effective mass, since it must carry its induced deformation with it as it moves through the lattice. The decrease in energy of the small polaron relative to that of the electron in the undistorted lattice is called the small-polaron binding energy. The reduction in energy of the small polaron is the net result of two opposite contributions: on the one hand a lowering due to the displacements of the surrounding atoms, and on the other hand an increase (half that of the decrease) due to the strain energy originating from the induced lattice distortion. *Holstein* [5.38] found that for a crystal the small-polaron states overlap sufficiently to form a polaron band, a feature analogous to the electronic band formation in an undeformed lattice. In a crystal, the small-polaron band is generally rather narrow and its width decreases exponentially with increasing temperature. In an ideal crystal a small polaron can move via two distinct mechanisms. At low temperatures, band conduction can take place. This process involves the motion of a small polaron without any change in the phonon population. At high temperatures the small polaron moves by "hopping" from one lattice position to an equivalent one, which can only be created by a similar distortion of the surrounding lattice. This deformation of the lattice requires energy which is delivered by the phonons and the motion can therefore be considered as a phonon-assisted tunneling between adjacent sites.

In his paper on small-polaron motion *Holstein* [5.38] introduced the fundamental concept called "coincidence event" which characterizes the hopping mechanism. The electronic energy level of a carrier, occupying a particular atomic site, is a function of the instantaneous positions of the atoms. Due to the atomic vibrations these positions are constantly changing so that the electronic energy of the trapped carrier also changes in time. Now, at a given moment a situation may occur in which the energy of the site containing the bound electron equals that of a nearest neighbor. Such a momentary occurrence is called a "coincidence event".

Two cases are generally distinguished:

I) the "adiabatic" regime, in which the electron goes backwards and forwards several times during the period that the two potential wells have equal depth. The carrier will possess a high probability to hop to the adjacent site.

II) the "nonadiabatic" regime, in which the electron cannot follow the lattice vibrations and the time required for an electron to hop is large compared to the duration of a coincidence event. In this case, the carrier will have many coincidence events before it hops, its probability for transfer being much smaller than in the adiabatic case.

In order to find the conductivity for hopping motion of small polarons, one can make use of the Einstein relation which relates the mobility to the diffusion coefficient D:

$$\sigma = ne\mu = ne\,\frac{eD}{kT}.$$

The diffusion coefficient can be written as the product of the hopping probability P with the square of the interatomic distance a. Thus the dc conductivity for small polarons, being proportional to the probability for hopping, is given by

$$\sigma = (ne^2 a^2 / kT)P. \tag{5.56}$$

As seen before, the physical picture of hopping is based on occasional fluctuations in the positions of the atoms, causing at particular moments equal deformations in the occupied and unoccupied lattice sites. At each coincidence event the carrier will have a certain probability to jump. The jump probability can therefore be written as a product of two terms: the probability p_1 for the occurrence of a coincidence event and the probability p_2 of charge transfer during this event. The probability for a configuration with equal potential can be expressed by

$$p_1 = \frac{\omega_0}{2\pi} \exp(-W/kT), \tag{5.57}$$

where $\omega_0/2\pi$ is an average phonon frequency and W is the minimum energy necessary to obtain two equivalent sites. The hopping energy W is related to the small-polaron binding energy E_b by $W = E_b/2$. The total probability P becomes

$$P = \frac{\omega_0}{2\pi} \exp(-W/kT)p_2. \tag{5.58}$$

In the adiabatic regime, where the carrier can follow the motion of the lattice, the probability for jumping during coincidence is high, and one can put $p_2 = 1$. In the nonadiabatic regime where the carrier is slow, one expects $p_2 \ll 1$. *Holstein* [5.38] derived the following expression for p_2:

$$p_2 = \frac{2\pi}{\hbar\omega_0}\left(\frac{\pi}{WkT}\right)^{1/2} J^2. \tag{5.59}$$

The electronic transfer integral J is a measure of the overlapping of the wave functions.

Using (3.59), one obtains for the conductivity mobility in the nonadiabatic regime

$$\mu_C = \frac{ea^2}{kT}\frac{1}{\hbar}\left(\frac{\pi}{WkT}\right)^{1/2} J^2 \exp\left(-\frac{W}{kT}\right). \tag{5.60}$$

The mobility will increase exponentially over a broad temperature range. This thermally activated behavior is one of the essential features of small-polaron theory. In the temperature range where kT becomes of the order of W, the preexponential term, which varies at $T^{-3/2}$, will become predominant.

The calculation of Hall mobility of the small polaron has proved to be more difficult. It was found that the local geometry of the atoms is extremely important and may lead to different results. For a triangular lattice, *Friedman* and *Holstein* [5.39] derived the following expression for the nonadiabatic case:

$$\mu_H = \frac{ea^2}{\hbar} J\left(\frac{\pi}{12kTW}\right)^{1/2} \exp\left(-\frac{W}{3kT}\right), \tag{5.61}$$

and

$$\frac{\mu_H}{\mu_C} = \frac{2}{3\sqrt{3}}\frac{kT}{J}\exp\left(\frac{2}{3}\frac{W}{kT}\right). \tag{5.62}$$

They find that the Hall mobility is thermally activated with an activation energy which is only one-third of that associated with the conductivity mobility. Since the preexponential term decreases with increasing temperature, the Hall mobility will reach a maximum at $kT = W/3$. Furthermore, since $J \gg kT$, the Hall mobility will generally be larger than μ_C. An important result is concerned with the sign of the Hall coefficient. Indeed, the theory predicts a sign anomaly when conduction is due to hopping of small-polaron holes: the sign of the thermopower is positive whereas the sign of the Hall effect is that which would correspond to carriers being electrons.

The calculation of the Hall mobility of small polarons in a square lattice has been carried out by *Emin* [5.40]. In the limit of high temperature, one gets

$$\mu_H \propto \left(\frac{\theta}{T}\right)^{3/2} \exp\left(-\frac{W}{3kT}\right), \tag{5.63}$$

where θ is the Debye temperature.

For this cubic geometry, the Hall mobility may vary slowly with temperature over a considerable temperature range.

Another transport property which has received special attention in the experimental studies on small polarons is the thermoelectric power S. For small-polaron hopping, S was found [5.41] to be expressed by a formula of the classical form:

$$S = -\frac{k}{e}\left(\frac{E}{kT} + A\right), \tag{5.64}$$

where E is the energy associated with the thermal generation of the carriers. It was shown that the kinetic term A could be very small if there is no transfer of vibrational energy associated with a small-polaron hop.

5.4 Chalcogenide Glasses

5.4.1 Preparation and Structure

In the past a considerable amount of experimental work has been published on the transport properties of covalent amorphous semiconductors. Most of the research has been carried out on two groups, namely the tetrahedrally bonded semiconductors, in particular amorphous Si and Ge, and the chalcogenide glasses. In this section we shall restrict ourselves to a discussion of some relevant pieces of information which have been obtained on the second group of materials. We shall not attempt to summarize all the published data on the chalcogenide glasses but we shall focus our attention on some results which have helped to produce a rather coherent picture of these materials. This section will be mainly devoted to a brief description of some results obtained from dc and ac conductivity, thermopower, Hall effect, and steady-state photoconductivity experiments. Use will be made of the transport equations presented in the preceding section.

Glasses based on the chalcogenides S, Se, and Te can be easily prepared on two different ways: thin-film deposition using evaporation or sputtering techniques, and quenching from the melt. The binary systems As–S, As–Se, and As–Te, which embrace the compounds As_2S_3, As_2Se_3, and As_2Te_3, have been studied extensively. Ternary and quarternary systems, containing the elements S, Se, Te, As, Ge, Tl, and others can also be prepared in the glassy state. The multicomponent systems, e.g., $As_{3.0}Te_{4.8}Si_{1.2}Ge_{1.0}$, are of great practical importance because of their switching and memory properties. In general, the chalcogenide glasses show good thermal stability: for example, As_2S_3 can be transferred from the solid to the liquid state without crystallization.

In the glassy state, the chalcogen atoms possess a well-defined coordination number. In the ground state the electronic configuration of the elements S, Se, and Te is represented by s^2p^4. In the glasses of the type $A_2^V B_3^{VI}$ (A = As, Sb; B = S, Se, Te) the structure consists of a network showing complete satisfaction of the two- and threefold covalent bonding requirements of the B^{VI} and A^V elements. Each chalcogen atom is linked to two As atoms, while each As atom is linked to three chalcogens. Because of the twofold coordination there remains in each chalcogen atom a nonbonding electron pair in a so-called lone-pair orbital. This free electron pair does not take part in the formation of a chemical bond. An inherent consequence of the way chalcogenide glasses are prepared is the presence of a large number of defects such as vacancies, dangling bonds, and nonbridging atoms. It may be expected that these defects will give rise to relatively well-defined energy levels in the mobility gap, leading to a peaked density-of-states distribution. In recent years, some ideas – some rather speculative – have been put forward concerning the nature of the gap states. Chapter 3 of this volume specifically addresses the question of gap states.

5.4.2 dc Conductivity, Thermopower, and Hall Effect

Combined measurements of dc conductivity, thermopower, and Hall effect have yielded valuable information about the transport mechanism in chalcogenide glasses. The composition of most binary and multicomponent alloys can be varied over a large range. In nearly all cases, this produces a gradual change in the electrical and optical properties. In many systems, some common experimental features appear, which can be summarized as follows: (I) When measured over a broad temperature range, the dc conductivity often shows slightly curved lines, which gradually level out with decreasing temperature. In this case the conductivity cannot be represented by one single exponential of the form $\sigma = C \exp(-E/kT)$. The conductivity activation energy, which varies slightly with temperature, is about half the magnitude of the optical gap (see Chap. 4 for a discussion of the meaning of the optical gap). This means that the Fermi level is located not far from the center of the gap. (II) The thermoelectric power is usually positive. The sign of the thermopower is believed to give a reliable indication of whether the conduction is dominated by holes or electrons. In the majority of cases the thermopower data, when plotted as a function of $1/T$, yield slightly curved lines. They exhibit, however, a less steep temperature dependence than the conductivity curves. In the whole range of measurement, a difference in slopes of the order of 0.1 to 0.2 eV is commonly observed. (III) The sign of the Hall coefficient is usually negative. There exists, however, one alloy of composition $Tl_2Te-As_2Te_3$ for which a positive sign has been reported [5.42]. The sign discrepancy between the thermopower and the Hall coefficient has received a great deal of attention. The Hall mobility calculated as the product of the conductivity times the Hall coefficient usually shows a thermally activated behavior. The activation energy associated with the Hall mobility curves is in the range 0.03–0.07 eV.

The results obtained by *Nagels* et al. [5.43, 44] on the amorphous system As_2Te_3 with a small addition of Si and on a quarternary Si–Te–As–Ge (STAG) glass give a nice illustration of these basic experimental facts. Their dc conductivity, thermoelectric power, and Hall mobility data are represented in Fig. 5.5. The authors have stressed that the σ and S data are best fitted by smooth curves with gradually varying slopes. For both transport coefficients, the deviation from a straight line is small, thus requiring data over an extended temperature range in order to detect the slight curvature in the plots. The difference in slopes between the conductivity and thermopower curves is of the order 0.1–0.15 eV. The experimental Hall mobility increases very slightly with increasing temperature, the activation energy being in the range 0.03–0.05 eV.

The two major features of the results are the weaker temperature dependence of the thermopower as compared with the conductivity and the thermally activated nature of the Hall mobility. The apparent discrepancy between the activation energies for S and σ has been observed in other chalcogenide systems such as As_2Se_3 [5.45]; $As_2(Se_{1-x}Te_x)$ [5.46]; As–Te–I and As–Te–Ge [5.47]. At the moment there are two consistent interpretations of these experimental observations. A possible resolution of the problem may be found through a two-channel model of conduction proposed by *Nagels* et al. [5.43]. In their analysis of the results represented in Fig. 5.5, the authors started from the Davis–Mott model for the energy distribution of the density of states and took into account simultaneous conduction of holes in extended states and in band-tail localized states. The positive sign of the thermopower was taken as an indication that holes are the most numerous carriers. It has been pointed out in the previous section that the conductivity due to holes excited into the extended states near the mobility edge can be represented by an equation of the form

$$\sigma_{ext} = \sigma_0 \exp[-(E_F - E_V)/kT].$$

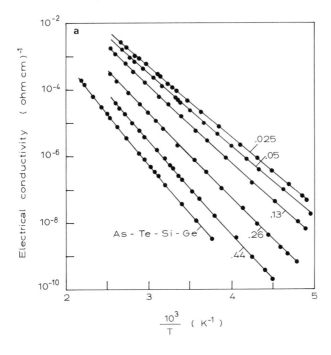

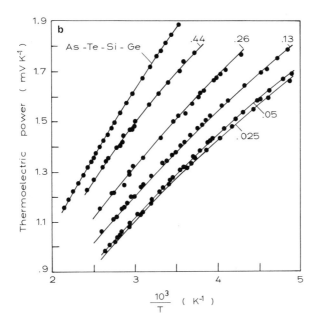

Fig. 5.5. Electrical conductivity, (**b**) thermopower, and (**c**) Hall mobility versus $10^3/T$ for five $AsTe_{1.5}Si_x$ glasses (the Si content is indicated) and a $As_{3.0}Te_{4.8}Si_{1.2}Ge_{1.0}$ glass [5.41,42]

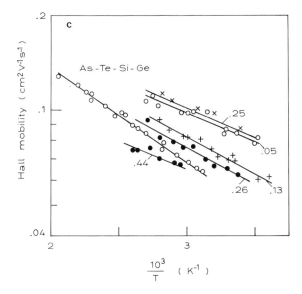

As - Te - Si - Ge

Fig. 5.5c. Caption see oposite page

If the conduction is due to holes excited into the tail of the valence band, then the conductivity behaves like

$$\sigma_{hop} = \sigma_{0\,hop} \exp[-(E_F - E_A + W)/kT],$$

where W is the hopping energy.

When both conduction channels contribute to the conduction, a kink in the conductivity curves will only be observed under the condition that the rate constants of both exponentials differ markedly. On the other hand, when the rate constants are similar in magnitude, a semilogarithmic graph of the conductivity data will not show any breakpoint. Indeed, the curve will change its slope only gradually and will not differ noticeably from a straight line. Both situations are schematized in Fig. 5.6a. A kink in the conductivity curve has clearly been observed in amorphous silicon by *Le Comber* et al. [5.48]. *Nagels* et al. were able to decompose their experimental conductivity curves into a sum of two exponentials functions, having the following values for the pre-exponential constants: $\sigma_0 \simeq 3000\,\Omega^{-1}\,cm^{-1}$ and $\sigma_{0\,hop} \simeq 90\,\Omega^{-1}\,cm^{-1}$. This large difference is consistent with *Mott*'s prediction that the mobility should drop drastically at the energy which demarcates the extended states of the valence band. From this analysis it followed that the conduction in the band tails contributes to the total conductivity for a nonnegligible amount even at the highest temperature of measurement.

As seen in Sect. 2, the thermopower in both conduction regimes is expected to follow an equation of the form

$$S = \frac{k}{e}\left(\frac{\Delta E}{kT} + A\right),$$

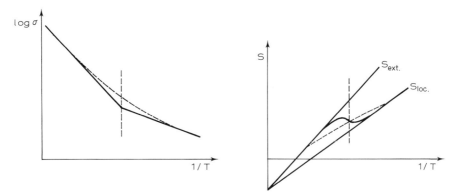

Fig. 5.6. Schematic plots of log σ and S versus $1/T$ in the range where conduction changes from band conduction above the mobility edge to hopping

with ΔE defined by $E_F - E_V$ for conduction in extended states and by $E_F - E_B$ for hopping in band-tail localized states. When a transition from hopping at E_A to conduction at E_V occurs by raising the temperature, the behavior of the thermopower will depend on the form of the conductivity curve. This is illustrated in Fig. 5.6b. If a sharp kink in the conductivity curve is observed, one expects a sudden rise in S for increasing temperature in the neighborhood of the kink. For a slowly changing conductivity, the thermopower is gradually transferred from its contribution associated with conduction in the localized states to that of the extended states and the S vs $1/T$ plot shows a decreased slope over a wide temperature range. *Nagels* et al. have fitted their thermopower data obtained on the As–Te–Si glasses with the help of these two contributions using the values of the parameters deduced from the conductivity curves. It is important to stress that, according to their model, the activation energy of the experimentally observed thermopower has no direct physical significance. The experimental Hall mobility was found to increase with temperature (see Fig. 5.5c). *Nagels* et al. used the two-channel model to interpret this behavior. For carriers conducting in the extended and localized states the measured Hall mobility will be given by

$$\mu_H = \frac{\mu_{H,\,ext}\sigma_{ext} + \mu_{H,\,hop}\sigma_{hop}}{\sigma_{ext} + \sigma_{hop}}.$$

The Hall mobility of holes conducting in extended states near the mobility edge E_V is given by *Friedman*'s equation [5.3, 45]. The authors furthermore assume that the Hall mobility in the hopping regime is much smaller than in the extended states. Under this assumption the observed Hall mobility is given by

$$\mu_H = \frac{\mu_{H,\,ext}}{\left\{1 + \dfrac{\sigma_{0\,hop}}{\sigma_0}\exp[(E_B - E_V - W/kT]\right\}}.$$

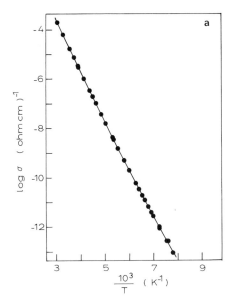

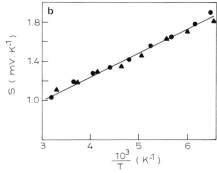

Fig. 5.7. (a) Conductivity and **(b)** thermopower against $10^3/T$ for amorphous As_2Te_3 [5.43]

The values of $\mu_{H, \text{ext}}$ evaluated from the experimental data are completely temperature independent, in agreement with *Friedman*'s theory. They are equal to $(1.1 - 1.4) \times 10^{-1}$ cm^2 V^{-1} s^{-1} for the different compositions of the As–Te–Si glasses and 1.9×10^{-1} cm^2 V^{-1} s^{-1} for the STAG glass. The proposed interpretation, based on the Davis–Mott model, may lend some support for the existence of narrow tails of localized states of the order of 0.1 eV at the band extremities. In a recent paper *Van der Plas* and *Bube* [5.49] have shown that the two-channel model, providing transport simultaneously in both extended and band-tail-localized states, is applicable to a wide variety of amorphous semiconductors.

An alternative model which can provide a straightforward description of the measured difference between the activation energies of the conductivity and the thermopower and, moreover, of the thermally activated Hall mobility is the small-polaron hopping model. *Emin* et al. [5.50] were first to present some arguments in support to the occurrence of self-trapping of charge carriers in noncrystalline solids and to point out that the predictions of the small-polaron theory are in agreement with the experimental observations on some chalcogenide glasses. *Seager* et al. [5.45, 47] reported measurements of dc conductivity, thermopower, and Hall mobility on a number of As–Te–I, As–Te–Ge, As–Te, As–Se, and As–S chalcogenide glasses, and concluded that their results lend considerable support to the suggested small-polaron transport mechanism. Figure 5.7a shows that dc conductivity and Fig. 5.7b the thermopower obtained by these authors on amorphous As_2Te_3. The slope of the conductivity curve changes smoothly from 0.38 to 0.42 eV with rising temperature. The thermopower varies linearly with $1/T$, the activation energy being equal to 0.25 eV. In the following we shall reproduce the main arguments used by *Emin* and co-workers to support the validity of their viewpoint. According to the small-polaron theory, the electrical conductivity will essentially be thermally activated: $\sigma = \sigma_0 \exp(-E/kT)$, the activation energy E being the sum of the energy required to generate the carriers and an activation energy W associated with the hopping of small polarons. The value of the preexponential σ_0 is given by $\sigma_0 = (Ne^2 a^2 kT \omega_0 / 2\pi)p_2$ [see (5.56) and (5.58)], where N is the number of equivalent sites per cm^3 and $p_2 \simeq 1$ in the adiabatic regime and $\ll 1$ in the nonadiabatic limit. The experimental values of σ_0 fall within the limits 10^2–10^3 Ω^{-1} cm^{-1}, which is consistent with the theoretical estimates deduced from the above equation. In the case of small-polaron hopping the thermopower varies linearly with reciprocal temperature [see (5.64)]. The value of its slope yields the activation energy for carrier creation so that $E_\sigma - E_S = W$. For the family of the As–Te glasses the deduced hopping energy W lies within narrow limits (0.17–0.20 eV) which led the authors to conclude that the holelike small polarons move between adjacent sites in a network of As–Te units. The Hall coefficient associated with the holelike motion is negative and the Hall mobility possesses an activation energy of the order of 0.06 eV for all the As–Te based glasses being roughly one-third of the hopping energy W. The theory of *Friedman* and *Holstein* [5.39] of the Hall mobility does in fact predict that the hopping of holelike small polarons will yield an n-type Hall effect if they move within a triangular-type geometry, which is believed to characterize the As–Te unit. The theory furthermore predicts that the Hall mobility should be low and characterized by a small activation energy, which should be equal to $\frac{1}{3}W$. It is clear that this second approach also provides a consistent physical basis for the essential features of the transport properties and cannot be definitively ruled out. It must be stressed that there exist at the moment no direct arguments to prove in a definite way the validity either of the two-channel model or of the small-polaron model.

In the past several attempts to enhance the electrical conductivity of chalcogenide glasses by doping were unsuccessful. This insensitivity to impurities was ascribed to the fact that in the disordered material the coordination environment can adjust to satisfy the valency requirements of each atom [5.51]. *Ovshinsky* and co-workers [5.52], however, recently succeeded in increasing the electrical conductivity of a large variety of these glasses by many orders of magnitude. Extrinsic conduction was achieved by "chemically modifying" the amorphous semiconductors with Ni or other transition metals. The authors used the term modifier instead of dopant because of the relatively large concentrations of impurities added. Their preparation technique consists in a deposition of thin films of the chalcogenide system plus desired amounts of the modifier by rf sputtering at low temperature. The essential feature of their method appears to be the fact that the chemical modifiers are incorporated at sufficiently low temperature so that the material cannot reach a structural or configurational equilibrium. Attempts to obtain similar results by adding the modifying elements to the melt of the chalcogenide material were unsuccessful.

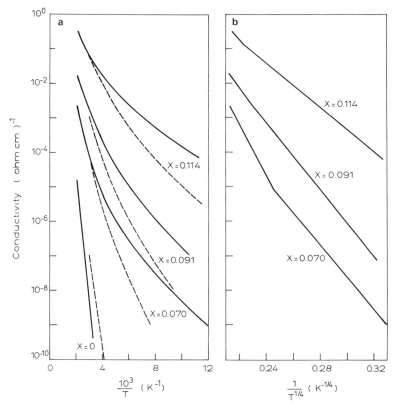

Fig. 5.8. (a) Conductivity versus $10^3/T$ and (b) conductivity as a function of $T^{-1/4}$ for a $(Ge_{0.32}Te_{0.32}Se_{0.32}As_{0.04})_{1-x}Ni_x$ glass [5.51]

Flasck et al. [5.53] reported measurements of dc electrical conductivity, thermoelectric power, and optical absorption in a glass of composition $Ge_{32}Te_{32}Se_{32}As_4$ modified by the addition of variable amounts of Ni. Their conductivity data, shown in Fig. 5.8, suggest that at low temperature the conduction is governed by variable range hopping as can be seen from the $\ln\sigma$ versus $T^{-1/4}$ dependence. Optical measurements over a limited range were interpreted as indicating that the energy gap remains essentially unaltered. The steady-state photoconductivity was found to be greatly enhanced by the addition of the modifier.

Prior to these findings, information about the occurrence of variable range hopping associated with localized gap states was very scarce in chalcogenide glasses. In materials prepared in the amorphous state by quenching the melt, hopping conductivity has not been observed. *Hauser* and *Hutton* [5.54] have recently found evidence for variable range hopping in films of As_2Te_3 deposited at 77 K by sputtering. The conductivity data of the as-deposited films could be fitted by the relationship $\sigma \propto \exp(-A/T^{1/4})$, characteristic for conduction in localized states at the Fermi level. The density of states at E_F deduced from the term A using Mott's equation (5.30) was found to equal $N(E_F) \simeq 2 \times 10^{17}$ eV^{-1} cm^{-3} with $\alpha^{-1} \simeq 5.5$ Å. After annealing at room temperature the hopping conductivity disappeared and σ varied exponentially with temperature.

5.4.3 ac Conductivity

It is recalled from the preceding section that the determination of the ac conductivity may provide information about the density of states at the Fermi level [see (5.47)]. Experimental results of the ac conductivity have been reported: on thin films of As_2Se_3 [5.55]; on bulk samples of As_2Se_3, As_2S_3, and As_2Se_3–As_2Te_3 [5.56–59]; on thin films of As_2Te_3 [5.60], and on bulk samples of $(As_2Te_3)_{95}Ge_5$ [5.61]. According to the Austin–Mott formula (5.48), a frequency-dependence of the ac conductivity $\sigma(\omega) \propto \omega^s$ with $s \simeq 0.8$ is expected to occur. In general, this relationship was observed at frequencies below 10^8 Hz, yielding s values in the range 0.7–1.0. Some investigators found, however, a sudden increase of $\sigma(\omega)$ to an ω^2 law in the megacycle frequency range. Using the data of the different authors, Davis and Mott [5.2] estimated the $N(E_F)$ values, which were found to be in the range 10^{18}–10^{21} eV^{-1} cm^{-3} (see Fig. 5.9). In a recent paper Eliott [5.62] made a critical analysis of the different difficulties which arise when conventional tunneling models are used to interpret the ac conductivity observed in chalcogenide glasses. One of these difficulties is the high magnitude of $N(E_F)$ evaluated with the help of the Austin-Mott formula. Instead of the thermally assisted tunneling process, he proposed a mechanism in which two electrons hop over a potential barrier between defect sites, the height of the barrier being correlated with the intersite separation. This model predicts a slightly sublinear frequency dependence of $\sigma(\omega)$ and a temperature dependence of the exponent tending to unity as the temperature is lowered. This feature has been observed in some chalcogenide glasses. In addition, the ac conductivity expression deduced using this model yields values for the density of states which lie in a range lower than those evaluated from other theories.

In most studies on the temperature dependence of $\sigma(\omega)$ it was observed that the ac component of the total conductivity becomes progressively less temperature dependent with increasing frequency and, moreover, that the measured ac conductivity approaches the dc value at high temperatures. Measurements of the temperature dependence of ac and dc conductivity carried out by Rockstad [5.60] on As_2Te_3 films, are illustrated in Fig. 5.10. The low-temperature part is proportional to T and can therefore be ascribed to hopping in localized gap states. In the figure the quantity $\sigma_1 \equiv \sigma(10^5 \text{ Hz}) - \sigma_{dc}$ is also plotted. Above 200 K, $\sigma_1(\omega)$ was found to rise much more rapidly with temperature and to be proportional to σ^s, where s was slightly less than unity. Rockstad attributed the component σ_1 to conduction in localized tail states.

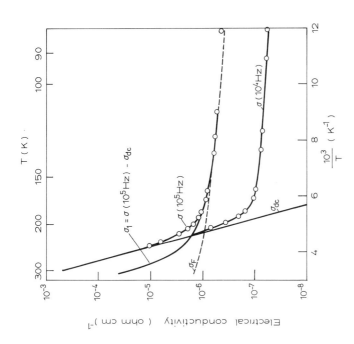

Fig. 5.10. Temperature dependence of ac and dc conductivity of an amorphous As$_2$Te$_3$ film. The dashed line represents a conductivity proportional to T [5.58]

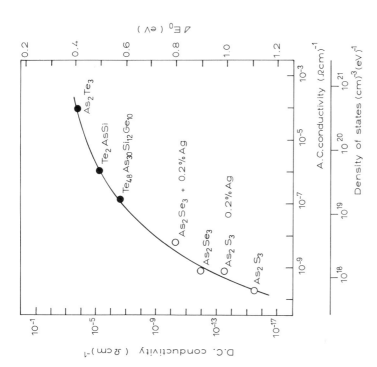

Fig. 5.9. ac Conductivity at $\omega = 10^6$ s^{-1} against dc conductivity in various chalcogenide glasses. Also shown are the density of states at E_F and the activation energies E for dc conduction. [5.2]

5.4.4 Photoconductivity

As a last point we shall turn to experiments on photoconductivity in chalcogenide glasses and see what information can be extracted from these measurements in connection with states in the gap acting as recombination centers. Figure 5.11 shows the results reported by *Main* and *Owen* [5.37] for the steady-state photocurrent of amorphous As_2Te_3 (a) as a function of the photon flux at two different temperatures and (b) as a function of temperature. These results are typical for the chalcogenide glasses and show the characteristics which have already been discussed in Sect. 3.5. A striking feature is that there are well-defined activation energies involved in the temperature dependence of the photoconductivity. This behavior suggests that the recombination centers are located at relatively discrete levels of localized states. *Main* and *Owen* have interpreted their data on the basis of a simple four-level model containing an electron trap and a hole trap. The problem remains, however, to identify the origin of the gap states which are responsible for recombination. In the past the description of localized gap states in chalcogenides has been complicated due to the observation of apparently conflicting data. Indeed, although ac conductivity measurements gave ample evidence for a large number of gap states, ESR and magnetic susceptibility measurements revealed the absence of significant densities of free spins. To explain this, *Street* and *Mott* [5.63] and *Mott* et al. [5.64] postulated the model of the so-called charged dangling bonds. Their first hypothesis was that the chalcogenide glasses contain a high number of frozen-in structural defects and that missing chalcogen atoms leaving broken or "dangling" bonds on other atoms could be the source of localized defect states. When neutral, a dangling bond contains an unpaired electron which, since the number of dangling bonds is supposed to be at least $10^{18} \, cm^{-3}$, would give rise to an observable ESR signal. The authors assumed, therefore, that the dangling bond states are either unoccupied or occupied by pairs of electrons so that positively D^+ and negatively charged D^- dangling bonds are formed. Here, the fundamental assumption is that the reaction,

$$2D^0 \rightarrow D^+ \rightarrow D^-$$

is exothermic. Here, D^0 represents the neutral dangling bond which contains an unpaired electron. In this picture all the states are positively or negatively charged and, hence, are diamagnetic. The neutral state can only be formed by excitation. *Mott* and co-workers argued furthermore that this unusual feature results from the electronic configuration of the chalcogen atoms. In their compounds these elements have a nonbonding lone-pair electron which can interact with a dangling bond. A positively charged dangling bond is attracted to a neighboring lone pair and forms a strong bond with it. This displacement of the two neighboring chalcogen atoms causes a distortion of the surrounding lattice. It is supposed that D^- does not form a bond with a lone pair of an adjacent chalcogen atom. *Mott* and co-workers assumed that the energy gained

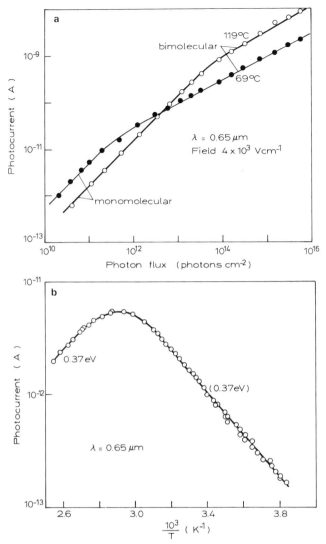

Fig. 5.11. (a) Photocurrent versus photon flux for amorphous As_2Te_3 showing monomolecular and bimolecular kinetics, **(b)** photocurrent as a function of reciprocal temperature for 5×10^{11} photons cm^2 s^{-1}. [5.35]

by lattice distortion exceeds the energy lost in the process that brings two electrons into D^-, yielding in this way a situation which is energetically more favorable than that of the neutral defect states. The electronic energy levels of the dangling bonds are strongly effected by the induced lattice distortion. The D^+ defect state yields a donor state in the upper half of the gap, whereas D^- forms an acceptor state in the lower half of the gap. The energy level of D^0 lies intermediate between that of D^+ and D^-.

Mott et al. pointed out that the model of charged dangling bonds can account for the photoconductivity data observed on chalcogenide glasses. The D^+ and D^- centers can act as discrete traps for electrons and holes generated by light excitation. This capture of photoexcited charge carriers leads to an excess concentration ΔN of D^0 centers. The trapped carriers can be released to the valence or conduction band and, hence, give rise to an excess conductivity $\Delta\sigma$. If holes dominate the conduction process, as is the case for most chalcogenic glasses, then

$$\Delta\sigma = \text{const} \cdot \Delta N \exp(-W/kT),$$

where W is the energy required to excite a hole from a dangling bond to the valence band. *Mott* pointed out, furthermore, that photogenerated holes, trapped in D^- centers, are lost when an electron tunnels between two D^0 centers forming a stable D^+ and D^-.

The model of charged dangling bonds has been successful in explaining a number of essential observations in lone-pair semiconductors. A detailed discussion is given in Chap. 3 of this volume.

5.5 Concluding Remarks

Although over the last five years considerable progress has been made in the understanding of the amorphous state, many problems still remain unsettled. The existence of sharp mobility edges at energies where the wave function changes its nature from localized to extended, which for a long time was only a matter of speculation, has found some grounds in some experiments on electrical transport. The main evidence comes from the studies of systems in which one was able to move to Fermi level through the mobility edge. As a consequence of this shift the conductivity can drastically change its nature, leading to some kind of insulator–metal transition called by *Mott* the "Anderson transition" [5.65]. As pointed out by *Mott* such a transition has been experimentally observed in systems where changes in composition can give rise to an increase of disorder and to a change in occupation of the band.

The extent of tailing of the bands also needs further investigation. It is now clear that the Cohen–Fritzsche–Ovshinsky model, which implies an overlapping of the band tails, does not describe the experimental situation for chalcogenide glasses. The essential feature of the proposed band structure contained in the concept of band tailing certainly requires more theoretical foundation. Once the band structure of amorphous semiconductors is settled on a firm basis, it will become possible to analyze the experimental results with some confidence.

Another problem which urgently demand clarification is the role played by lattice deformation in noncrystalline solids. It must be emphasized that at present the plausibility of small-polaron formation cannot be completely ruled out. Many experimental data have been gathered, especially in the field of the transport properties, but a consistent analysis proved to be possible starting from the Davis–Mott model or using the small-polaron theory. The considerable amount of experimental work has not helped to solve this controversary in a definite way.

In a short review of this type, it is inevitable that various other techniques such as thermally stimulated currents and field effect, which also yielded valuable information about the amorphous state, cannot be included. Furthermore, even in the restricted field of the chalcogenide glasses only a minor part of all the experimental work published in the literature has been incorporated in this article. In recent years much more information on the properties of other amorphous materials, such as arsenic, the III–V compounds and other tetrahedral glasses (e.g., $CdGeAs_2$) has become available but these results have not been presented here because lack of space.

Concluding, one can say that there is still a whole series of challenging problems.

Acknowledgement. Gratitute is due to my colleague R. Colson who has helped me in the preparation of this article.

References

5.1   M. H. Cohen, H. Fritzsche, S. R. Ovshinsky: Phys. Rev. Lett. **22**, 1065 (1969)
5.2   E. A. Davis, N. F. Mott: Philos. Mag. **22**, 903 (1970)
5.3   D. Emin: "Aspects of the Theory of Small Polarons in Disordered Materials," in *Electronic and Structural Properties of Amorphous Semiconductors*, ed. by P. G. Le Comber, J. Mort (Academic Press, London, New York 1973) p. 261
5.4   P. W. Anderson: Phys. Rev. **109**, 1492 (1958)
5.5   N. F. Mott: Philos. Mag. **22**, 7 (1970)
5.6   W. E. Spear, P. G. Le Comber, A. J. Snell: Philos. Mag. B **38**, 303
5.7   M. H. Cohen: J. Non-Cryst. Solids **4**, 391 (1970)
5.8   J. M. Marshall, A. E. Owen: Philos. Mag. **24**, 1281 (1971)
5.9   A. E. Owen, W. E. Spear: Phys. Chem. Glasses **17**, 174 (1976)
5.10 N. F. Mott: "Electrons in Non-crystalline Materials," in *Electronic and Structural Properties of Amorphous Semiconductors*, ed. by P. G. Le Comber, J. Mort (Academic Press, London, New York 1973) p. 1
5.11 N. K. Hindley: J. Non-Cryst. Solids **5**, 17 (1970)
5.12 L. Friedman: J. Non-Cryst. Solids **6**, 329 (1971)
5.13 N. F. Mott: Philos. Mag. **19**, 835 (1969)
5.14 H. Overhof: "Hopping Conductivity in Disordered Solids," in *Festkörperprobleme* (Advances in Solid State Physics), Vol. XVI, ed. by J. Treusch (Vieweg, Braunschweig 1976) p. 239
5.15 H. Fritzsche: Solid State Commun. **9**, 1813 (1971)
5.16 M. Cutler, N. F. Mott: Phys. Rev. **181**, 1336 (1969)
5.17 I. P. Zvyagin: Phys. Status Solidi (b) **54**, 733 (1972)
5.18 V. Čápek: Phys. Status Solidi (b) **57**, 733 (1973)
5.19 H. Overhof: Phys. Status Solidi (b) **67**, 709 (1975)
5.20 W. Brenig: In *Proc. 5th Int. Conf. Amorphous and Liquid Semiconductors*, ed. by J. Stuke, W. Brenig (Taylor and Francis, London 1973) p. 31
5.21 T. Holstein: Philos. Mag. **27**, 225 (1973)
5.22 P. G. Le Comber, D. Jones, W. E. Spear: Philos. Mag. **35**, 1173 (1977)
5.23 H. Mell: To be published

5.24 R. Callaerts, M. Denayer, F. H. Hashmi, P. Nagels: Discuss. Faraday Soc. **50**, 27 (1970)
5.25 E. Mytilineou, E. A. Davis: In *Proc. 7th Int. Conf. Amorphous and Liquid Semiconductors*, ed. by W. E. Spear (G. G. Stevenson, Dundee 1977) p. 632
5.26 D. Emin: Philos. Mag. **35**, 1189 (1977)
5.27 I. G. Austin, N. F. Mott: Adv. Phys. **18**, 41 (1969)
5.28 G. Pfister: Phys. Rev. Lett. **36**, 271 (1976)
5.29 G. Pfister: Phys. Rev. Lett. **33**, 1474 (1974)
5.30 W. D. Gill: In *Proc. 5th Int. Conf. Amorphous and Liquid Semiconductors*, ed. by J. Stuke, W. Brenig (Taylor and Francis, London 1973) p. 901
5.31 H. Scher, E. W. Montroll: Phys. Rev. B **12**, 2455 (1975)
5.32 G. Pfister, H. Scher: Phys. Rev. B **15**, 2062 (1971)
5.33 K. Weiser, R. Fischer, M. H. Brodsky: *Proc. Tenth Int. Conf. Physics of Semiconductors* (U.S. Atomic Energy Commission, Oak Ridge 1970) p. 667
5.34 T. C. Arnoldussen, R. H. Bube, E. A. Fagen, S. Holmberg: J. Appl. Phys. **43**, 1798 (1972)
5.35 R. Fischer, U. Heim, F. Stern, K. Weiser: Phys. Rev. Lett. **26**, 1182 (1971)
5.36 J. G. Simmons, G. W. Taylor: J. Non-Cryst. Solids **8–10**, 947 (1972)
5.37 C. Main, A. E. Owen: "Photoconductivity and Noise in Chalcogenide Glasses," in *Electronic and Structural Properties of Amorphous Semiconductors*, ed. by P. G. Le Comber, J. Mort (Academic Press, London, New York 1973) p. 527
5.38 T. Holstein: Ann. Phys. (N.Y.) **8**, 343 (1959)
5.39 L. Friedman, T. Holstein: Ann. Phys. (N.Y.) **21**, 494 (1963)
5.40 D. Emin: Ann. Phys. (N.Y.) **64**, 336 (1971)
5.41 K. D. Schottke: Z. Phys. **196**, 393 (1966)
5.42 P. Nagels, R. Callaerts, M. Denayer, R. De Coninck: J. Non-Cryst. Solids **4**, 295 (1970)
5.43 P. Nagels, R. Callaerts, M. Denayer: In *Proc. 5th Int. Conf. Amorphous and Liquid Semiconductors*, ed. by J. Stuke, W. Brenig (Taylor and Francis, London 1973) p. 867
5.44 P. Nagels: "Electronic Properties of Amorphous Semiconductors," in *Linear and Nonlinear Electron Transport in Solids*, ed. by J. T. Devreese, V. E. Van Doren (Plenum Press, New York, London 1976) p. 435
5.45 C. H. Seager, R. K. Quinn: J. Non-Cryst. Solids **17**, 386 (1975)
5.46 P. Nagels, R. Callaerts, M. Denayer: *Proc. 11th Int. Conf. Physics of Semiconductors*, ed. by M. Miasek (PWN – Polish Scientific Publishers, Warsawa 1972) p. 549
5.47 C. H. Seager, D. Emin, R. K. Quinn: Phys. Rev. B **8**, 4746 (1973)
5.48 P. G. Le Comber, A. Madan, W. E. Spear: J. Non-Cryst. Solids **11**, 219 (1972)
5.49 A. Van der Plas, R. H. Bube: J. Non-Cryst. Solids **24**, 377 (1977)
5.50 D. Emin, C. H. Seager, R. K. Quinn: Phys. Rev. Lett. **28**, 813 (1972)
5.51 N. F. Mott: Adv. in Phys. **16**, 49 (1967)
5.52 R. Ovshinsky: *Proc. 7th Int. Conf. Amorphous and Liquid Semiconductors*, ed. by W. E. Spear (G. G. Stevenson, Dundee 1977) p. 519
5.53 R. Flasck, M. Izu, K. Sapru, T. Anderson, S. R. Ovshinsky, H. Fritzsche: In *Proc. 7th Int. Conf. Amorphous and Liquid Semiconductors*, ed. by W. E. Spear (G. G. Stevenson, Dundee 1977) p. 524
5.54 J. J. Hauser, R. S. Hutton: Phys. Rev. Lett. **37**, 868 (1976)
5.55 A. I. Lakatos, M. Abkowitz: Phys. Rev. B **3**, 1791 (1971)
5.56 A. E. Owen, J. M. Robertson: J. Non-Cryst. Solids **2**, 40 (1970)
5.57 M. Kitao: Jpn. J. Appl. Phys. **11**, 1472 (1972)
5.58 E. B. Ivkin, B. T. Kolomiets: J. Non-Cryst. Solids **3**, 41 (1970)
5.59 H. Segawa: J. Phys. Soc. Jpn. **36**, 1087 (1974)
5.60 H. K. Rockstad: J. Non-Cryst. Solids **8–10**, 621 (1972)
5.61 X. Le Cleac'h, J. F. Palmier: J. Non-Cryst. Solids **18**, 265 (1975)
5.62 S. R. Eliott: Philos. Mag. **36**, 1291 (1977)
5.63 R. A. Street, N. F. Mott: Phys. Rev. Lett. **39**, 1293 (1975)
5.64 N. F. Mott, E. A. Davis, R. A. Street: Philos. Mag. **32**, 961 (1975)
5.65 N. F. Mott: *Metal–Insulator Transitions* (Taylor and Francis, London 1974) p. 30

6. Luminescence in Amorphous Semiconductors

R. Fischer

With 25 Figures

Radiative recombination can be observed in amorphous semiconductors as luminescence with high quantum efficiency. In discussing the various results, the usual distinction is made between the tetrahedrally coordinated materials (chiefly silicon) on the one hand and the chalcogenides, arsenic, and selenium on the other. The luminescence results of optimally prepared amorphous silicon are presented first. Then the influence of network defects and doping on the luminescence is discussed. Also included are results from Si_xC_{1-x}. As_2Se_3, As_2S_3 and selenium are treated as model substances of the second group of materials. In this case, and also for As and $GeSe_2$, the specific models for interpreting the results are based on the assumption of large Stokes shifts.

6.1 Introductory Remarks

6.1.1 Survey of Materials

Luminescence, or radiative recombination, may or may not be observed in a semiconductor. This is because there always exist radiationless recombination mechanisms which can bypass the radiative process. Radiationless recombination in semiconductor crystals is often connected with the presence of localized states in the forbidden gap. In an amorphous semiconductor the density of localized states in the gap is quite high [6.1, 2] which should make radiationless recombination rather easy. Nevertheless luminescence is observed from various amorphous semiconductors out of both groups: the glasses, and the tetrahedrally coordinated ones. Quite surprisingly, the quantum efficiency is commonly quite high and reaches the order of unity in some favorable cases.

In general, the luminescence spectra of amorphous semiconductors lie at energies well below the optical gap. The luminescence bands are rather broad and without any fine structure. The absorption and luminescence spectra of silicon [6.3, 4] and As_2Se_3 [6.5–10] are shown in Fig. 6.1. The corresponding spectra of the crystals are also shown [6.5, 7, 11–13]. Silicon and Si_xC_{1-x} alloys [6.14] are to date the only amorphous semiconductors with tetrahedral coordination that exhibit measureable luminescence. On the other hand, there are many examples of glassy semiconductors with efficient luminescence, and As_2Se_3 is thought to be a typical representative of the entire group.

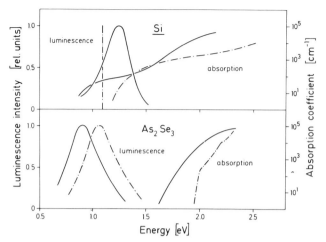

Fig. 6.1. Luminescence and absorption spectra of the amorphous and crystalline forms of silicon [6.3,4,11,12] and As$_2$Se$_3$ [6.5,10,13] (amorphous: solid lines)

6.1.2 Information from Luminescence Measurements

For luminescence to be observed, excess carriers must first be excited in the sample (Fig. 6.2). In amorphous semiconductors this is done almost exclusively by incident light (photoluminescence). One paper reports the observation of electroluminescence [6.15], i.e., excitation by carrier injection. Light emission has also been observed during electrical switching in chalcogenide glass semiconductors [6.16]. The excess carriers are in most cases generated above the band edges in a region of high density of states. There, it is more likely that an excess carrier emits phonons and thermalizes within the band than that it recombines with a carrier of opposite sign. Thermalization will not stop until, by virtue of a decreasing density of states, the phonon emission rate becomes smaller than the recombination rate [6.17, 18]. In pure crystals this point is reached abruptly at the excitonic states immediately below the band edges.

It is a peculiarity of amorphous and disordered [6.19] semiconductors that recombination takes over gradually because of the fuzzy band edges. Therefore, the excess carriers in general occupy localized states before recombination. Depending on the degree of localization and on the strength of the electron–lattice coupling, the network may deform (relax) around an occupied site, thereby lowering the total energy. This network relaxation (Fig. 6.2) produces a Stokes shift of the luminescence band with respect to the absorption edge. The chalcogenide glasses represent a clear case of a Stokes-shifted luminescence band, since practically no optical absorption is measurable in the luminescence region [6.20, 21]. In general, however, unambiguous evidence for Stokes shifts in amorphous semiconductors is rather difficult to obtain because of the lack of any sharp structure in the luminescence and absorption spectra. After these

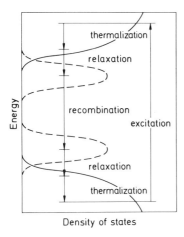

Density of states

Fig. 6.2. Density-of-states model for an amorphous semiconductor

preceding steps the excess carriers may recombine with or without emission of radiation.

The information obtained from luminescence results in amorphous substances cannot be so detailed as in crystals [6.22]. There the observed sharp lines permit the investigation of the influence of external perturbations such as electric and magnetic fields, uniaxial strain, hydrostatic pressure, and so on. These relatively small effects are very likely not to be observed with the broad luminescence bands of amorphous semiconductors. However, because of thermalization, luminescence measurements probe the band tail states and other localized states in the first place. It is these states that are probably most interesting here and also most important considering technical applications. Luminescence results are therefore most informative when regarded in connection with absorption, photoconductivity, electron spin resonance, and related results.

6.1.3 Experimental Note

For the luminescence measurements reported in this chapter, the experimental setup was rather standard. In many cases, the spectral range of the observed luminescence prohibits the use of sensitive, fast photomultipliers. Instead rather insensitive and slow photovoltaic and photoresistive devices are employed. On the other hand these cells have the advantage of an essentially flat spectral distribution of quantum efficiency.

The variation of the spectral sensitivity of the detection system introduced by the dispersion of the monochromator is in general not taken into account. This may generate changes in the shape of the spectrum of minor nature. Differences on the order of 0.04 eV in the positions of the luminescence maxima as observed by different authors might be due to this effect [6.23].

6.1.4 Scope of the Chapter

In the presentation of the various results the usual distinction is made between tetrahedrally coordinated semiconductors, on the one hand, and chalcogenide glass and related semiconductors, on the other. This distinction is not necessary in connection with the recombination mechanisms alone. Electron–hole correlation, Stokes shifts, and even light-induced changes of the electronic properties must be taken into account for both groups. It is nevertheless believed that the many results may be presented in a clearer way if the two groups are treated separately. Silicon is discussed first since the results do seem to be more closely related to the luminescence results of conventional crystalline semiconductors and can be discussed by analogy. This includes topics as excitonlike electron–hole pairs and the possibility of doping. Then the chalcogenide glasses, selenium, and arsenic are discussed together. Their discussion includes the less familiar valence-alternation pairs and large Stokes shifts.

To conclude this introduction I would like to expressly mention the review articles by *Mott* et al. [6.24], *Street* [6.23], *Bishop* et al. [6.25], and *Mott* [6.26], which cover a similar or closely related field. The general principles of luminescence are contained in the articles by *Gershenzon* [6.27] and by *Bebb* and *Williams* [6.28].

6.2 Amorphous Silicon

6.2.1 Influence of Preparation Conditions

Until recently, photoluminescence could only be found in amorphous silicon prepared by the decomposition of silane (SiH_4) in a glow discharge (GD–Si). Now there also exist luminescence results from hydrogenated sputtered [6.15, 29], and vacuum evaporated [6.30] silicon films. Also evaporated films subsequently exposed to a hydrogen plasma exhibit luminescence [6.31]. The quantum efficiencies of luminescence are, however, highest in optimally prepared GD–Si films [6.4]. In the case of GD–Si, both the shape of the luminescence spectrum and the luminescence intensity depend on the deposition parameters, the most important of which are the substrate temperature T_s during deposition and the deposition rate. In the range of deposition rates normally used ($1–20\,\text{Å s}^{-1}$), the photoluminescence properties practically do not change. The luminescence signal decreases, however, when the samples are prepared at higher rates [6.32] (up to $350\,\text{Å s}^{-1}$). The variation of the spectrum with T_s is seen [6.4] from Fig. 6.3: the spectrum moves to higher energy with increasing substrate temperature. There is no change in the spectrum for $250\,°C < T_s < 400\,°C$. This apparently indicates that some kind of intrinsic state of the sample is reached. The same conclusion can be drawn from the corresponding dependence of the luminescence intensity on T_s

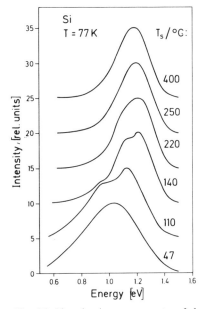

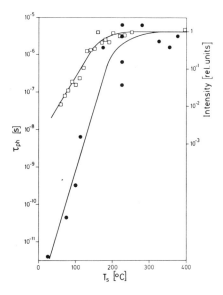

Fig. 6.3. Photoluminescence spectra of glow-discharge produced amorphous silicon [6.4] deposited at various substrate temperatures T_s

Fig. 6.4. Photoluminescence intensity [6.4] (□) and photocarrier life-time [6.3] τ_{ph} (●) of GD-Si deposited at various substrate temperatures T_s.

τ_{ph} is proportional to the density of photo-carriers

(Fig. 6.4): The intensity also saturates above approximately 200 °C, and quantum efficiencies of 0.1–1 are estimated in that range of substrate temperatures [6.4]. GD–Si films deposited at a temperature of 250–400 °C and at moderate deposition rates (1–20 Å s^{-1}) are therefore termed intrinsic. It is not intended to preclude the presence of hydrogen in intrinsic films.

There is a certain scatter in the position of the luminescence maximum of intrinsic samples from 1.2 eV to about 1.3 eV. A similar range of peak energies is found if results from different laboratories are compared [6.4, 33–35]. This probably indicates the influence of other deposition parameters (just to mention one example: the distance of the bright area in the gas discharge from the substrate) which are difficult to control and which are therefore in general not specified. Those changes in deposition parameters which degrade luminescence are also those which increase the residual spin density [6.35a].

6.2.2 Intrinsic Amorphous Silicon

The photoluminescence spectrum of intrinsic GD–Si, measured at low temperature ($T \leq 100$ K), consists of a single broad band centered at around 1.25 eV

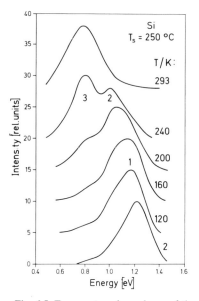
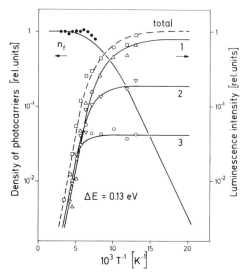

Fig. 6.5. Temperature dependence of the photoluminescence spectrum of GD-Si [6.36]

Fig. 6.6. Temperature dependence of the density of photocarriers [6.37, 38] and of the luminescence intensity [6.36] of GD–Si. The numbers refer to the three luminescence bands discussed in Sect. 6.2.4a

with a half-width of about 0.25 eV [6.4]. When the temperature is increased, the luminescence spectrum changes its shape [6.36] (Fig. 6.5) and the luminescence intensity decreases [6.36] (Fig. 6.6). Two additional bands appear at elevated temperature, and the low-temperature maximum 1 disappears (see also the discussion in Sect. 6.2.4a). The intensities of the three bands are plotted in Fig. 6.6 versus reciprocal temperature. It can, nevertheless, be seen that all three bands decrease at high temperature in a similar fashion. This suggests in the first place that there is a competing nonradiative recombination process which is temperature activated. It is possible to assign an activation energy ΔE (0.13 eV in this case) to the activated decrease at high temperature ($T \leq 150$ K) which seems to be the same for all three luminescence bands.

It is essential for the following discussion that photoconductivity has a completely different temperature dependence, though it also depends on the generation of excess carriers as luminescence. In the expression

$$\sigma_{\text{ph}} = e n_f \mu \tag{6.1}$$

for the photoconductivity, the mobility μ and σ_{ph} are known [6.37, 38], and it is possible to extract a quantity which is proportional to the excess free carrier concentration n_f. This quantity (Fig. 6.6, black circles) is constant in a temperature range where the photoluminescence intensity falls off steeply. The radiationless competing process, which causes the decrease of photolumines-

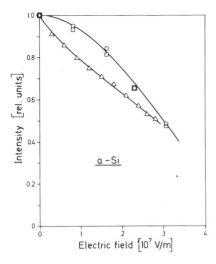

Fig. 6.7. Luminescence intensity of GD-Si at 77 K as a function of an externally applied electric field, for near-band-gap excitation (□,○: [6.39], △: [6.40])

cence, apparently does not affect the excess carriers responsible for photoconduction. Another result points in the same direction: the fact that photoconductivity decays with time constants [6.3] of the order of 1 ms, whereas photoluminescence decays much faster [6.32], time constants down to 20 ns being reported [6.39]. A possible clue to the solution of this problem is given by the result that an externally applied electric field quenches luminescence [6.15, 39, 40] (Fig. 6.7). This suggests that there exists a correlation between the photoexcited electrons and holes which is destroyed by the electric field. Analogous results are indeed found for semiconductor crystals in the case of the radiative decay of excitons: the luminescence intensity falls off with increasing temperature in an activated way, which allows one to deduce the exciton binding energy [6.41]. Also, photoconductivity and luminescence exclude each other: the neutral excitons do not conduct, and the free carriers responsible for photoconduction recombine in most cases nonradiatively. Furthermore, luminescence is quenched by an electric field through exciton dissociation [6.42].

We conclude from this analogy that in amorphous silicon the photoexcited electrons and holes form pairs. This, however, does not necessarily imply that these pairs are excitons, though this is certainly one possibility. Other explanations for this pairing are   that the electron and hole are trapped at the same center giving a bound exciton [6.15], or that both particles are trapped close to each other in the respective band tail [6.39]. The additional though natural assumption in this case is that the band tail states have the same charge character as the respective band, i.e., a carrier in the conduction band tail has a negative charge. If the bond between the electron and the hole of a pair is Coulombic, the binding energy is

$$\Delta E = \frac{1}{4\pi\varepsilon\varepsilon_0} \frac{e^2}{d},$$ (6.2)

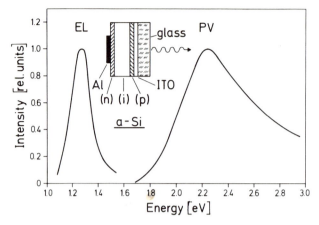

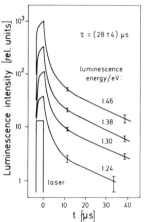

<figure>Fig. 6.8. Spectra of electroluminescence (EL) and photovoltaic collection efficiency (PV) of GD-Si at 78 K [6.15]</figure>

Fig. 6.9. Time decay of the luminescence intensity after pulsed excitation, at four different luminescence energies [6.32]

where d is the electron–hole distance. Identifying ΔE with the experimental value of about 0.13 eV [6.36, 39], (6.2) gives (with $\varepsilon = 10$) $d = 10$ Å. This is a short distance, and if the pairs can be identified with (bound) excitons this means that they are rather of the Frenkel than of the Wannier-Mott type [6.43]. It is interesting to note that this pairing occurs also when the excess carriers are injected, as in electroluminescence [6.15] (Fig. 6.8).

An important consequence of pairing is that radiative recombination obeys first-order (monomolecular) kinetics, i.e., the recombination rate is proportional to the density of pairs n_{p}. This yields the rate equation

$$\dot{n}_{\mathrm{p}} = g - \frac{n_{\mathrm{p}}}{\tau_{\mathrm{r}}}, \tag{6.3}$$

where g is the generation rate, and τ_{r} is the radiative lifetime. Monomolecular kinetics were directly confirmed by *Tsang* and *Street* [6.32], who measured the decay time of radiative recombination [(28 ± 4) μs, Fig. 6.9] and found no dependence on the excitation intensity. Bimolecular and higher-order kinetics, on the other hand, would yield decay times dependent of the excess carrier concentration [6.44].

6.2.3 Electron–Hole Pairs

a) Recombination Versus Dissociation

Equation (6.3) is not generally valid; it holds only at low temperature, where radiationless recombination plays no role [6.36] (Fig. 6.6). The results of Fig. 6.6, however, indicate a temperature-activated radiationless process, which comes into play above about 100 K. The competing process apparently generates photoconduction, and so it is assumed that this process is the dissociation of the pairs into uncorrelated photocarriers of densities n_f and p_f. The dissociation rate is proportional to the density of bound pairs, so that now the rate equation reads, neglecting the reformation of pairs from uncorrelated carriers [6.39],

$$\dot{n}_p = g - \frac{n_p}{\tau_r} - \frac{n_p}{\tau_d}. \tag{6.4}$$

Because of the exponential decrease of the luminescence intensity at high temperature, an exponential expression is assumed [6.39] for the dissociation coefficient τ_d^{-1}:

$$\tau_d^{-1} = \tau_{d,0}^{-1} e^{-\Delta E/kT}. \tag{6.5}$$

In this model, ΔE is identified with the pair binding energy. In steady state ($\dot{n}_p = 0$), one obtains for the luminescence intensity [6.39]

$$I_L(T) \propto \frac{n_p(T)}{\tau_r} = g \left(1 + \frac{\tau_r}{\tau_{d,0}} e^{-\Delta E/kT} \right)^{-1}. \tag{6.6}$$

This is in good agreement with the experimental results (Fig. 6.6).

In order to obtain a similar fit to the photoconductivity results and a prediction of the behavior of n_f at low temperature, the rate equation for n_f is needed. The pair dissociation rate enters this equation as a generation rate. According to photoconductivity results by *Spear* et al. [6.38], who find an approximate square-root behavior of photoconductivity with excitation intensity (at high generation rate and low temperature), the recombination of the photocarriers (electrons [6.38]) is second order (bimolecular):

$$\dot{n}_f = + \frac{n_p}{\tau_d} - b n_f^2, \tag{6.7}$$

where b is the second-order recombination coefficient. In steady state ($\dot{n}_f = 0$),

$$n_f(T) = \left(\frac{g}{b}\right)^{1/2}\left(1 + \frac{\tau_{d,0}}{\tau_r}e^{+\Delta E/kT}\right)^{-1/2}. \tag{6.8}$$

So the predicted behavior for $T \to 0$ is an exponential decrease with an activation energy of $\Delta E/2$. At high temperature, the free carrier concentration will be constant (Fig. 6.6). It is expected for an amorphous semiconductor that there is a distribution of binding energies ΔE around an average value. The effect of a distribution of ΔE will be to make the transition from constant n_p or n_f to the activated decrease according to Eqs. (6.6) and (6.8) less sharp, and this is indeed found [6.36].

The assumption that the dissociation of pairs is the generation rate for photoconductivity is also able to explain a very different result, namely the dependence of photoconductivity on substrate temperature T_s (Fig. 6.4) [6.3]. Photoconductivity depends much more strongly on this parameter than luminescence. The low-T_s samples are very likely to contain more defects than the high-T_s samples, and a strong nonradiative recombination mechanism with monomolecular kinetics dominates Eqs. (6.4) and (6.7) [6.36]. The nonradiative rate will be proportional to the concentration N_R of the defects, which form recombination centers. We then have

$$\dot{n}_p = g - rN_R n_p, \tag{6.9}$$

$$\dot{n}_f = \frac{n_p}{\tau_d} - rN_R n_f. \tag{6.10}$$

Combining (6.9) and (6.10) for steady-state conditions yields

$$I_L \propto \frac{n_p}{\tau_r} = \frac{g}{\tau_r r}\frac{1}{N_R}, \tag{6.11}$$

and

$$\sigma_{ph} = en_f\mu = e\mu\frac{g}{\tau_d rr'}\frac{1}{N_R^2}. \tag{6.12}$$

The quadratic dependence of σ_{ph} on N_R explains why σ_{ph} depends much more strongly on preparation conditions than photoluminescence. In fact, the slope of the solid curve in Fig. 6.4 for photoconductivity is, at low T_s, twice as steep as the corresponding curve for photoluminescence.

b) Stokes Shifts

In the preceding section a close correlation between an electron and a hole in amorphous silicon was assumed, but it was not necessary to specify the nature of this correlation further. It seems that at present no real decision is possible whether the electron–hole pairs are pairs of trapped particles or rather excitons trapped at some defect or impurity. Possible candidates for exciton trapping sites would be Si–H bonds [6.15]. In any case, however, the excited pair is confined to rather a small volume, if the identification of the experimentally found activation energy with a Coulombic binding energy is correct [see Sect. 6.2.2 and (6.2)]. Excited states of small spatial extension possess a strong tendency to lower their energy by deforming the surrounding network so that small polarons will form [6.45]. Therefore Stokes shifts of the luminescence band with respect to the absorption edge have to be taken into account.

There are indeed indications for polaronic effects in GD–Si. Drift mobility [6.37] and photoconductivity [6.38] are activated with about 0.15 eV, and a similar energy (0.1 eV) is found as the difference between the activation energies of dark conductivity and thermopower [6.46]. *Street*, following detailed-balance arguments, estimates a Stokes shift of around 0.4 eV [6.35]. Photostructural changes indicating strong electron–lattice coupling are also found [6.47]. However, direct and unambiguous evidence for Stokes shifts is hard to obtain. Luminescence in amorphous silicon, in contrast to the chalcogenide semiconductors, occurs in an energy region, where still measurable optical absorption exists. Unfortunately the luminescence spectrum, the excitation spectrum, and the absorption edge lack the sharp structure by whose aid Stokes shifts can unambiguously be identified.

c) Radiationless Recombination

There are two kinds of radiationless mechanisms: those that compete directly with luminescence, that is *after* the electron–hole pairs have formed (an electron–hole pair, after having formed, has the choice either to emit photons, or to recombine nonradiatively); and those processes that compete in an indirect way *before* the pairs have formed (excess electrons and holes either form pairs or recombine nonradiatively). The indirectly competing processes are likely to occur, even at low temperature, in samples with a high defect concentration. The indirect mechanism was assumed in (6.9) and (6.10), for example (see Fig. 6.4). The directly competing processes can also occur in samples without defects: an example is the dissociation of electron–hole pairs which comes into play at elevated temperatures [see Fig. 6.6 and (6.4)]. These processes influence the decay times, as can be seen from (6.4),

$$\dot{n}_p = g - n_p \left(\frac{1}{\tau_r} + \frac{1}{\tau_d} \right) = g - \frac{n_p}{\tau_{eff}}. \tag{6.13}$$

According to *Tsang* and *Street* [6.32], an external electric field does not influence the decay constant τ_{eff}, and therefore induces an indirectly competing radiationless process, possibly by impeding the formation of electron–hole pairs from free excess carriers.

6.2.4 Network Defects

a) Network Defects in Glow-Discharge Produced Silicon

Network defects affect the luminescence by changing the spectrum and/or by changing the intensity. Which influence appears in each particular case depends on the kind of the electronic energy levels introduced by the defect. If the energy levels lie within either band, they will hardly have an effect, since they lie in a region where thermalization dominates. If the energies of the defect levels fall between or in the vicinity of the band edges, it depends on the charge state and on the particular electron–lattice coupling whether they act as radiative or nonradiative recombination centers [6.44].

In the case of radiative centers, the defects shift the luminescence spectrum to lower energy, or generate an extra luminescence band at lower energy. Defects of this kind are present in intrinsic GD–Si without special treatment. They are responsible for the slight asymmetry of the luminescence band at low temperature and dominate the spectrum at high temperature [6.36] (Fig. 6.5). For each temperature the spectra were fitted with three Gaussians centered at about 0.8 eV, 1.1 eV, and 1.2 eV. In this way the temperature dependence of the intensity of each of the three bands could be determined (Fig. 6.6), and very similar dependencies were found. This means in the electron–hole-pair picture that the binding energy of the pairs does not, or only little, depend on whether the excess carriers are in deep or shallow traps.

Luminescence bands at similar energies can already be detected at low temperature if the GD–SI films are deposited on low-T_s substrates [6.4] (Fig. 6.3), or if high-T_s layers have been irradiated with Ar^+ ions [6.48] (Fig. 6.10). These extra bands vanish in both cases at elevated substrate temperature (Fig. 6.3) or when the samples are annealed (Fig. 6.10). This confirms the initial impression that the additional luminescence bands are due to network defects. In order to account for the defect energy levels, a density-of-states model has been developed on the basis of the luminescence results [6.4] (Fig. 6.11). This model is in very good qualitative agreement with the density-of-states distribution which was deduced by *Madan* et al. from field effect results [6.2, 49].

In general, a density of states as in Fig. 6.11 gives rise to three extra bands instead of the two that are observed. Transitions between the two defect levels may be excluded since transitions between two deep defect states are unlikely because of insufficient overlap of the two wave functions. In addition, following *Spear*'s suggestion [6.2], we may identify the defects with the amorphous analog of a double silicon vacancy. The positively and negatively charged states of a double vacancy in crystalline silicon [6.50] are in fair agreement with the density of states diagram of Fig. 6.11. Optical transitions between the two levels of a double vacancy are, however, not possible [6.50].

One example of the second kind of defects (centers for nonradiative recombination) has already been described: the luminescence intensity decreases when the substrate temperature is lowered (Fig. 6.4). Very similar results

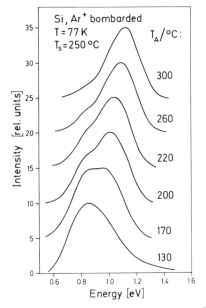

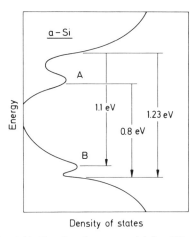

Fig. 6.10. Photoluminescence spectra of ion-bombarded GD-Si [6.48] after annealing at temperature T_A

Fig. 6.11. Density-of-states model for GD-Si [6.4]

are obtained when a high-T_s film is irradiated with noble gas ions (He$^+$, Ar$^+$) [6.48]. Then, depending on the dose, the luminescence intensity usually vanishes below the level of detection, but can be annealed back to the initial value (Fig. 6.12). Annealing also works in the case of low-T_s material [6.51]. In this case, however, the intensity level of high-T_s GD–Si is not reached.

The exact nature of these centers is not yet known. They could be related to localized states near midgap which are charged because of overlapping density-of-states tails. It was suggested by *Anderson* and *Spear* [6.52] that these centers also determine the magnitude of the photoconductivity.

b) The Effect of Hydrogen and Oxygen

Amorphous silicon evaporated in normal high vacuum presents an interesting problem: The luminescence intensity is in general (depending on the evaporation parameters) more than two orders of magnitude lower than in GD–Si. One could assume that the presence of oxygen enhances the radiationless rate through high-energy local network vibrations [6.53]. However, GD–Si implanted with oxygen ions anneals faster than implanted with helium ions [6.48] (Fig. 6.12). Apparently, oxygen helps in compensating centers for radiationless recombination, probably by saturating dangling bonds. Furthermore, the luminescence signal in ultra-high-vacuum evaporated silicon is similarly low as in silicon prepared in normal high vacuum [6.31]. This suggests that the low

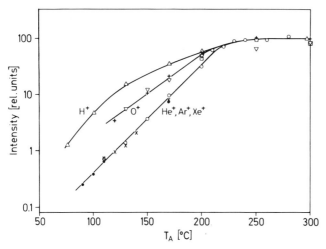

Fig. 6.12. Luminescence intensity of ion-bombarded GD-Si after annealing at temperatures T_A [6.48]

luminescence signal is caused by the evaporation process. It is well known that Si vapor contains a significant fraction of clusters [6.54]. Therefore, a large number of network defects will be introduced into evaporated films.

The role of hydrogen, which is present in GD-Si in large amounts [6.55, 51], has naturally received great attention. In principle, hydrogen in silicon may influence luminescence in several ways: (I) It may be present as interstitial ions and turn amorphous silicon n type. There is, however, no evidence for this at present (the corresponding effect of lithium in GD–Si [6.56] is discussed in Sect. 6.2.5). (II) It may form covalent bonds with silicon. These Si–H bonds may again influence luminescence in at least three ways: through their vibrational modes, through their electronic levels, or by providing trapping sites for electron–hole pairs (excitons).

The vibrational modes have been clearly identified [6.57, 58]. It does, however, not seem as if these localized phonons provided an efficient channel for radiationless recombination: the principal effect of hydrogen is to improve the luminescence intensity. This has been shown again by hydrogen implantation and annealing [6.48]: hydrogen-implanted samples anneal faster than those implanted with helium and oxygen (Fig. 6.12). This effect has also been demonstrated by the influence of hydrogen effusion on luminescence [6.51] (Fig. 6.13), and by hydrogenation during sputtering [6.29, 30]. According to *Pankove* and *Carlson* [6.51], the effect of hydrogen on the luminescence in GD–Si is either to saturate dangling bonds and thus remove radiationless centers, or to provide luminescent centers, for example, as trapping sites for electron–hole pairs.

The electronic levels of the Si–H bond in GD–Si (the bonding and antibonding states) probably merge with the valence and conduction bands,

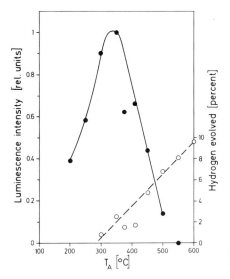

Fig. 6.13. Luminescence intensity (●) and hydrogen evolved [6.51] (○) from GD-Si after a 30-min anneal at temperatures T_A

respectively. This is plausible since the Si–H binding energy is larger than the Si–Si binding energy [6.59]. Another argument concerning the electronic levels of the Si–H bond will be discussed in Sect. 6.2.6 (on Si_xC_{1-x}).

6.2.5 Doped Amorphous Silicon

A separate chapter in this book (Chap. 9) is devoted to doped amorphous semiconductors (chiefly silicon), so that it is sufficient here to summarize the effects of doping which are relevant to the luminescence properties: (I) Doping creates donor and acceptor levels in the vicinity of the respective bands. These additional electronic states could generate additional luminescence bands. (II) The presence of a continuum of localized states throughout the quasi-forbidden gap has the effect that doping fills these localized states with electrons or holes. Thereby charged impurities and oppositely charged states near midgap are produced [6.60]. Thus, doping generates internal electric fields. (III) Only part of the impurity atoms are built into the network as substitutional or interstitial donors or acceptors [6.60]. The nondoping atoms within the amorphous silicon films are very likely to create network defects.

Quite as expected from the discussion in Sect. 6.2.4a, doping influences both the luminescence spectrum [6.61] and the intensity [6.32, 61] (Figs. 6.14, 15). The case of phosphorus doping will be discussed first (Fig. 6.14, right-hand side). An additional band at 0.8 eV appears with increasing phosphorus content, which is the only one to remain at high phosphorus concentration. Very similar results are obtained for GD–Si interstitially doped with lithium [6.62]. It is probably not accidental that the band introduced by doping, and the defect bands appearing after Ar^+ bombardment [6.48] (Fig. 6.10) or at high

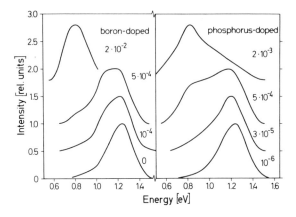

Fig. 6.14. Luminescence spectra of doped GD-Si [6.61] at 77 K. The numbers indicate the SiH_4/PH_3 and SiH_4/B_2H_6 mixing ratios

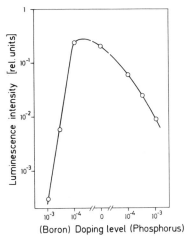

Fig. 6.15. Luminescence intensity of doped GD–Si [6.32]

temperature [6.36] (Fig. 6.5) lie at practically the same energy – they might indeed be of common origin. This can be understood in the following ways: (I) The electronic states leading to the 0.8 eV band are present anyway, even in intrinsic samples, since they appear at high temperature (Fig. 6.5). The appearance of this band at low temperature by doping is due to the occupation of the states at A (Fig. 6.11) by shifting the Fermi level across them [6.63]. This will drastically enhance the radiative recombination of electrons from these levels to photoexcited holes in the valence-band tail. (II) Doping creates or stabilizes the defects whose energy levels lead to the 0.8 eV band. This explanation has been proposed by *Nashashibi* et al. [6.40], and this would also explain the results for boron doping [6.61] (Fig. 6.14, left half): there is an indication that boron doping generates a shoulder at 1.1 eV (which could be related to band 2 in Fig. 6.3, and to the levels B in Fig. 6.11), but it nevertheless also produces a band at 0.8 eV. Preliminary results on Li-doped films [6.62] also support the interpretation that doping creates defects: varying the degree of Li doping above a certain level practically does not influence the position of the Fermi level, but it strongly affects the additional luminescence band.

The decay of the luminescence intensity upon doping may be explained in the following ways: either doping adds centers for radiationless recombination to the network, or the luminescence intensity decreases in response to the internal electric fields of ionized umpurities [6.40, 61, 64]. *Tsang* and *Street* [6.32] argue that, if the lowering of the intensity is via an electric field, this electric field must come into play before the electron–hole pairs are formed, since doping does not affect the radiative lifetime.

6.2.6 The System Si_xC_{1-x}

There are technical and practical reasons why one might think of extending or shifting the photoluminescence spectrum of amorphous semiconductors to higher energies in a suitable alloy system. The technical reason is that visible luminescence is more relevant to applications (in displays, etc.) than infrared luminescence. The practical reason is very simply that radiation at energies higher than about 1.2–1.3 eV can be investigated using extremely sensitive photomultipliers.

One such system is Si_xC_{1-x} produced in a glow-discharge of SiH_4 and C_2H_4 (and therefore inevitably containing hydrogen in addition to Si and C). The optical absorption edge was shown by *Anderson* and *Spear* [6.65] to shift linearly to higher energies in the silicon-rich region. The photoluminescence spectra consist of two bands [6.14] (Fig. 6.16), one of which increases in height and moves to higher energy similar to the optical gap. The other practically does not shift but stays rather fixed at the energy of the luminescence maximum in pure GD–Si.

The energy shift of the one luminescence maximum as well as of the optical gap E_g with x can readily be understood in terms of the virtual-crystal model [6.66]. The coherent overlap of the wave functions of neighboring Si and C atoms causes a monotonic dependence of E_g on x which is close to linear. The luminescence maximum with almost fixed position may be understood if one recollects that luminescence, as a result of the thermalization of excess carriers, always overemphasizes electronic states at the bottom of the bands or below. It is suggested that, as this luminescence band is energetically related to the luminescence band in GD–Si, these lower states have to do with Si clusters, which are to be expected in a material of this kind. If the bound-exciton model is adopted, the size of the clusters need not be very large, but can be limited to the spatial extension of a bound exciton.

There is one more consequence of these results. As the electronic states of the Si–H bond in GD–Si are still unknown (see Sect. 6.2.4b), one might speculate that the efficient luminescence of this material originates at the Si–H antibonding state. This view is not supported by the results on Si_xC_{1-x}. Here both luminescence bands are likely to have common origin since their energetic

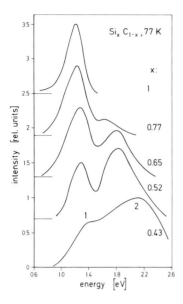

Fig. 6.16. Photoluminescence spectra of GD-Si_xC_{1-x} at 77 K for various values of x [6.14]

position extrapolates to the same value for $x = 1$. Band 2, however, cannot be due to an increasing number of C–H bonds. In this case there would be no linear shift of this luminescence band. Its energy would instead remain rather fixed at the C–H antibonding energy. It is, however, still conceivable that luminescence comes from electron–hole pairs or excitons bound to Si–H or C–H bonds.

6.3 Amorphous Chalcogenide and Related Semiconductors

6.3.1 Influence of Preparation Conditions Survey

The principal difference, from an experimental point of view, between amorphous chalcogenide and related semiconductors, and amorphous silicon, is the fact that the former materials can be prepared in bulk form by rapid cooling from the melt, and this is why they are termed glasses. As *Kastner* has pointed out [6.67, 68], this property is closely linked to the chemical bonding and the electronic structure of these substances: the prerequisite for glass formation seems to be the presence, in the network, of atoms with a coordination number of less than four. In this case, only little energy is required for the two or three bonds of a single atom to arrange in a way that a three-dimensional amorphous network can be built. These principles are realized in their purest form in amorphous SiO_2, which is the most typical glass. It has turned out that a-As_2Se_3 is the typical glassy semiconductor. Its bonding scheme [6.69] is shown in Fig. 6.17.

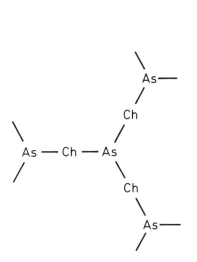

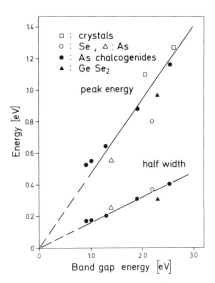

Fig. 6.17. Simplified structure of As$_2$Ch$_3$ (Ch: chalcogen) [6.69]

Fig. 6.18. Peak energies and half-widths of the luminescence spectra of glassy semiconductors vs optical gap [6.23]

Quenching from the melt is, however, not the only way of preparation. Amorphous semiconductors of this group can be obtained by evaporation and sputtering as well. Generally the preparation conditions play only a minor role in this group, and no extra luminescence bands are observed as in the case of a-Si when the samples are prepared in different ways. Therefore, the luminescence and several other electronic properties can be termed intrinsic properties.

The discussion here is based on the results for As$_2$Se$_3$ and As$_2$S$_3$ as the typical representatives of the group, and for amorphous selenium as an example of an elemental glassy semiconductor. The principal reason for this limitation is seen in Fig. 6.18, which shows the positions of the luminescence bands and their half-widths as a function of the optical gap [6.23]. Figure 6.18 demonstrates the similarity of the luminescence properties of the glassy semiconductors. It demonstrates further that the As–chalcogenide alloy system obeys the virtual-crystal model [6.66]: there is a single luminescence band in this system, which changes its energetic position and half-width along with the composition, suggesting that the atoms in this alloy system are built into the network completely at random, and no clustering occurs as in the Si$_x$C$_{1-x}$ system (Sect. 6.2.6).

The common features of the luminescence results of the semiconductor glasses are briefly listed as follows: (I) The luminescence spectrum lies at very low energy, at about half the optical gap. (II) The luminescence signal decreases strongly with increasing temperature. (III) Apart from selenium, the spectra of the amorphous and crystalline forms are very similar. (IV) In general the luminescence signal decreases, at low temperature, during exposure to the exciting light ("fatiguing") [6.9, 25, 70, 71]. This effect is correlated with photoinduced absorption and electron spin resonance [6.72–75].

6.3.2 Arsenic Chalcogenides and Selenium

The amorphous chalcogenides differ from amorphous silicon in that their luminescence spectra lie in an energy region where no detectable optical absorption is found [6.5–10] (Fig. 6.1). The straightforward interpretation is that strong electron–lattice interaction leads to large Stokes shifts of these levels. This is commonly described in a configuration coordinate diagram [6.24, 76, 77] (Fig. 6.19). By optical excitation, starting from the ground state B, point A is reached on the energy curve of the exited state. The configuration at A, described by the coordinate q_1, is not stable since at C, on the same curve, the total energy of the electron (hole) in a deformable network is lower. Absorption of a photon therefore changes the configuration of the network from q_1 to q_2 (at C). In the recombination process the ground state is not reached directly, but via D at the same network configuration q_2. From D, the network relaxes back to the configuration most favorable for the ground state. In Fig. 6.19 the Franck–Condon principle [6.78, 79] is used; this states that optical excitation or deexcitation is a faster process than network relaxation. Optical transitions, therefore, have to be drawn vertically in configuration coordinate diagrams.

The temperature dependence of photoluminescence in the chalcogenides [6.17, 70, 80] is qualitatively similar to that in amorphous silicon: the luminescence intensity decreases steeply towards high temperature (Fig. 6.20). A fixed activation energy cannot, however, be attributed to the decrease. Instead, the luminescence intensity follows the empirical law [6.81]

$$I_L = I_{L,0} e^{-T/T_0}. \tag{6.14}$$

As in the case of silicon (Sect. 6.2.2, 3), this decay of the luminescence with temperature is attributed to a (non-Coulombic) correlation between the excited electrons and holes yielding bound excitons [6.32]. The particular form of the empirical law (6.14) is, according to *Street* et al. [6.10, 81], due to the electron in the bound exciton tunneling away from the hole through a potential barrier.

There is further support for the electron–hole-pair or bound-exciton concept from a comparison of the photoluminescence and photoconductivity decay times. Photoconductivity in As_2Se_3 decays, at 250 K, with a time constant larger than 1 ms [6.82, 83], with a tendency to longer times at lower temperature. The decay times of photoluminescence at 4.2 K are as low as 25 μs [6.81, 84, 85]. This suggests different decay mechanisms for photoconductivity and luminescence. The strongest support for the pairing model comes from the observation by *Ivashchenko* et al. [6.85] that the luminescence decay time does not depend on the excitation intensity. This result can only be understood by monomolecular recombination (see also Sect. 6.2.2).

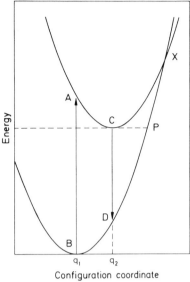
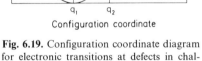

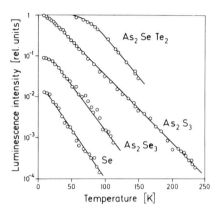

Fig. 6.19. Configuration coordinate diagram for electronic transitions at defects in chalcogenide semiconductors [6.77]

Fig. 6.20. Temperature dependence of the luminescence intensity for chalcogenide semiconductors [6.23]

The first indication that luminescence is related to defects comes from excitation spectra [6.8, 70, 86, 87], which show the luminescence intensity as a function of the energy of the exciting light (Fig. 6.21). At lower energy the excitation efficiency follows the absorption edge [6.87]. With increasing energy the excitation spectrum reaches a maximum and then drops off. This has been interpreted by *Street* et al. [6.86] as follows: Radiative recombination can only occur if the exciton is excited in the vicinity of specific defects or localized states. These localized levels are energetically close to the conduction or valence band edges, and are responsible for the exponential absorption tail. Higher-energy light creates the electrons and holes higher up in the respective bands, and the probability decreases that this excitation results in excitons trapped at radiative recombination centers – which explains the peaked excitation spectrum. The more common explanation by radiationless surface recombination [6.89] does not work for several reasons [6.23]; the clearest one is probably that the excitation spectrum does not depend on temperature as would be expected from the dependence of surface recombination on the diffusion length.

The results on the fatiguing of the luminescence signal during illumination [6.9, 25, 70, 71] seem to present further evidence that defect states play the central role in luminescence. The fatiguing is reversible after heating the samples up to room temperature [6.71] or irradiating them with low-energy light [6.9, 70]. These results are closely related to the occurrence of photoinduced absorption and electron spin resonance [6.25, 72–75] (Fig. 6.22).

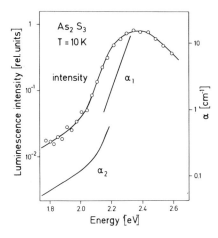

Fig. 6.21. Absorption and excitation spectra of amorphous As_2S_3 at 10 K [6.87] (α_1:[6.88], α_2:[6.20])

Finally, we shall compare the luminescence and absorption spectra of the crystalline and amorphous forms. In general, the spectra of both modifications are very similar (Fig. 6.1) with the exception of selenium (Fig. 6.23). Crystalline selenium [6.90–92] behaves as the standard crystalline semiconductor, silicon. No Stokes shifts occur between absorption and emission; the energy difference between the absorption edge and the luminescence peaks is due to the fact that the band edge excitons are indirect, and both absorption and emission occur, at low temperature, with the aid of phonon emission [6.91]. The results for a-Se [6.10, 93], on the other hand, are very similar to those for a-As_2Se_3, and exhibit large Stokes shifts. Furthermore, the luminescence signal of a-Se is very weak and can be increased by adding arsenic [6.10]. The specific models which are able to account for these results will be described next.

6.3.3 Specific Models

a) Defects

The following discussion of specific models is based on the concept by *Street* and *Mott* [6.77] (which in turn leads back to a paper by *Anderson* [6.94]). A more detail discussion appears in Chap. 3. These authors assume that the basic unit responsible for radiative recombination is a defect level (for example that of a dangling bond) which is charged in the ground state according to

$$2D^0 \rightarrow D^+ + D^- . \tag{6.15}$$

The energy to form a D^- is expected to be overcompensated by the deformation energy released by the D^+ strongly coupling to the surrounding network. As D^+ and D^- possess no unpaired electrons, this model accounts

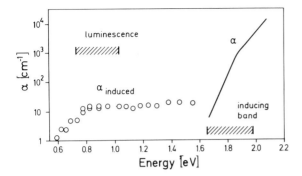

Fig. 6.22. Spectrum of the photoinduced absorption [6.72] of amorphous As$_2$Se$_3$ at 6 K

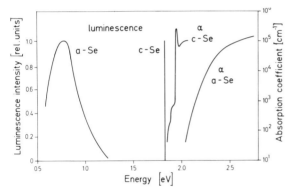

Fig. 6.23. Absorption and luminescence spectra of the amorphous [6.10,93] and crystalline [6.90–92] forms of selenium

well for the absence of a spin signal. The D$^-$ energy is assumed close to the valence band edge, since the electron configuration is similar to a chalcogen lone pair. The absorption tail is due to transitions of electrons from D$^-$ states to the conduction band, forming loosely bound e–D^0 pairs. The total energy of these pairs is lowered by electron–lattice interaction (relaxation from A to C in Fig. 6.19). The luminescence band originates from transitions from the relaxed e–D^0 state to the D$^-$ state as shown in Fig. 6.19. The small non-Coulombic binding energy of the excited e–D^0 pair before relaxation is responsible for the temperature dependence of photoluminescence since this pair will dissociate at elevated temperature.

As the incident light also generates neutral defect states, the luminescence signal will decrease as a results of the decreasing density of D$^-$ centers. The reaction in (6.15) leading back to thermal equilibrium is not possible at low temperature, which makes the D^0 states metastable. This explains metastable photoinduced absorption and spin resonance. Photoconductivity is attributed to different processes,

$$D^0 \to D^- + h,$$

or   (6.16)

$$D^0 \to D^+ + e.$$

Here the $D^-(D^+)$ states are considered to be trapping sites for photocarriers. The different processes occurring with photoconduction and luminescence explain the different decay times.

The *Street–Mott* model [6.77] does not account for the intrinsic nature of photoluminescence in chalcogenides (Sect. 6.3.1) as it relates the luminescence properties with defect states. This difficulty has been solved by *Kastner* et al. [6.95, 96] by specifying the nature of the defects. They postulated the formation of valence-alternation pairs which need so little energy of formation that even crystals cannot be grown without them. This concept also explains the great similarity between the amorphous and crystalline luminescence spectra. These valence-alternation pairs occur, when an atom of valence v happens to be on a site with coordination number $v + 1$, and at the same time an atom with valence w is on a site with coordination $w - 1$. For example (the coordination number appears in parentheses):

$$2Se^0(2) \to Se^+(3) + Se^-(1),$$ (6.17)

$$As^0(3) + Se^0(2) \to As^-(2) + Se^+(3).$$ (6.18)

b) Small Polarons

The *Street–Mott* and related models need additional assumptions to explain the intrinsic nature of luminescence in glassy semiconductors (valence-alternation pairs). In the polaron model of *Emin* [6.97], the intrinsic nature of the luminescence properties is the basic assumption. Immediately after excitation, the excess carriers form small polarons, which leads to the appearence of polaron bands in the density of states (Fig. 6.2). This concept automatically implies Stokes shifts, since the total energy of the excess carriers is lowered by network relaxation (formation of small polarons). The basic electronic properties of the chalcogenide semiconductors can be explained, if those ideas are combined with the models by *Knights* and *Davis* [6.93] and *Pai* and *Enck* [6.98] for photoconduction and recombination in amorphous selenium: luminescence occurs if electron and hole polarons are created spatially close to each other (by near-band-gap light). Radiationless recombination dominates if both particles are generated at greater distances (by light with energy larger than the band gap). A small polaron at a large distance from polarons of opposite sign is metastable at low temperature – resulting in luminescence fatigue as well as photoinduced spin resonance and absorption. The induced absorption band in this model is due to photon-induced hopping of small polarons.

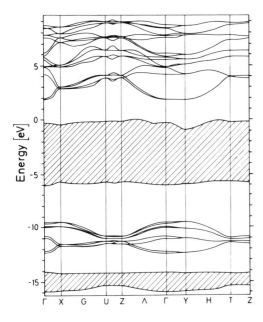

Fig. 6.24. Simplified band structure of c-As$_2$Se$_3$ [6.99]. The top of the valence band is at $E = 0$

c) Band Structure of Chalcogenides

These two models (defects versus polarons) seem to exclude each other. There is, however, the possibility of a compromise, if the following results from amorphous and crystalline selenium are considered: The luminescence signal of a-Se is small compared with that of the chalcogenides and increases with the addition of arsenic [6.10]. Furthermore, the luminescence in c-Se [6.91] is not Stokes-shifted as in c-As$_2$Se$_3$. The consequence which one has to draw on the basis of the defect model is that selenium itself does not form valence-alternation pairs. The explanation probably has to do with the two major differences between selenium and As$_2$Se$_3$: (I) Selenium is an elemental semiconductor, whereas As$_2$Se$_3$ is a compound. This means in terms of the defect model that the defects relevant for luminescence must be of a kind that can occur only in the compound, for example according to (6.18). (II) The uppermost valence band in c-Se (probably also in a-Se) is a separate band made up exclusively of the Se $4p$ lone-pair electrons. According to recent band structure calculations by *Althaus* et al. [6.99], this is not valid for c-As$_2$Se$_3$ (probably neither for a-As$_2$Se$_3$). Their band structure results are reproduced in Fig. 6.24. In the highest occupied band the As–Se bonding states and the Se lone pairs cannot be separated from each other [6.100], as a result of the ionicity of the As–Se bond. A consequence of this band structure is that incident light may break bonds by transitions from bonding to antibonding states [6.100]. These photostructural effects are indeed observed in crystalline and amorphous As$_2$Se$_3$ [6.101, 102]. On this basis, the defect model and the

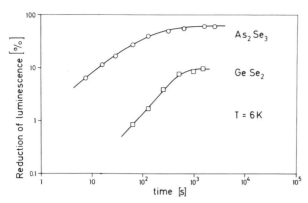

Fig. 6.25. Time dependence of the photoluminescence fatigue in glassy As_2Se_3 and $GeSe_2$ [6.104]

polaron model might be reconciled, since a photoinduced broken bond can be considered as an extreme case of electron–lattice coupling of a fundamental excitation.

d) Radiationless Recombination

There is still one problem to be dealt with: So far all the defects mentioned are centers for radiative recombination. It is, however, evident from luminescence and photoconductivity (for example Fig. 6.20), that at high temperature almost all recombination processes are radiationless. This implies that there are also centers or mechanisms for nonradiative recombination. Instead of introducing ad-hoc radiationless centers, *Mott* and *Stoneham* [6.76] proposed a model according to which radiative and nonradiative recombination can occur at the same center. They suggest an energy diagram as in Fig. 6.19, with the cross-over point X lying sufficiently low so that the excited relaxed state C may tunnel to state P on the ground-state parabola. Thermal excitation across the barrier at X is also possible, which explains the temperature dependence of the luminescence intensity. A modification of this concept has been used by *Street* [6.103] to account for the dominating nonradiative recombination in amorphous selenium.

6.3.4 Related Materials

The study of materials related to the chalcogenide glass family is not only interesting in itself but may also give additional clues to the solution of problems encountered there. The study of As and $GeSe_2$ has proved to be of particular interest. Both As [6.75] and $GeSe_2$ [6.104] fit very well into the picture established for chalcogenide glasses (see, e.g., Fig. 6.18, and Sect. 6.3.1). They also exhibit photoluminescence fatigue and photoinduced spin resonance [6.25, 75, 104]. Interestingly, the luminescence in a-As fatigues faster than in

As_2Se_3 or As_2S_3, and it is hardly possible to establish an unfatigued excitation spectrum. On the other hand, $GeSe_2$ shows a slower fatiguing rate than As_2Se_3 and a higher saturated luminescence signal (Fig. 6.25). *Bishop* et al. [6.104] suggest that this is correlated with the dimensionality of the local order. Arsenic and As_2Se_3 possess locally more or less layered structures, while $GeSe_2$ consists of a three-dimensionally coordinated network, as, for example, SiO_2. Another suggestion is, following Sect. 6.3.3c, that the varying strength of the photoinduced properties is related to the character of the valence band edge. In As, the valence band is a band of As–As bonding states [6.67]; in As_2Se_3 it contains Se lone pairs in addition [6.99, 100]. If, in $GeSe_2$, the highest valence band consists primarily of Se lone-pair states, then we have from As via As_2Se_3 and $GeSe_2$ to Se a decreasing probability of breaking bonds by optical excitation. Based on these considerations, one may predict photoinduced changes of electronic properties also for a-Si, if one recollects that the valence and conduction bands of silicon have largely bonding and antibonding character, respectively [6.105]. These effects have been found just recently [6.47].

Acknowledgements. I would like to thank W. Fuhs, H. Mell, J. Stuke, and G. Weiser for helpful discussions.

References

6.1   N.F.Mott, E.A.Davis: *Electronic Processes in Non-Crystalline Materials* (Clarendon Press, Oxford 1971)

6.2   W.E.Spear: In *Amorphous and Liquid Semiconductors*, ed. by J.Stuke, W.Brenig (Taylor and Francis, London 1974) p. 1

6.3   R.J.Loveland, W.E.Spear, A.Al-Sharbaty: J. Non-Cryst. Solids **13**, 55 (1973)

6.4   D.Engemann, R.Fischer: In *Proc. 12th Int. Conf. Physics of Semiconductors*, ed. by M.H.Pilkuhn (Teubner, Stuttgart 1974) p. 1042

6.5   R.Zallen, R.E.Drews, R.L.Emerald, M.Slade: Phys. Rev. Lett. **26**, 1564 (1971)

6.6   J.T.Edmond: Br. J. Appl. Phys. **17**, 979 (1966)

6.7   B.T.Kolomiets, T.N.Mamontova, A.A.Babaev: J. Non-Cryst. Solids **4**, 289 (1970)

6.8   S.G.Bishop, D.L.Mitchell: Phys. Rev. B **8**, 5696 (1973)

6.9   J.Cernegora, F.Mollot, C.Benoit à la Guillaume: Phys. Status Solidi (a) **15**, 401 (1973)

6.10 R.A.Street, T.M.Searle, I.G.Austin: Philos. Mag. **30**, 1181 (1974)

6.11 W.C.Dash, R.Newman: Phys. Rev. **99**, 1151 (1955)

6.12 J.R.Haynes, M.Lax, W.F.Flood: J. Phys. Chem. Solids **8**, 392 (1959)

6.13 R.A.Street, I.G.Austin, T.M.Searle, B.A.Smith: J. Phys. C **7**, 4185 (1974)

6.14 D.Engemann, R.Fischer, J.Knecht: Appl. Phys. Lett. **32**, 567 (1978)

6.15 J.I.Pankove, D.E.Carlson: Appl. Phys. Lett. **29**, 620 (1976)

6.16 G.Z.Vezzoli, P.J.Walsh, P.J.Kisatsky, L.W.Doremus: J. Appl. Phys. **45**, 4534 (1974)

6.17 R.Fischer, U.Heim, F.Stern, K.Weiser: Phys. Rev. Lett. **26**, 1182 (1971)

6.18 F.Stern: J. Non-Cryst. Solids **8–10**, 954 (1972)

6.19 D.Redfield: Phys. Rev. B **2**, 1830 (1970)

6.20 J.Tauc, A.Menth, D.L.Wood: Phys. Rev. Lett. **25**, 749 (1970)

6.21 D.L.Wood, J.Tauc: Phys. Rev. B **5**, 3144 (1972)

6.22 E.W.Williams, H.B.Bebb: In *Semiconductors and Semimetals*, Vol. 8, ed. by R.K.Willardson, A.C.Beer (Academic Press, New York 1972) p. 321

6.23 R.A.Street: Adv. Phys. **25**, 397 (1976)

6.24 N.F.Mott, E.A.Davis, R.A.Street: Philos. Mag. **32**, 961 (1975)

6.25 S.G.Bishop, U.Strom, P.C.Taylor: In *Proc. 7th Int. Conf. Amorphous and Liquid Semiconductors*, ed. by W.E.Spear (University of Edinburgh 1977) p. 595

6.26 N.F.Mott: Adv. Phys. **26**, 363 (1977)

6.27 M.Gershenzon: In *Semiconductors and Semimetals*, Vol. 2, ed. by R.K.Willardson, A.C.Beer (Academic Press, New York 1966) p. 289

6.28 H.B.Bebb, E.W.Williams: In *Semiconductors and Semimetals*, Vol. 8, ed. by R.K.Willardson, A.C.Beer (Academic Press, New York 1972) p. 181

6.29 M.H.Brodsky, J.J.Cuomo, F.Evangelisti: In *Proc. 7th Int. Conf. Amorphous and Liquid Semiconductors*, ed. by W.E.Spear (University of Edinburgh 1977) p. 397

6.30 D.Engemann, R.Fischer, H.Mell: In *Proc. 7th Int. Conf. Amorphous and Liquid Semiconductors*, ed. by W.E.Spear (Edinburgh 1977) p. 387

6.31 D.Kaplan, N.Sol, G.Velasco, P.A.Thomas: Appl. Phys. Lett. **33**, 440 (1978)

6.32 C.Tsang, R.A.Street: Philos. Mag. **37**, 601 (1978)

6.33 D.Engemann, R.Fischer: In *Amorphous and Liquid Semiconductors*, ed. by J.Stuke, W.Brenig (Taylor and Francis, London 1974) p. 947

6.34 T.S.Nashashibi, I.G.Austin, T.M.Searle: Philos. Mag. **35**, 831 (1977)

6.35 R.A.Street: Philos. Mag. B **37**, 35 (1978)

6.35a R.A.Street, J.C.Knights, D.K.Biegelsen: Phys. Rev. B **18**, 1880 (1978)

6.36 D.Engemann, R.Fischer: Phys. Status Solidi (b) **79**, 195 (1977)

6.37 P.G. Le Comber, W.E.Spear: Phys. Rev. Lett. **25**, 509 (1970)

6.38 W.E.Spear, R.J.Loveland, A.Al-Sharbaty: J. Non-Cryst. Solids **15**, 410 (1974)

6.39 D.Engemann, R.Fischer: In *Structure and Excitations of Amorphous Solids*, ed. by G.Lucovsky, F.L.Galeener (American Institute of Physics, New York 1976) p. 37

6.40 T.S.Nashashibi, I.G.Austin, T.M.Searle: In *Proc. 7th Int. Conf. Amorphous and Liquid Semiconductors*, ed. by W.E.Spear, (University Edinburgh 1977) p. 392

6.41 V.S.Bagaev, L.I.Paduchikh: Sov. Phys.-Solid State **13**, 389 (1971)

6.42 D.L.Smith, D.S.Pan, T.C.McGill: Phys. Rev. B **12**, 4360 (1975)

6.43 R.S.Knox: *Theory of Excitons*, Solid State Physics, Supplement 5 (Academic Press, New York 1963)

6.44 J.S.Blakemore: *Semiconductor Statistics* (Pergamon Press, Oxford 1962) p. 2

6.45 P.W.Anderson: Nature (London) Phys. Sci. **235**, 163 (1972)

6.46 W.Beyer, H.Mell, H.Overhof: In *Proc. 7th Int. Conf. Amorphous and Liquid Semiconductors*, ed. by W.E.Spear (University of Edinburgh 1977) p. 328

6.47 D.L.Staebler, C.R.Wronski: Appl. Phys. Lett. **31**, 292 (1977)

6.48 D.Engemann, R.Fischer, F.W.Richter, H.Wagner: In *Proc. 6th Int. Conf. Amorphous and Liquid Semiconductors*, ed. by B.T.Kolomiets (Nauka, Leningrad 1975) p. 217

6.49 A.Madan, P.G.Le Comber, W.E.Spear: J. Non-Cryst. Solids **20**, 239 (1976)

6.50 G.D.Watkins, J.W.Corbett: Phys. Rev. **138**, A 543 (1965)

6.51 J.I.Pankove, D.E.Carlson: Appl. Phys. Lett. **31**, 450 (1977)

6.52 D.A.Anderson, W.E.Spear: Philos. Mag. **36**, 695 (1977)

6.53 H.J.Hrostowski, R.H.Kaiser: Phys. Rev. **107**, 966 (1957)

6.54 H.Mell, M.H.Brodsky: Thin Solid Films **46**, 299 (1977)

6.55 C.C.Tsai, H.Fritzsche, M.H.Tanielian, P.J.Gaczi, P.D.Persans, M.A.Vesaghi: In *Proc. 7th Int. Conf. Amorphous and Liquid Semiconductors*, ed. by W.E.Spear (University of Edinburgh 1977) p. 339

6.56 W.Beyer, R.Fischer: Appl. Phys. Lett. **31**, 850 (1977)

6.57 M.H.Brodsky, M.Cardona, J.J.Cuomo: Phys. Rev. B **16**, 3556 (1977)

6.58 J.C.Knights, G.Lucovsky, R.J.Nemanich: Philos. Mag. B **37**, 467 (1978)

6.59 L.Pauling: *The Nature of the Chemical Bond* (Cornell University Press, Ithaca 1960)

6.60 W.E.Spear, P.G.LeComber: Philos. Mag. **33**, 935 (1976)

6.61 W.Rehm, D.Engemann, R.Fischer, J.Stuke: In *Proc. 13th Int. Conf. Physics of Semiconductors*, ed. by F.G.Fumi (Rome, 1976) p. 525

6.62 H.W.Spies, D.Zangmeister, R.Fischer, W.Beyer: Verh. Dtsch. Phys. Ges. (VI) **13**, 44 (1978); to be published

6.63   R. Fischer: In *Festkörperprobleme*, Advances in Solid State Physics, Vol. 17, ed. by J. Treusch (Vieweg, Braunschweig 1977) p. 85
6.64   D. Redfield: Phys. Rev. **130**, 916 (1963)
6.65   D. A. Anderson, W. E. Spear: Philos. Mag. **35**, 1 (1977)
6.66   J. C. Philips: *Bonds and Bands in Semiconductors* (Academic Press, New York 1973) Chap. 8
6.67   M. Kastner: Phys. Rev. Lett. **28**, 355 (1972)
6.68   M. Kastner: Phys. Rev. B **7**, 5237 (1973)
6.69   A. L. Renninger, B. L. Averbach: Acta Crystallogr. B **29**, 1583 (1973)
6.70   R. A. Street, T. M. Searle, I. G. Austin: J. Phys. C **6**, 1830 (1973)
6.71   F. Mollot, J. Cernegora, C. Benoit à la Guillaume: Phys. Status Solidi (a) **21**, 281 (1974)
6.72   S. G. Bishop, U. Strom, P. C. Taylor: Phys. Rev. Lett. **34**, 1346 (1975)
6.73   S. G. Bishop, U. Strom, P. C. Taylor: Phys. Rev. Lett. **36**, 543 (1976)
6.74   S. G. Bishop, U. Strom, P. C. Taylor: Phys. Rev. B **15**, 2278 (1977)
6.75   S. G. Bishop, U. Strom, P. C. Taylor: Solid State Commun. **18**, 573 (1976)
6.76   N. F. Mott, A. M. Stoneham: J. Phys. C **10**, 3391 (1977)
6.77   R. A. Street, N. F. Mott: Phys. Rev. Lett. **35**, 1293 (1975)
6.78   J. Franck: Trans. Faraday Soc. **21**, 536 (1925)
6.79   E. U. Condon: Phys. Rev. **32**, 858 (1928)
6.80   B. T. Kolomiets, T. N. Mamontova, E. A. Smorgonskaya, A. A. Babaev: Phys. Status Solidi (a) **11**, 441 (1972)
6.81   R. A. Street, T. M. Searle, I. G. Austin: In *Amorphous and Liquid Semiconductors*, ed. by J. Stuke, W. Brenig (Taylor and Francis, London 1974) p. 953
6.82   C. Main: Thesis (Edinburgh 1974)
6.83   W. Fuhs, D. Meyer: Phys. Status Solidi (a) **24**, 275 (1974)
6.84   Yu. N. Ivashchenko, B. T. Kolomiets, T. N. Mamontova, E. A. Smorgonskaya: Phys. Status Solidi (a) **20**, 429 (1973)
6.85   Yu. N. Ivashchenko, B. T. Kolomiets, T. N. Mamontova: Phys. Status Solidi (a) **24**, 401 (1974)
6.86   R. A. Street, T. M. Searle, I. G. Austin: Philos. Mag. **29**, 1157 (1974)
6.87   R. A. Street, T. M. Searle, I. G. Austin: Philos. Mag. **32**, 431 (1975)
6.88   R. A. Street, T. M. Searle, I. G. Austin, R. S. Sussmann: J. Phys. C **7**, 1582 (1974)
6.89   H. B. De Vore: Phys. Rev. **102**, 86 (1956)
6.90   H. Zetsche, R. Fischer: J. Phys. Chem. Solids **30**, 1425 (1969)
6.91   R. Fischer: Phys. Rev. B **5**, 3087 (1972)
6.92   S. Tutihasi, I. Chen: Phys. Rev. **158**, 623 (1967)
6.93   J. C. Knights, E. A. Davis: J. Phys. Chem. Solids **35**, 543 (1974)
6.94   P. W. Anderson: Phys. Rev. Lett. **34**, 953 (1975)
6.95   M. Kastner, D. Adler, H. Fritzsche: Phys. Rev. Lett. **37**, 1504 (1976)
6.96   M. Kastner: In *Proc. 7th Int. Conf. Amorphous and Liquid Semiconductors*, ed. by W. E. Spear (University of Edinburgh 1977) p. 504
6.97   D. Emin: In *Proc. 7th Int. Conf. Amorphous and Liquid Semiconductors*, ed. by W. E. Spear (Edinburgh 1977) p. 261
6.98   P. M. Pai, R. C. Enck: Phys. Rev. B **11**, 5163 (1975)
6.99   H. L. Althaus, G. Weiser, S. Nagel: Phys. Status Solidi (1978) in press
6.100   I. Chen: Phys. Rev. B **8**, 1440 (1973)
6.101   J. Feinleib, J. de Neufville, S. C. Moss, S. R. Ovshinsky: Appl. Phys. Lett. **18**, 254 (1971)
6.102   J. S. Berkes, S. W. Ing, W. J. Hillegas: J. Appl. Phys. **42**, 4908 (1971)
6.103   R. A. Street: In *Proc. 7th Int. Conf. Amorphous and Liquid Semiconductors*, ed. by W. E. Spear (University of Edinburgh 1977) p. 509
6.104   S. G. Bishop, U. Strom, P. C. Taylor: In *Proc. 13th Int. Conf. Physics of Semiconductors*, ed. by F. G. Fumi (Rome 1976) p. 563
6.105   D. R. Penn: Phys. Rev. **128**, 2093 (1962)

7. Spin Effects in Amorphous Semiconductors

I. Solomon

With 14 Figures

In this chapter, we discuss electron spin resonance and related spin-dependent effects almost exclusively in tetrahedrally coordinated amorphous semiconductors, and particularly in amorphous silicon. This is not only to limit the chapter to a reasonable size, but also because many of the new and exciting applications of magnetic resonance in the recent years have been obtained in these materials. This chapter is more an attempt to highlight some new developments related to magnetic resonance methods than an exhaustive review of the subject.

7.1 Survey

Since the classical work of *Feher* [7.1], the application of magnetic resonance techniques to semiconductor crystals has provided very detailed information on the properties of impurities and structural defects in semiconductors. In particular, a large part of the information on the structure of radiation-induced defects in crystalline silicon is due to the electron-spin-resonance studies of *Watkins*, *Corbett*, and co-workers [7.2, 3].

It was quite natural to attempt to use the same techniques for the study of localized gap states postulated to explain the distribution of electronic states in amorphous semiconductors. Such an attempt in chalcogenide glasses [7.4] gave both a surprising and disappointing result: the number of unpaired spins turned out to be much smaller than expected and the magnetic resonance in thermal equilibrium undetectable. This result, important in itself, has been explained by *Anderson* [7.5] and *Street* et al. [7.6] in terms of the "lone-pair semiconductors." It is assumed that a negative effective correlation energy, brought about by local lattice distortion, exists between two electrons in the same gap state. The existence of two broken bonds D_A and D_B is unstable against the migration of an electron from one of the bonds to the other. In other words, the reaction

$$D_A^0 + D_B^0 \rightarrow D_A^+ + D_B^-$$

is believed to be exothermic, the energy gained in the deformation of the lattice compensating the repulsion energy of the two electrons on the same center D_B^-. The net result is the transformation of two paramagnetic centers into a pair of

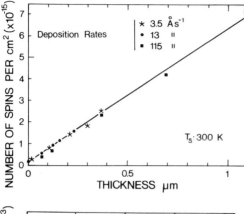

Fig. 7.1. Number of spins in amorphous silicon prepared by evaporation in ultra-high vacuum on a substrate at a temperature $T_S = 300\,\mathrm{K}$. The density of spins is independent of the deposition rate. [7.22]

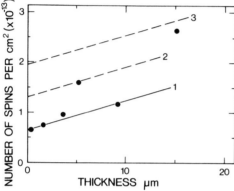

Fig. 7.2. Number of spins in amorphous silicon prepared by decomposition of silane in a glow discharge. Points marked 2 and 3 have been obtained by two and three successive runs. The results are interpreted as a surface (or interface) number of 7×10^{12} spins·cm^{-2} and a volume density of 6×10^{15} spins·cm^{-3}. [7.13]

stable diamagnetic centers, and therefore no magnetic resonance signal can be detected in this type of material. Chapter 3 discusses the defect gap states in further detail.

It has been possible, however, to restore the unstable paramagnetic states by optical injection of electron–hole pairs [7.7]. These light-induced paramagnetic centers are metastable at low temperature in chalcogenide glasses such as As_2S_3 and in amorphous arsenic, even after suppression of the exciting light. The study of these centers by spin-resonance techniques in correlation with optical properties has provided information concerning the density of optically induced states at different temperatures, the local bonding configurations, and the distribution of the paramagnetic centers on the different atoms of the solid. A well-documented review on this type of study has been recently given by *Bishop* et al. [7.8].

The situation is quite different, and apparently simpler, in tetrahedrally coordinated pure amorphous semiconductors, where the unpaired electrons of the broken bonds are quite stable, and give a strong spin-resonance signal [7.9, 10]. It appears that the ideal continuous random network [7.11] where all the normal covalent bonds are satisfied is not stable, and that paramagnetic dangling bonds are necessary to stabilize the disordered structure. The magnetic resonance signal, directly proportional to the number of the localized centers, provides a powerful tool for the experimental study of the local microstructure of these materials: the results will be presented in some detail in the next section. In particular, electron-spin-resonance techniques have been essential in the determination of the structure of hydrogenated silicon produced by glow-discharge decomposition of silane gas [7.12, 13]. In this material it is believed that the hydrogen forms covalent bonds with the broken bonds of the amorphous material, thus restoring the chemical continuity. Spectacular and direct evidence of the neutralization of the dangling bonds by covalent interaction with the hydrogen atoms is given by the almost complete disappearence of the magnetic resonance signal compared to the pure amorphous material.

The novel and original studies in the past two years concern the influence of the spin statistics on electronic processes in amorphous semiconductors, in particular, in hydrogenated silicon. Spin-dependent photoconductivity [7.14] and spin-dependent luminescence [7.15, 16] not only provide an extremely sensitive method of spin-resonance detection, but also give precise information on the physics of recombination in this material, and on the nature of the electronic states in the gap. Also the surprising low-magnetic-field effects on the luminescence [7.17] and on the transport [7.18–20] in amorphous silicon and germanium can only be explained by spin effects, and illustrate the importance of the electron spins in the physics of amorphous materials. These spin-dependent effects are discussed in detail in Sect. 7.3.

7.2 Electron Spin Resonance in Tetrahedrally Coordinated Amorphous Semiconductors

A very large amount of data has been accumulated on group-IV tetrahedrally coordinated amorphous semiconductors since the first discovery of a large electronic spin resonance by *Brodsky* and *Title* in Ge, Si, and SiC [7.9]. This technique has proved particularly useful in amorphous silicon where large differences are found in the different kinds of materials. The most striking result is a variation of more than four orders of magnitude in the numbers of spins between the pure and "contaminated" material (Figs. 7.1, 2).

For pure evaporated or sputtered material, the signal is always very large, showing the existence of 10^{19}–10^{20} paramagnetic centers per cm^3, with a g value close to $g = 2.0055$ (Fig. 7.1). At the other extreme, silicon deposited in the presence of atomic hydrogen [glow-discharge decomposition of silane (SiH_4) or sputtering in the presence of hydrogen] gives a much weaker signal, showing a very large decrease in the number of paramagnetic centers in the bulk of the material (Fig. 7.2). There are also variations in the shapes and saturation properties of the resonance lines in the different samples, but the interpretation is not so straightforward, and only special cases will be discussed below.

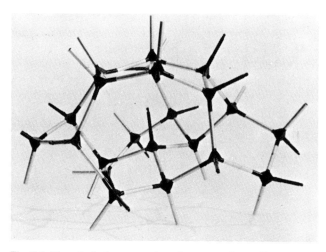

Fig. 7.3. Hypothetical construction of an amorphous silicon lattice showing the existence of a "lone broken bond." Such a center does not exist in crystalline silicon where missing atoms (vacancies) produce several adjacent broken bonds. [7.11]

7.2.1 Magnetic Properties of Centers in Silicon

For the study of the very large variety of defects or centers to be found in amorphous materials, the large amount of data obtained in crystalline materials provides useful landmarks. In particular, extremely valuable guidelines are provided by the monumental work of *Watkins* and co-workers [7.2, 3] on paramagnetic centers in irradiated crystalline silicon (see Appendix A). The situation however is expected to be more complicated in amorphous materials. In contrast with crystals, the disorder in the environment of a center will give a spread in its energetic and magnetic properties. Certain type of centers can exist only in disordered material [7.11]. A lone broken bond (Fig. 7.3), for example, cannot exist in a crystal where a vacancy produces several broken bonds at once.

The qualitative picture proposed for the gap states in amorphous group-IV material is that of broken bonds or defects with a wide spread in energy due to disorder. As far as magnetic properties are concerned, we can predict two limiting cases (Figs. 7.4, 5):

a) Very Localized Centers ("Hard Centers")

An elementary result of quantum mechanics relates the energy spacing ΔE between quantum levels to a characteristic spatial extent λ of the electronic wave function. This is given, in order of magnitude, by

$$\Delta E \simeq (2/\lambda)^2, \tag{7.1}$$

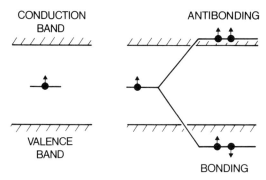

Fig. 7.4. A localized center giving a single level in the gap ("hard center"). Because of the small extension of the wave function, the level can accomodate only one electron, and the center is thus paramagnetic. The addition of an other electron, either coming from an other center or from a contaminating gas (e.g., atomic hydrogen) makes a bonding or antibonding level which lies outside of the gap

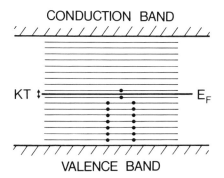

Fig. 7.5. Quasi continuum of levels in the gap due to a distribution of extended centers ("soft centers") like the multivacancies described in [7.2]. Each level can accomodate 0, 1, or 2 electrons. The states above the Fermi level E_F are empty, and the states below E_F are populated by 2 electrons of opposite spin directions so that these "soft centers" are not magnetic, except for a very small region of about kT around E_F

where ΔE is expressed in eV and λ in Å. Thus, for an electron localized on one atomic site ($\lambda \simeq 2$ Å), there will be at most one state in the gap. Besides, we can expect the Hubbard-type repulsion U to be large so that the doubly occupied states will be outside the gap. Thus, such a center will be paramagnetic, independently of the position of the Fermi level in the gap.

b) Electrons with a Wavefunction Spread on Several Atomic Sites ("Soft Centers")

The same (7.1) shows that the energy spacing for a soft center can be a small fraction of an eV and that such a center can have many possible levels in the gap. We can also expect, because of the extension of the wave function, and perhaps because of the polaronic distorsion of the lattice [7.21], the Hubbard correlation energy U between two electrons to be small, so that a quantum level can be singly as well as doubly occupied. Such a center can be visualized as one of the multivacancies described by *Watkins* et al. [7.2].

In sharp contrast with the very localized "hard centers," the "soft centers," as shown in Fig. 7.5, will not in general be magnetic. In the quasi-continuum distribution of states, the states above the Fermi level will be empty, and the states below this level will be doubly occupied with two spins of opposite directions. Only a small range of the order of kT around the Fermi level will be singly occupied, giving a very small contribution to the ESR signal. This assumes that the effective correlation energy U for two electrons in a gap state is very small ($|U| < kT$), as an intermediate situation between localized states in silicon where U is large and positive, and gap states in chalcogenide glasses where U is negative. Further studies are necessary to confirm this point, but so far this assumption is compatible with the experimental magnetic resonance results.

7.2.2 Electron Spin Resonance in Very Pure Amorphous Silicon

As for other techniques, it is often difficult to deduce a clear-cut picture or model in disordered materials from magnetic resonance results alone. Firstly, the sensitivity to preparation conditions can lead to inconsistant results between different authors. Secondly, one almost always obtains signals which are clearly a superposition of lines having a wide distribution of properties. This is particularly the case for the positions (g values), the linewidths, and the relaxation times.

In that respect, the efforts of *Thomas* et al. [7.22] to obtain a close control of the experimental conditions have proved very efficient in clarifying some aspects of spin resonance in pure amorphous silicon. Their novel idea, which had a marked influence on the results, was to prepare the samples by evaporation under ultrahigh vacuum, and keep them under the same vacuum conditions for the measurements: the number of uncontrolled impurities deposited in (or on) the film was at least one order of magnitude lower than the number of spins detected. The first conclusion is thus that the paramagnetic centers are due to structural defects and not to impurities in the material. The density of spins obtained in these conditions was remarkably constant. The number of spins deposited at room temperature was $6 \times 10^{19}/\text{cm}^{-3}$. The effect of increasing the temperature of the substrate T_S or the effect of subsequent annealing at a temperature T_A is shown in Fig. 7.6. It can be seen that the increase of T_S or T_A, up to crystallization temperature, decreases the density of spins by, at most, a factor of 3. This is in sharp contrast with previous studies [7.10], where variations of several orders of magnitude were found, illustrating the important effect, to be discussed below, of contamination on the density of spins.

The conclusion to be drawn from the other properties of the resonance line (width, position, relaxation time, exchange interactions), is much less clear. There is little doubt that these properties result from an average of a range of values, and we refer the reader to the original article [7.22] for details. It can be

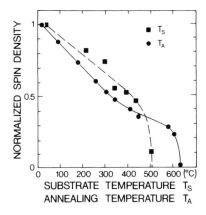

Fig. 7.6. Variation of spin density in evaporated amorphous silicon versus deposition temperature T_S or annealing temperature T_A. When these operations are performed in ultrahigh vacuum to avoid contamination, the decrease of spin density is not larger than a factor of 3. The drop in spin density at high temperature ($T_S \geq 500\,°C$, $T_A \geq 600\,°C$) is due to the onset of crystallisation. [7.22]

remarked also that, in contradiction with other studies [7.23], the susceptibility in this pure material follows a well-defined Curie law, and the magnetic average interaction between the spins is experimentally less than 1 K.

7.2.3 The Effect of Contamination

A most useful result of the paper [7.22] on ultrahigh vacuum conditions has been to draw attention to the importance of contamination. It is clear from the discussion in Sect. 7.2.1 that the paramagnetic centers found in evaporated amorphous silicon are very localized defects ("hard centers"), probably of the broken-bond type. Spin resonance thus indicates unambiguously that in the preparation of evaporated amorphous films, one bond is not satisfied for about 10^3 silicon atoms, i.e., about one broken bond every 10 atomic neighboring sites. Such a broken bond is expected to be very reactive. Covalent bonding with a contaminating gas, such as oxygen or hydrogen, is energetically very favorable, the Si–H bond, for example, having an energy of several eV.

Magnetic resonance is an excellent and unambiguous monitor of such a reaction, since the pairing of the electron with the electron of the reacting atom removes the center from the electronic resonance line. Evidence of contamination is found in the following cases:

During preparation. For very slow deposition rates, of the order of $1\,\text{Å s}^{-1}$, the amount of uncontrolled gas is not negligible compared with the number of spins. *Thomas* et al. [7.22] found, at high deposition temperature ($T_S = 380\,°C$), that very slowly deposited samples have a spin density about 20 % lower than films at deposited at higher rates. The extreme case of neutralization of the paramagnetic centers during deposition will be found in the preparation of hydrogenated amorphous silicon to be discussed in the next section.

By atmospheric exposure after preparation. The centers in the bulk are not affected, but there is a small surface contamination and for normal samples, a decrease of about 10 % of the signal can be found after long air exposure. However for very thin films, or material having a large specific area, the effect can be much larger. This is the case in particular for films deposited at oblique incidence which are known to be very porous [7.24]. For films deposited at 60° incidence, the decrease upon exposure to air of the magnetic resonance signal is at least one order of magnitude larger than for normal-incidence films deposited in the same conditions.

Our conclusion is that *pure* amorphous silicon is *a well-defined material.* We have seen that it is so for pure evaporated silicon, but we believe that it is also the case for material prepared by sputtering or ion implantation in controlled conditions, where contamination by active gases such as hydrogen or oxygen is kept below a critical level. The material so obtained is amorphous, with a large density of states in the gap (of the order of $10^{20} \, cm^{-3} \, eV^{-1}$) due to very localized centers. These states are paramagnetic (2 to 6×10^{19} centers/cm^3) and probably also dominate the electrical properties of the material.

We believe also that the wide scatter of properties obtained previously by some authors, in particular at slow deposition rates, as well as some erratic annealing effects, are spuriously due to contamination, especially for very thin or porous samples.

In fact this material appears in several other conditions. It has been established [7.25], for example, that magnetic resonance of the damaged surface of crystalline silicon is very similar to that of the amorphous material. It is then probable that, except in very special cases of ultrahigh-vacuum cleaving, or shortly after certain surface etches, the actual surface of crystalline silicon/(to a depth ranging from 100 to 1000 Å) is the amorphous material that we have been discussing in this section.

7.2.4 Hydrogenated Amorphous Silicon

A quite different magnetic resonance signal is obtained when the material is prepared in presence of a large amount of hydrogen. This was first discovered in amorphous silicon prepared by glow-discharge decomposition of silane gas (SiH$_4$) deposited at high temperature ($T_s > 150 \, °C$) [7.12, 13, 26], but quite similar results are obtained in material prepared by sputtering in the presence of ionized hydrogen [7.27–30]. As expected from the previous section, all localized centers are involved in covalent bonding with hydrogen and the spin resonance signal becomes vanishingly small.

However, direct information about the density of states in the material can be gained by the measure of the intensity of the remaining signal. If we take this material as being free of paramagnetic "hard centers," all bound to hydrogen, we are left only with "soft centers" (Fig. 7.5) forming a quasicontinuum of energies in the gap. We assume that each level can accomodate 0, 1, or 2 electrons, and that the ensemble of these levels is responsible for most of the density of states $\varrho(E)$ in the gap of this material. In this picture, we have a Pauli-type paramagnetism [7.31], and only a range of the order of αkT ($\alpha \simeq 3$ for uncorrelated spins 1/2) around the Fermi level is paramagnetic and participates in the magnetic resonance signal. Thus, from the intensity of the signal, we can measure the value

$$N_S = \alpha kT \varrho(E_F) \tag{7.2}$$

which gives an unambiguous determination of the density of states provided the correlation energy U between two electrons on a gap state is negligible compared with kT.

The experimental difficulty of a quantitative measurement is not so much the small value of the signal, but rather large errors induced by spurious signals due to substrate, cavity, surface states, etc. By a careful study of the signal as a function of thickness it is, in principle, possible to sort out the spin density of the bulk (Fig. 7.2). In a typical case, *Knights* et al. [7.13] have obtained a value of $N_S = 6 \times 10^{15}$ spins cm^{-3} (and also an unexplained surface spin density of 7×10^{12} cm^{-2}). From (7.2) we obtain for this particular case a density of states $\varrho(E_F) \simeq 8 \times 10^{16}$ cm^{-3} eV^{-1}, a value which is somewhat in the low part of the range given by other methods [7.32].

Because of the experimental difficulties already mentioned, the method is quite tedious and has not been employed systematically. It is possible, however, as for chalcogenide glasses [7.7, 8] to increase the signal artificially by electron–hole injection. This is accomplished [7.11, 13, 33, 34] by off-equilibrium populating the levels by light irradiation. In contrast with the case of chalcogenide glasses, the photoinduced paramagnetic states are not stable, even at low temperature, and they disappear when the light is switched off.

The principle of this light-induced paramagnetism is shown in Fig. 7.7. High intensity of extrinsic light generates a large number of electron–hole pairs. Electrons are captured by empty states above the Fermi level, these states thus becoming paramagnetic, and similarly for holes captured on states below the Fermi level. A light-induced signal is thus obtained, which has the practical advantage of being much larger than the signal in the absence of light, and also that of being easily distinguished from any parasitic or background line by on–off modulation of the light. It is expected that the normally empty levels will be populated, around the Fermi level, up to two energy limits (Fig. 7.7). These limits are defined as the energy levels where the probability of thermal reexitation into the bands is of the same order as the rate of capture of the electrons and holes.

Since the probability of the reexcitation varies exponentially with the distance to the band edge, one expects a logarithmic, or at least rather weak, dependence of the signal with the light intensity. It is indeed found that above a certain light intensity ($\simeq 50$ mW cm^{-2}), the magnetic resonance signal levels off, at a value corresponding to 6×10^{16} spins cm^{-3} [7.33]. If we assume that the populated levels are spread on a 1 eV range, we deduce a density of states

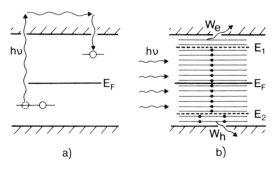

Fig. 7.7a, b. Spin resonance signal induced by light irradiation. The levels in the gap of this material (hydrogenated silicon) are due to "soft centers" and are not magnetic in the absence of light. (a) The light excites an electron from a state below the Fermi level, which is subsequently trapped by a center normally empty. Both centers become paramagnetic. (b) For high intensity illumination, most of the centers become paramagnetic. Only the levels close to the bands (above E_1 and below E_2) are not populated by the light because of the high probabilities W_e and W_h of reexcitation in the bands of the trapped electrons or holes

$\varrho(E_F) \simeq 6 \times 10^{16}\,\mathrm{cm}^{-3}\,\mathrm{eV}^{-1}$, quite compatible with the results of spin resonance in the dark. It has been found recently [7.11] that the light-induced signal at a given light intensity $(5 \times 10^{17}\,\mathrm{photons\,s}^{-1})$ is remarkably constant in undoped amorphous silicon obtained by glow-discharge decomposition of silane. This is obtained in a series of films of variable hydrogen content, showing a large variation of optical and transport properties [7.35]. This result seems to indicate that although the amount of hydrogen in the material has a strong influence on the gap structure, it has little effect on the defects that remain after elimination of the unpaired dangling bonds. Thus, if the proposed model for the photoinduced magnetic resonance signal is valid, it provides a practical tool for the study of the density of states in amorphous silicon. Unfortunately, at very high light excitation, of the order of $1\,\mathrm{W\,cm}^{-2}$, the magnetic resonance signal increases again [7.11, 34], an effect which is as yet unexplained. It shows that the actual processes are probably more complicated than the simple model presented in Fig. 7.7.

7.2.5 Hydrogenation and Dehydrogenation of Amorphous Silicon

We are thus led, from magnetic as well as other physical properties of the materials, to consider the existence of two well-defined amorphous silicons:

Pure silicon, having a density of states of some $5 \times 10^{19}\,\mathrm{cm}^{-3}\,\mathrm{eV}^{-1}$ mostly due to localized states of the broken-bond type; and

Silicon–hydrogen alloys, with a density of states three orders of magnitude lower, with no paramagnetic localized states in the gap.

In this picture of neutralization of the broken bonds by covalent bonding with atomic hydrogen, it should be possible to transform one material into the other by high-temperature decomposition or by chemical reaction techniques. This has indeed been realized in both directions, and electron spin resonance has been a powerful and direct tool for studying this type of reaction.

When heated to sufficiently high temperature (350–600 °C), but below the crystallization temperature, hydrogenated silicon loses most of its hydrogen. As expected, paramagnetic dangling bonds are left behind, and a large increase of the magnetic resonance signal is observed [7.36]. Recently, the reverse process has been obtained by *Kaplan* et al. [7.37]. They have shown that it is possible to produce a complete chemical reaction between the broken bonds of the pure material with external hydrogen, thus transforming the first material into low density-of-states hydrogenated silicon, the decrease of the spin resonance signal being a convenient and direct monitor of the reaction. Of course, this cannot be achieved with the molecular gas, since the breaking of the H–H bond would require too much energy. The reaction is therefore performed in a hydrogen plasma containing a large proportion of atomic hydrogen. The size of the atom is quite small, and at a temperature of 500 °C, it can easily diffuse into the bulk of the pure amorphous silicon and make a stable covalent bond with the localized defects of the material. The experiment shows that, to obtain the complete neutralization of the localized paramagnetic centers, it is necessary to incorporate a much larger number (at least one order of magnitude) of hydrogen atoms than the number of spins. This proportion has also been found in the preparation of low-density-of-states hydrogenated germanium by sputtering in the presence of hydrogen [7.27]. This seems to indicate that other nonparamagnetic defects (like paired broken bonds of the type described in Appendix A) exist in the material and can also chemically react with the atomic hydrogen. In very similar experiments, both reactions of dehydrogenation and hydrogenation have been performed by *Pankove* et al. [7.38] on the *same samples*. After removal, by heat treatment, of the hydrogen from amorphous silicon produced by glow-discharge decomposition of silane, they have been able to restore the electrical properties of the material by exposure to a hydrogen plasma. Magnetic resonance studies, which were not performed during these experiments, could be useful to obtain precise information about the reversibility of the rehydrogenation process.

7.2.6 Low-Temperature-Deposited Hydrogenated Silicon

When deposited at low temperature ($T_s < 100\,°C$) in a glow-discharge reactor or by reactive sputtering in presence of hydrogen, another type of material is produced [7.39] which is characterized by a much lower density than the other amorphous materials and, in general, a much higher hydrogen content ($\sim 30\%$ or more). Recent extensive studies of ir vibrational spectra [7.40, 41] have established that this material contains, besides the expected Si–H bonds, an excess of SiH_3 groups, and also a somewhat larger amount of hydrogen bond in polymeric form $(SiH_2)_n$ than in the high-temperature material.

This is not sufficient to explain why this material should exhibit a strong resonance signal [7.42], since, whatever the silicon–hydrogen groups, all bonds are satisfied and no unpaired spins should be present.

Recently *Knights* et al. [7.43] have shown that the paramagnetic defects are related to the morphology of the films rather than to the composition of the material. They have found that the low-temperature-deposited material contains voids that grow perpendicular to the film surface, and they have been able to explain the electron-spin-resonance signal on the basis of the dangling bonds on the internal surface of these voids.

7.2.7 Conclusion

A large amount of data on the properties of the spin resonance signals in the different amorphous materials has been accumulated in the past few years. The results on the shapes and positions of the lines are always very complex. This is not surprising if we consider that even in crystalline silicon (see Appendix A) a large variety of centers can be found, all with anisotropic g tensors. In the amorphous material, one obtains a superposition of powder patterns which makes the interpretations rather hazardous, although trends can be found in some cases. Exchange interactions and relaxation studies also suffer from the same spread of parameters, although significant progress has been made in the understanding of the relaxation processes correlated with transport properties [7.44]. The recent most significant results can be found in a well documented review given by *Stuke* [7.45].

We have therefore restricted our discussion to the intensity of the signal which, simple as it is, gives a clearcut picture of two well-defined materials, one being a pure disordered silicon with a large, but stable, number of broken bonds, the other being a silicon–hydrogen alloy with a low, and rather stable, density of states in the gap. We believe that the large variety of materials produced in different experimental conditions can always be described as an admixture, in variable proportions, of these two materials.

7.3 Spin-Dependent Effects in Amorphous Silicon

7.3.1 The Existence of Spin-Dependent Effects in Semiconductors

It was first recognized by *Honig* [7.46] and *Schmidt* et al. [7.47] that the spin properties of carriers can have a large influence on transport in semiconductors. Following the pioneering work of *Lepine* [7.48] on the spin-dependent photoconductivity in silicon, a wide range of studies have been published illustrating the importance of spin statistics on the physics of recombination in semiconductors. These studies cover recombination on surfaces [7.49], dislocations [7.50, 51], space-charge region of a junction [7.52], amorphous material [7.14] with a few theoretical models to explain the magnitude of the effect [7.53, 54].

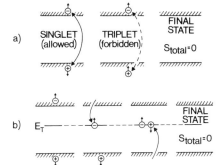

Fig. 7.8a, b. Principle of the spin-dependent recombination. (a) Direct electron–hole recombination. (b) *SHOCKLEY-READ* model [7.56] of a recombination by a deep trap. It both cases, the final state has an angular momentum of zero and, if the probability of a spin flip in the recombination is small, the recombination of the singlet spin state is much more probable than that of the triplet state

7.3.2 The Basis of Spin-Dependent Recombination

Because of the conservation of angular momentum in an electron–hole collision, it is expected that the recombination will depend on the spin state of the carriers. This is illustrated in Fig. 7.8a in the hypothetic case of direct electron–hole recombination. For low-energy carriers, the collision is mostly S-like (orbital angular momentum $L=0$) and the recombination of a triplet spin state is forbidden since the total spin of one cannot disappear in the electron–hole annihilation.

The argument remains the same for the more realistic case of a recombination center (Fig. 7.8b). In the *Shockley–Read* model [7.55], an electron, for example, is first trapped on a deep center. The trapped electron then captures a free hole, completing the recombination. Here, also, the trapped electron can only recombine with a hole of opposite spin (singlet spin state) to conserve the total angular momentum in the recombination.

Thus, in the general case, the triplet recombination rate W_T will be zero, or at least much smaller than the singlet recombination W_S. In zero applied magnetic field, the spin directions of the injected carriers are completely random (spin polarizations zero) and the probabilities of a singlet or triplet electron–hole spin state are respectively 1/4 and 3/4. The recombination rate is then

$$W_{\text{random}} = \tfrac{1}{4} W_S + \tfrac{3}{4} W_T \tag{7.3}$$

with $W_T \ll W_S$.

If the spins of the carriers have finite polarizations P_e and P_h as will be the case, for example, in an applied magnetic field, the probabilities of singlet or triplet electron–hole spin states depart slightly from the values 1/4 and 3/4. They are given, for spin 1/2, by the average values of the projection operators

$$\pi_S = 1/4 - S_e \cdot S_h$$
$$\pi_T = 3/4 + S_e \cdot S_h, \tag{7.4}$$

where S_e and S_h are the spins of the recombining electron and hole.

For uncorrelated spins, the average values are simply given by

$$\langle S_e \rangle = P_e/2, \langle S_h \rangle = P_h/2 ; \tag{7.5}$$

so that the recombination rate is now

$$W = \frac{W_S}{4}(1 - P_e \cdot P_h) + \frac{W_T}{4}(3 + P_e \cdot P_h) + W', \tag{7.6}$$

where a hypothetic spin-independent process W' had been added for the sake of generality. This can also be expressed as

$$W = W_0(1 - \alpha P_e \cdot P_h), \tag{7.7}$$

with $W_0 = (W_S + 3W_T + 4W')/4,$

$$\alpha = \frac{W_S + W_T}{W_S + 3W_T + 4W'}.$$

In this model, we remark that $\alpha \leq 1$, the maximum value $\alpha = 1$ being obtained when the "leakage" processes W_T and W' are zero.

7.3.3 Experimental Detection of the Spin-Dependent Recombination

Equation (7.7) shows that a change of the spin polarization of the electron (or hole) will result in a change of the recombination rate which can be detected, for example, by a small variation of photoconductivity.

This is usually accomplished by "saturation" of the electron spin resonance [7.56]. The principle of the saturation is described as follows: in the absence of a microwave field, the polarization relaxes towards its equilibrium value $P_0 = \hbar\omega/2kT$ at a rate $1/T_1$ (T_1 is the longitudinal relaxation time and $\hbar\omega$ is the spin splitting in the applied magnetic field). At resonance, the microwave field induces transitions at a rate W_{RF}, thus tending to equalize the populations of the spin states and to reduce the spin polarization. The two competing effects result in a steady-state polarization

$$P = P_0 \frac{1}{1 + W_{RF} T_1}. \tag{7.8}$$

The quantity W_{RF} is proportional to the microwave power, so, by extrapolation to infinite power, one has effectively a way of "nulling" the polarization of the resonant spins, the other experimental parameters remaining unchanged.

The experimental arrangement is shown in Fig. 7.9. The sample is illuminated by visible light, and the variation of the photoconductivity is detected

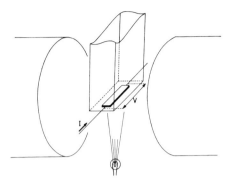

Fig. 7.9. Experimental arrangement for the detection of spin-dependent photoconductivity in silicon. The light reaches the sample through a series of slits in the bottom of the cavity. The microwave power is on–off modulated and the corresponding voltage variations across the sample are detected by a lock-in amplifier

phase sensitively by a lock-in amplifier. The polarization of the spins is periodically decreased by saturation with the resonant microwave power, on–off modulated at the lock-in frequency.

The arrangement provides an extremely sensitive detection of a magnetic resonance signal, without the usual limitations of classical resonance. Provided the recombination is dominated by the spin-dependent process, which is almost always the case for silicon, and that the resonance line can be saturated with the available microwave power, the signal is independent of the number of spins and many orders of magnitude can be gained in sensitivity for thin films or surface studies.

Figure 7.10 shows a typical signal obtained by this technique in a sample of amorphous silicon produced by glow-discharge decomposition of silane [7.14].

7.3.4 Recombination by Trapped Electron–Hole Pairs

Equation (7.7) gives the magnitude of the spin-dependent effect when the thermal equilibrium spin polarizations are the only correlation between the spins of the recombining electrons and holes. The maximum variation of the recombination rate upon complete saturation is then

$$(\Delta R/R) \leqq P_e \cdot P_h \tag{7.9}$$

At room temperature, in a field $H_0 = 3300\,\text{G}$ (resonance frequency 9.5 GHz), this predicts a rather small effect of the order of $\Delta R/R \simeq 10^{-6}$.

Now, it is found experimentally that the effect can be much larger [7.14]. In particular, in undoped hydrogenated amorphous silicon we have recently found at room temperature a spin-dependent effect as large as 10^{-3}, three orders of magnitude larger than the value given by (7.9).

We are thus led to a model of recombination where a large spin correlation must exist between the recombining electron–hole pairs [7.57]. Such a correlation has been found, for example, in the case of donor–acceptor recombination in crystalline material [7.58]. In hydrogenated silicon, all the

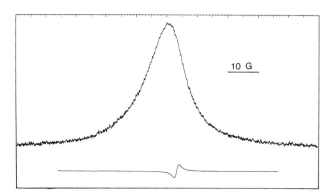

Fig. 7.10. Resonance signal obtained with the arrangement shown in Fig. 7.9 in a film of amorphous silicon slightly boron doped. The bottom line is a DPPH marker [7.53]

levels of the "soft centers" in the gap can play the role of donor or acceptor according to their position with respect to the Fermi level. An electron and a hole trapped on these levels in a situation of proximity ("trapped electron–hole pair") will have the possibility of recombining directly. We shall see that because this recombination depends strongly on the spin state of the pair, a large correlation, much larger than (7.7), is obtained between the spins of the trapped pairs. A striking result is the existence, in steady state, of more triplets and less singlets than in a random distribution.

This process of recombination through pairs of trapped electron and hole is illustrated in Fig. 7.11. Two "soft centers" A and B, of the type described in the last section, have a quantum level in the gap, below the Fermi level for center A and above for center B. At thermal equilibrium center A is normally full and center B empty. We now offset the populations of A and B by light illumination or some other injection method, resulting in one electron on each level (see Fig. 7.7a). It is understood in this example that center B is populated through the conduction band and not directly with the electron removed from A, so that its spin state, when trapped on B, is completely uncorrelated with the spin of A^1.

Let us assume that the equilibrium is restored by direct return of the electron from B to A (direct trapped electron–hole recombination). This requires that the two centers are not too far apart so that there is a small, but finite, overlap of the wave functions resulting, for example, in a finite hopping probability from B to A.

Before we proceed in the discussion of the spin properties of such a pair, it is worth remarking that it can provide a very efficient recombination mechanism. Each center can be optimized as an efficient trap for each of the carriers, centers

---

1 It is usually said that A and B have trapped respectively a hole and an electron. This language is somewhat misleading here: it appears, for example, that the spin attributed to the hole must be *opposite* to the spin that it replaces if we want the electron–hole singlet recombination to be allowed.

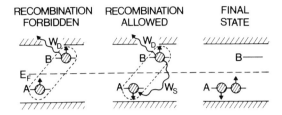

Fig. 7.11. Recombination through electron–hole trapped pairs. Light irradiation has removed one electron from center A and put one electron on center B (see Fig. 7.7a). If the two centers are within hopping distance, equilibrium can be restored by the direct return of the electron from B to A, provided the two electrons are in a singlet spin state (rate W_s). The competing process to this recombination mechanism is the dissociation of the pair (rate W_D) by reexcitation of the electron, for example, to the conduction band

A for holes and centers B for electrons. This is in contrast with the *Shockley–Read* model [7.55], where a single recombination center has to be placed, as a compromise, deep in the center of the gap which results in a rather small trapping cross section of the carriers.

The spin-dependent effect will result from the fact that the direct recombination of the electron–hole trapped pair is possible only if the two electrons on A and B are in a singlet spin state (Fig. 7.11). Indeed, the final state, with 2 electrons on the same level, has to be a singlet spin state and since in silicon the spin–orbit interaction is very weak, the electron and hole of the pair recombine without spin flip. This provides a strong correlation between the spins of a pair : pairs created in singlet configurations have a shorter lifetime than pairs created in triplet configurations. Therefore their steady-state population is smaller than in the random situation and we are left with a net surplus of triplet configurations.

If we now saturate the electron spin resonance, we restore the random distribution which contains more singlets so that the recombination rate in enhanced. As remarked in [7.57], the pair concept is central in this model because the spin orientation of the system as a whole remains random and only the two spin orientations in the *same pair* are correlated.

The magnitude of the effect results from a competition between the recombination rate W_S of a pair in a pure singlet state and its dissociation rate W_D. In the present model, the dissociation is due to the actual motion of the carriers, like the reexcitation of an electron into the conduction band or a hole into the valence band. We assume for the time being that relaxation mechanisms that flip one spin with respect to the other and thus change the spin state of a pair can be neglected. We shall discuss the effect of spin relaxation later.

An explicit calculation of the effect is given in [7.57] with some simplifying assumptions, but sufficient to illustrate the physical features of the model. It is assumed that pairs are created by light irradiation at a generation rate G, and that they disappear both by direct recombination at a rate W_S when they are in a singlet spin state (we neglect completely the triplet recombination W_T), and by dissociation at a rate W_D.

In these conditions, it is shown in [7.57] that the total recombination rate R is given by

$$R = G\left[1 - \frac{2W_D}{W_S}\log\left(1 + \frac{W_S}{2W_D}\right)\right]. \tag{7.10}$$

If we destroy the spin correlation between pairs by a strong resonant saturation microwave field, the singlet and triplet proportions recover their random values $1/4$ and $3/4$, and the recombination rate R_{sat} is simply given by

$$R_{sat} = G\frac{W_S}{4W_D + W_S}. \tag{7.11}$$

The experimentally measured quantity is the variation of photoconductivity upon saturation of the resonance line. Extrapolated to complete saturation, this is equal to the relative change of the recombination rate

$$\frac{\Delta R}{R} = \frac{R - R_{sat}}{R}. \tag{7.12}$$

This quantity is plotted in Fig. 7.12 as a function of the ratio $\lambda = W_D/W_S$.

As expected, the spin-dependent effect disappears when the dissociation rate is very large: $W_D \gg W_S$. The extreme case of a large W_D would be the recombination of free carriers. It shows that the large experimental spin-dependent effect can only be explained if there is a finite "sitting time" of the recombining electron–hole pair, which is another way of stating the existence of trapped electron–hole pairs.

In the case of a small dissociation rate, $W_D < W_S$, the system possesses some interesting physical features. Figure 7.13 shows, as a function of $\lambda = W_D/W_S$, the value of the ensemble average among the pairs of $\langle S_e \cdot S_h \rangle$, which is a direct measure of the correlation between the spins of a pair. In the limit of $\lambda \to 0$, this average $\langle \overline{S_e \cdot S_h} \rangle$ increases slowly (logarithmically) towards the value $1/4$, showing a "pumping" of all the pairs in a triplet state.

In the simplified model presented here, we have assumed that there is no spin relaxation. However, when the lifetime of a pair becomes very long (W_D and W_S small), longer than a relaxation process flipping one spin with respect to the other, relaxation can no longer be neglected. Such a relaxation process, which can be longitudinal (T_1 relaxation) or transverse (T_2 effect), changes the spin state of a pair and tends to restore a random distribution. For fast relaxation, the effect is the same as the saturation of resonance, the correlation between the spins is lost and the spin-dependent effects disappear. The intermediate case of a relaxation of the same order as W_D or W_R is complicated and will not be discussed here. It is interesting to notice, however, that for a small probability of dissociation ($W_D \to 0$), the recombination is dominated by

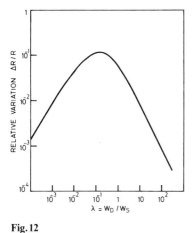

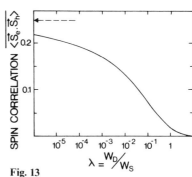

Fig. 12   **Fig. 13**

Fig. 7.12. Relative variation of the recombination rate $\Delta R/R$ upon saturation of the spin resonance, as a function of the dissociation to singlet-recombination rate ratio $\lambda = W_D/W_S$. For a wide range of values of λ, the effect is much larger than the value given by (7.7). [7.58]

Fig. 7.13. Correlation between the spins of a trapped electron–hole pair expressed as the ensemble average of the scalar product of their spins. In the limit of a dissociation rate W_D much smaller than the singlet recombination rate W_S, this average tends (slowly) towards 1/4 and all the pairs are in a triplet spin state. In practice this correlation is decreased by the spin relaxation processes

the spin-relaxation times, which shows the importance of the spin physics of the carriers in a domain where it is usually neglected.

In an actual material, there is a distribution of the distances between the trapped electron and hole, so that one expects a large distribution of the ratio $\lambda = W_D/W_S$. One can see however that for many orders of magnitude of λ the effect is much larger than the value predicted by (7.7), which is quite sufficient to explain the large experimentally observed value of $\Delta R/R$.

A crucial experimental verification of the model is provided by the magnetic field dependence of the effect. Previous theories, where the effect is proportional to the product of the electron and hole polarizations [see (7.7)], predict a spin-dependent effect varying like the square of the applied field. In the present model, the correlation between spins is independent of the spin polarizations so that the effect should not depend upon the applied magnetic field. This is well verified by experiment [7.57]: A variation of a factor of five in the applied magnetic field does not change the magnitude of the spin-dependent effect in a sample of undoped hydrogenated amorphous silicon within experimental error.

7.3.5 Spin-Dependent Luminescence

Since the electron–hole recombination is affected by the spin statistics, spin-dependent effects are observable in luminescence experiments and provide a useful tool for the study of luminescence centers [7.58, 59]. Recently large spin-dependent effects have been found independently by *Biegelsen* et al. [7.15] and *Morigaki* et al. [7.16] in hydrogenated amorphous silicon.

In the high-temperature range of the luminescence experiments (around liquid nitrogen temperature) the situation is relatively simple. Microwave saturation of a spin resonance line at 80 K in a field of 3000 G results in a *decrease* of the luminescence by a factor of about $\Delta L/L = -10^{-3}$. A single line is thus obtained which is very similar to the spin resonance line induced by light irradiation (Sect. 7.2.4) or to the line obtained in the spin-dependent photoconductivity experiments discussed above, with a g value close to 2.006 and a linewidth of about 8 G.

In the framework of the spin-dependent recombination discussed previously, the explanation of the effect is straightforward. The luminescence intensity results from a competition between the direct radiative recombination and the nonradiative recombination though recombination centers [7.60]. At high temperature, the nonradiative recombination dominates and above, 200 K the luminescence disappears. Since the spin saturation of the recombination centers increases the nonradiative recombination as described above, the result is a decrease of the radiative recombination and thus of the luminescence intensity. In this model, the resonance line is obviously the same as the light-induced or the spin-dependent photoconductivity lines of the previous sections.

Another possible model has been proposed by *Biegelsen* et al. [7.15] to explain this spin-dependent luminescence, based on the concept of "geminate pairs." In an amorphous material, the mobility of the carriers may be sufficiently low that the photocreated electron–hole pairs (geminate pairs) remain bound. Since only singlet optical transitions are allowed, such a pair will be created with antiparallel spins. If the radiative lifetime τ is shorter than the spin-thermalization time T_1, the luminescence will result from the direct recombination of the pair in an antiparallel spin state. Now the application of a resonant microwave field induces spin transitions which allows the spins of a pair to relax to triplet states. Since only the singlet states are radiative, the luminescence is reduced. In this model, the correlation of the spins is provided by the photocreation, and the spin-dependent effect results from the destruction of this correlation by the spin saturation.

We remark that for both models, the effect is independent of the thermal equilibrium polarizations of the spins, and thus should be independent of the ratio H/T. This is well verified experimentally: the effect is independent of temperature [7.15] and the same order of magnitude of the spin-dependent luminescence is found at a very low field of 50 G for a resonant microwave frequency of 150 MHz (G. Lampel, M. Rosso, I. Solomon: Unpublished).

At low temperature (below 40 K), the situation is more complicated and new resonance lines appear, which can be a decrease as well as an increase of the luminescence. A complete discussion of the low-temperature lines is difficult since the results differ according to the different authors and depend upon the experimental conditions and probably on the nature of the samples. A common feature is the existence of a broad line ($\Delta H \simeq 200\,G$) centered close to $g = 2.0$ and detected as an *increase* of the luminescence under microwave saturation. From the temperature dependence of these lines [7.15], which vary like T^{-2}, it is suggested that the effect is proportional to the polarizations of the electrons and holes [see (7.7)]. The effect of saturation is to destroy the polarizations, thus slightly increasing the number of pairs in singlet states resulting in an enhancement of radiative recombination. The crucial assumption in this model is that the spins thermalize before the carriers recombine. For the low-temperature lines, dynamical effects are important and the intensities of the lines are very sensitive to the field modulation [7.15, 16].

7.3.6 Magnetic Field Effects Due to Spins

It has been found recently that magnetic fields as small as a few gauss can significantly change the properties of amorphous germanium and silicon, which is quite unusual in the physics of semiconductors. This effect is found in dark conductivity [7.18–20], luminescence [7.17], and photoconductivity [7.62] (T. Dietl, I. Solomon: To be published).

It is quite clear that only spin properties can be responsible for the effect of such small fields. This is best understood in the case of luminescence and photoconductivity where recombination mechanisms are involved. We saw in Sect. (7.3.4) that recombining electrons are in strongly correlated spin states and any spin relaxation process destroying this correlation will affect the recombination.

The following relaxation processes are field dependent and thus can explain the small-field dependence of luminescence and photoconductivity:

a) The longitudinal spin relaxation time T_1. If the magnetic field fluctuations responsible for the longitudinal relaxation time T_1 contain a low-frequency process, then T_1 is much shorter in zero or low fields than in high magnetic fields. This is the process invoked by *Street* et al. [7.17], who observe a luminescence increase in hydrogenated silicon of about 1% above a field of 30 G.

b) The effect of spin dephasing. We saw in Sect. 7.3.4 that when the relaxation time T_1 is longer than the recombination time τ, a strong correlation exists between the spins of a recombining pair, with a surplus of triplet states. However, if there is a distribution δg of g values, in an external applied magnetic field H, the two spins of a pair will not precess at the same rate and the spin correlation of the spins which are not quantized along the applied field H will be lost in a time of the order of

$$T_2^* = \left(\frac{\delta g\,\mu_B H}{\hbar}\right)^{-1}. \tag{7.13}$$

The correlation is thus destroyed in high fields when $T_2^* < \tau$, in opposition with the previous process which decreases the spin correlation in low magnetic fields. Both mechanisms have been invoked by Mell, Movaghar and Schweitzer [7.62] to explain a very anomalous field dependence of the photoconductivity.

c) Another effect can be very efficient in destroying the correlation between spins: when the external magnetic field becomes of the same order of magnitude as the spin resonance linewidth (i.e., of the order of the internal fields: dipole–dipole or exchange interactions), one is in the "spin-temperature" regime [7.63] where the Zeeman energy is strongly coupled with the spin–spin interaction reservoir. In this situation, a fast spin diffusion between spins of different pairs can occur and the spin correlation between the spins of a pair can be destroyed in a time of the order of the transverse relaxation time T_2, which is in general much shorter than T_1. This effect, which has not been discussed in the literature, offers an alternative possible explanation of the very-low-field effects found in luminescence and photoconductivity.

Spin statistics play also an important role in the anomalous dark magneto-resistance in the hopping regime, as observed by *Mell* et al. [7.18]. Although more complicated, this effect is basically quite similar to the case of spin-dependent recombination. The starting point, as in the case of recombination centers, is that a level in the gap near the Fermi level already containing one electron is a hopping center only for electrons having spins in the opposite direction. This induces a correlation between neighboring spins and the field relaxation processes discussed above will also affect the hopping transport. The theory of the effect, which is somewhat complicated, has been given, using a random walk approach, by *Movaghar* and *Schweitzer* [7.64].

Acknowledgements. It is a pleasure to thank Dr. D. Kaplan for many useful discussions during the preparation of this chapter.

Appendix A:
Summary of the *g* Values of Paramagnetic Centers in Irradiated Crystalline Silicon

The general trend in the large number of *g* values measured by the Schenectady Group has been compiled by *Lee* and *Corbett* [7.3] and can be summarized as follows (Fig. 7.14):

a) *Broken bond.* This is a very localized center and, in crystalline silicon, it has to be stabilized by a nearby donor state (phosphorus vacancy pair). The *g* tensor is axially symmetric around the [111] broken bond direction with values

$$g_{||} \simeq 2.0005 \qquad g_{\perp} \simeq 2.0100.$$

b) *Positively charged center.* The two broken bonds of two neighboring silicons are not stable and they couple to form a "weak bond" (see Fig. 7.14).

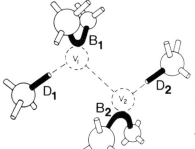

Fig. 7.14. Example of a center produced by electron irradiation in crystalline silicon (divacancy). The two broken bonds (making an angle of 109°) of two neighboring silicons couple to form a "bent bond" (B_1 and B_2). If the two broken bonds D_1 and D_2 are not too far apart, they can also couple to form a "parallel bond" [7.3]. These are examples of "weak bonds" discussed in the text

This neutral center contains two paired electrons and is not paramagnetic. However, the positively charged center, having lost one electron ("trapped hole") is paramagnetic, and has a g tensor which happens, by accidental compensation, to be very nearly spherically symmetric with a g value close to $g = 2.002$.

c) *Negatively charged center.* Similarly, a weak bond can capture an electron (trapped electron) and the negatively charged center is paramagnetic. The g tensor of such a trapped electron is highly anisotropic with components of the g tensor differing by as much as $\delta g = 1.5 \times 10^{-2}$. The angularly averaged g value of this type of center is close to the value $g_{av} \simeq 2.01$.

Several attempts have been made to associate the spin resonance lines in amorphous silicon to this type of center [7.11, 45]. In particular, it is quite natural to assign the line in pure amorphous silicon, which has the proper average g value, to a broken bond (center of type a). In hydrogenated silicon, two types of lines are generally found. By reference to the magnetic resonance properties of weak bonds (center of type b and c), *Kaplan* [7.11] has suggested assigning the high-field line ($g \simeq 2.0055$) to positively charged centers, i.e., to holes trapped somewhere in the lower part of the gap. This is not only supported by the similarities of the g values, but the spherical symmetry of the g tensor of this type of center could explain the comparatively small value of the linewidth. Similarly, the broader low-field line ($g = 2.012$) could be associated with trapped electrons lying on levels not far from the conduction band. Only correlated optical and resonance studies, as in the case of chalcogenide glasses [7.8], could verify these assumptions.

References

7.1   G. Feher: Phys. Rev. **114**, 1219 (1959)
7.2   G. D. Watkins, J. W. Corbett: Phys. Rev. **138**, A543 (1965)
7.3   Y. H. Lee, J. W. Corbett: Phys. Rev. **B8**, 2810 (1973)
7.4   S. C. Agarwal: Phys. Rev. **B7**, 685 (1973)
7.5   P. W. Anderson: Phys. Rev. Lett. **34**, 953 (1975)
7.6   R. A. Street, N. F. Mott: Phys. Rev. Lett. **35**, 1293 (1975)

7.7   S.G.Bishop, U.Strom, P.C.Taylor: Phys. Rev. B15, 2278 (1977)

7.8   S.G.Bishop, U.Strom, P.C.Taylor: In *Proc. 7th Int. Conf. Amorphous and Liquid Semiconductors*, ed. by W.E.Spear (University of Edinburgh, 1977) pp. 595–606

7.9   M.H.Brodsky, R.S.Title: Phys. Rev. Lett. 23, 581 (1969)

7.10   M.H.Brodsky, R.S.Title, K.Weiser, G.D.Pettit: Phys. Rev. B1, 2632 (1970)

7.11   D.Kaplan: In *Proc. 14th Int. Conf. Physics of Semiconductors* (Edinburgh 1978) Inst. Phys. Conf. Ser. 43, 1129 (1979)

7.12   U.Voget-Grote, J.Stuke, H.Wagner: In *Structure and Excitation of Amorphous Solids*, ed. by G.Lucovsky, F.L.Galeener (A.I.P., New York 1976) p. 91

7.13   J.C.Knights, D.K.Biegelsen, I.Solomon: Solid State Comm. 22, 133 (1977)

7.14   I.Solomon, D.K.Biegelsen, J.C.Knights: Solid State Comm. 22, 505 (1977)

7.15   D.K.Biegelsen, J.S.Knights, R.A.Street, C.Tsang, R.M.White: Philos. Mag. B37, 477 (1978)

7.16   K.Morigaki, D.J.Dunstan, B.C.Cavenett, P.Dawson, J.E.Nicholls: Solid State Comm. 26, 981 (1978)

7.17   A.Street, D.K.Biegelsen, J.C.Knights, C.Tsang, R.M.White: In *Proc. Int. Conf. Recombination in Semiconductors* (Southampton 1978) Solid State Electron. 21, 1461 (1978)

7.18   H.Mell, J.Stuke: J. Non-Cryst. Solids 4, 304 (1970)

7.19   J.Kubelik, A.Triska: Czech. J. Phys. B23, 123 (1973)

7.20   A.H.Clark, M.M.Cohen, M.Campi, H.P.D.Lanyon: J. Non-Cryst. Solids 16, 117 (1974)

7.21   P.W.Anderson: Nature (London) 235, 163 (1973)

7.22   P.A.Thomas, M.H.Brodsky, D.Kaplan, D.Lépine: Phys. Rev. B, 18, 3059 (1978)

7.23   H.Fritzsche, S.J.Hudgens: In *Proc. 6th Int. Conf. Amorphous Semiconductors*, ed. by B.T.Kolomiets (Nauka, Leningrad 1975) p. 6

7.24   N.G.Nakhodkin, A.I.Shaldervan: Thin Solid Films 10, 109 (1972)

7.25   D.Kaplan, D. Lépine, Y.Petroff, P.Thirry: Phys. Rev. Lett. 35, 1376 (1975)

7.26   P.G.Le Comber, R.J.Loveland, W.E.Spear, R.A.Vaughan: In *Proc. 5th Int. Conf. Amorphous and Liquid Semiconductors*, ed. by J.Stuke, W.Brenig (Taylor and Francis, London 1974) p. 245

7.27   G.A.N.Connell, J.R.Pawlik: Phys. Rev. B13, 787 (1976)

7.28   P.Gaezi, M.Tanielan, H.Fritzsche: Bull. Am. Phys. Soc. 21, 335 (1976)

7.29   J.J.Hauser: Solid State Comm. 19, 1049 (1976)

7.30   W.Paul, A.J.Lewis, G.A.N.Connell, T.D.Moustakas: Solid State Comm. 20, 969 (1976)

7.31   C.Kittel: *Introduction to Solid State Physics*, 4th ed. (Wiley, New York 1971) Chap. 15

7.32   A.Madan, P.G.Le Comber, W.E.Spear: J. Non-Cryst. Solids 20, 239 (1976)

7.33   D.K.Biegelsen, J.C.Knights: In *Proc. 7th Int. Conf. Amorphous and Liquid Semiconductors*, ed. by W.E.Spear (University of Edinburgh 1977) pp. 429–432

7.34   A.Friederich, D.Kaplan: In *Proc. 20th Electronic Material Conference*, Santa Barbara, 1978, J. Electron. Mater. 8, 79 (1979)

7.35   I.Solomon, J.Perrin, B.Bourdon: In *Proc. 14th Int. Conf. Physics of Semiconductors* (Edinburgh 1978) Inst. Phys. Conf. Ser. 43, 689 (1979)

7.36   C.C.Tsai, H.Fritzsche, M.H.Tanielian, P.J.Gaezi, P.D.Persans, M.A.Vesaghi: In *Proc. 7th Int. Conf. Amorphous and Liquid Semiconductors*, ed. by W.E.Spear (University of Edinburgh 1977) pp. 339–342

7.37   D.Kaplan, N.Sol, G.Velasco, P.A.Thomas: Appl. Phys. Lett. 33, 440 (1978)

7.38   J.I.Pankove, M.A.Lampert, M.L.Tarng: Appl. Phys. Lett. 32, 439 (1978)

7.39   J.C.Knights: In *Structure and Excitations of Amorphous Solids*, ed. by G.Lucovsky and F.L.Galeener (A.I.P., New York 1976) p. 296

7.40   M.H.Brodsky, M.Cardona, J.J.Cuomo: Phys. Rev. B16, 3556 (1977)

7.41   J.C.Knights, G.Lucovsky, R.J.Nemanich: Philos. Mag. B37, 467 (1978)

7.42   R.A.Street, J.C.Knights, D.K.Biegelsen: Phys. Rev. B18, 1880 (1978)

7.43   J.C.Knights, G.Lucovsky, R.J.Nemanich: J. Non-Cryst. Solids 32, 393 (1979)

7.44   B.Movaghar, L.Schweitzer: Phys. Status Solidi 80, 491 (1977)

7.45   J.Stuke: In *Proc. 7th Int. Conf. Amorphous and Liquid Semiconductors*, ed. by W.E.Spear (University of Edinburgh 1977) pp. 406–418

7.46 A. Honig: Phys. Rev. Lett. **17**, 186 (1966)
7.47 J. Schmidt, I. Solomon: C. R. Acad. Sci. Paris **263**, 169 (1966)
7.48 D. Lépine: Phys. Rev. B **6**, 436 (1972)
7.49 J. Caplan, J. N. Helbert, B. E. Wagner, E. H. Poindexter: Surf. Sci. **54**, 33 (1976)
7.50 T. Wosinski, T. Fifielski: Phys. Status Solidi **71**, 73 (1975)
7.51 D. Lépine, V. A. Grazhulis, D. Kaplan: In *Proc. 13th Int. Conf. Phys. Semiconductors*, ed. by F. G. Fumi (Tipografia Marves, Roma 1976) p. 1081
7.52 I. Solomon: Solid State Comm. **20**, 215 (1976)
7.53 R. M. White, J. F. Gouyet: Phys. Rev. B **16**, 3596 (1977)
7.54 V. S. L'vov, O. V. Tretyak, I. A. Kolomiets: Fiz. Tekh. Poluprovodn. **11**, 1118 (1977) [English trans.: Sov. Phys. Semicond. **11**, 661 (1977)]
7.55 W. Schockley, W. T. Read: Phys. Rev. **87**, 835 (1952)
7.56 A. Abragam: *Principles of Nuclear Magnetism* (Oxford University Press, Oxford 1961) Chap. 3
7.57 D. Kaplan, I. Solomon, N. F. Mott: J. Phys. (Paris) Lett. **39**, L51 (1978)
7.58 B. C. Cavenett: *Luminescence Spectroscopy*, ed. by M. D. Lumb (Academic Press, London 1978) Chap. 5
7.59 G. Lampel: In *Proc. 12th Int. Conf. Physics of Semiconductors* (Teubner, Stuttgart 1974) p. 743
7.60 C. Tsang, R. A. Street: Philos. Mag. B **37**, 601 (1978)
7.61 D. Engemann, R. Fisher: In *Proc. 5th Int. Conf. Amorphous and Liquid Semiconductors*, ed. by J. Stuke, W. Brenig (Taylor and Francis, London 1974) p. 947
7.62 H. Mell, B. Movaghar, L. Schweitzer: Phys. Stat. Sol. (b) **88**, 531 (1978)
7.63 M. Goldman: *Spin Temperature and Nuclear Resonance in Solids* (Clarendon Press, Oxford 1970)
7.64 B. Movoghar, L. Schweitzer: J. Phys. C **11**, 125 (1978)

8. Short-Range Order in Amorphous Semiconductors

G. Lucovsky and T. M. Hayes

With 15 Figures

The central issue in describing the structure of an amorphous solid is a specification of the short-range order, and the topological rules which determine it. The most important aspects of the short-range order are the number and type of immediate neighbors, and their spatial arrangement about a given reference atom. A second question concerns the connectivity of the solid and the extent to which structural correlations may then prevail over "several" interatomic dimensions. In this chapter, we emphasize the short-range order, and consider the network connectivity only when it bears directly on the interpretation of experimental results. There are a number of excellent review articles which treat structure in amorphous solids from a traditional point of view: Fourier inversion of X-ray, electron, or neutron diffraction data [8.1–4]. These procedures yield a radial distribution function, or RDF, which is the average density of atoms at a given distance from any other reference atom.

In this chapter, we develop the discussion along different lines with an emphasis on ultimately connecting local order to the electronic properties. The basis for this alternative approach resides in a realization that many of the interesting electronic properties of amorphous semiconductors are controlled by well-defined active chemical centers just as they are in crystalline semiconductors. These centers can arise in two ways: first, as characteristic defects that are determined by the bonding chemistry of the particular amorphous semiconductor [8.5, 6]; and second, as centers produced by adding small amounts of dopants [8.7] or "chemical modifiers" [8.8] to an elemental, compound, or alloy host. It is also necessary to consider the interactions between the dopants and the host-system defects. The paramount issue is then the local order in the immediate vicinity of these centers, and its relation to the resultant electronic properties. Other types of defects include dangling bonds; however, these will not be considered explicitly.

Any discussion of locally active centers must necessarily begin with some idealized model for the structure of the host material. One class of idealized structural models is based on continuous random networks in which all bonds are satisfied and in which each of the atomic species displays a coordination which satisfies the normal covalent bonding requirements. This type of structural model was first used by *Mott* to explain the dark dc conductivity in chalcogenide glasses [8.9]. The dc conductivity could be related to an activation energy that was approximately one-half of the optical band gap and could be varied in a continuous way by alloying. The model was found to be applicable to bulk-quenched glasses and well-annealed films, in which doping was not believed to be possible since the local environment could always adapt to accommodate the valence bonding requirements of a given atomic species. Films produced by quenching from the vapor phase onto low-temperature substrates are then expected to display different electronic properties, since they are inherently more susceptible to defects, doping, and chemical modification.

The organization of this chapter reflects this point of view. We first discuss the elements of short-range order in idealized random network structures and indicate the way in which real systems can display different types of local order. We then discuss two important probes of the local order: RDFs and vibrational spectra. Finally, we consider the application of these ideas to real systems, including as examples, doped hydrogenated amorphous Si and defects in a-As and representative chalcogenide glasses.

8.1 Elements of the Short-Range Order

Short-range order, hereafter designated as SRO, is defined as the local atomic arrangement about a given reference atom. A specification of the SRO must necessarily include: (I) the number of immediate neighbors, and their type; (II) their separation from the reference atom; and (III) their angular distribution. The specification of these aspects of the SRO does not completely or uniquely determine the structure of the amorphous solid. A complete description of the structure requires a specification of the network topology that defines the way the atomic sites are interconnected with each other. For the covalently bonded amorphous semiconductors, the most important aspect of the overall topology relates to the distribution of dihedral angles, and the resultant ring statistics [8.1].

8.1.1 Structural Models

There have been many discussions of the types of structural models that can be used to describe amorphous solids. These are identified in the review article of *Wright* and *Leadbetter* [8.4]. The two primary classes of models are the microcrystallite and random network models. Microcrystallite models are plagued by a serious shortcoming that relates to the relative fractions of interior and surface atoms [8.1]. To be meaningful, the average microcrystallite size must exceed a few unit cell dimensions; however, for a crystallite size equal to three unit cell spacings, about one-half of the atoms reside on the microcrystallite boundaries. Thus any "connective tissue" which bridges these boundaries is then as important as the crystallites themselves. This in effect means that in the limit of very small crystallite size, the microcrystallite model becomes indistinguishable from a random network model. We illustrate the important aspects of the local atomic structure by considering the random network model of *Polk* [8.10] which applies to the tetrahedrally bonded amorphous semiconductors a-Si and a-Ge.

The *Polk* model is a "ball and stick" model containing 440 atoms. Other workers, for example, *Schevchik* and *Paul* [8.11] and *Henderson* and *Herman* [8.12], have studied similar models generated by computer techniques. *Polk* has also refined his model using numerical techniques [8.13]. In the idealized random network structure, every atom is bonded to a number of immediate neighbors that satisfies its chemical valence. In an elemental system these atoms are all of the same type; whereas, in an alloy or compound, it is also necessary to specify the particular atomic species bonded at each reference site. The average nearest-neighbor interatomic distances are set equal to the values they have in an appropriate crystalline phase, or more generally to the sum of the respective covalent radii. The distribution of these bond lengths is kept small, on the order of $\pm 1\%$ of their average value. In the Polk model for a-Si or a-Ge, each atom then has four immediate neighbors arranged in a tetrahedral geometry. Bond angle deviations are allowed, so that the tetrahedra are generally distorted from an idealized geometry. In the Polk model, the bond angles are distributed about their average value of $109°28'$ with an average deviation of $\pm 10°$ and a maximum deviation of $\pm 20°$. The random character of the network results from a statistical distribution of dihedral angles. All of the atoms in the Polk model are included in either five- or six-membered rings. This ratio is 1:4 in the Polk model, whereas in the diamond-structure crystalline phases of Si and Ge, all of the atoms are in six-membered rings. The average bond angle and its distribution determine the average second-neighbor spacing and its distribution, respectively. The larger variation in the dihedral angle in effect removes a characteristic third-neighbor spacing.

The application of random network models to elemental systems containing either twofold- or threefold-coordinated atomic species raises interesting questions about the network topology. Models for a network of threefold-coordinated atoms, as in a-As, have been constructed by *Greaves* and *Davis* [8.14], and for a network of twofold-coordinated atoms, as in a-Se, by *Long* et al. [8.15]. Locally the network topology is fully three dimensional for a random network of fourfold-coordinated atoms; however for twofold- and threefold-coordinated atoms, the local topologies are respectively one and two dimensional. The lower coordination also allows the real solid to deviate from a completely interconnected network through the incorporation of discrete molecular species. The resultant solid can then be described in terms of a mixture of molecular and polymeric species. Discrete molecular species have been identified in a-Se as Se_8 ring molecules [8.16–18], in a-As as As_4 pyramidal molecules [8.19], and in "S-rich" glasses in the As–S and Ge–S binary alloys as S_8 ring molecules [8.20–23]. Random network models have also been constructed for compound amorphous semiconductors. The most notable of these is the *Bell* and *Dean* model for a-SiO_2 [8.23]. Other models have also been considered for chalcogenide glasses; however, they have not been studied in the same detail as the models for the elemental amorphous semiconductors or a-SiO$_2$.

The amorphous semiconductors we emphasize in this chapter will be derived principally from elements in the fourth, fifth, and sixth columns of the periodic table: Si and Ge; As; and S, Se, and Te respectively. We consider in some detail elements, compounds, and their alloys. We shall also extend our discussion to describe the way in which other types of atoms may be incorporated into network structures. We choose to approach the general subject of SRO by describing a methodology for treating the local order in binary alloy systems.

8.1.2 Local Order in Binary Alloys

Consider a binary alloy system A$_{1-x}$B$_x$, where A and B are two different atomic species and x is a normalized concentration variable. Let a and b be the columns of the periodic table from which the respective A and B atoms are chosen. A network structure can be specified in four stages: (I) the local coordination of each atomic species; (II) the distribution of bond-types, A–A, A–B, and B–B; (III) the specification of characteristic local molecular environments; and (IV) the topological rules for the interconnection of the molecular building units [8.21, 24]. We restrict our discussion to the first three stages that define the SRO.

a) Local Atomic Coordination

Consider first the local atomic coordinations. We assume that these satisfy the 8-N rule for normal covalent bonding, so that the coordinations of the A and B atoms, y_a and y_b, are given by 8–a and 8–b, respectively. This type of coordination dominates in bulk-quenched glasses and well-annealed films. Recently there has been considerable interest in "wrong atomic coordinations" or so-called valence alternation in chalcogenide glasses [8.5]. A small number of these wrong-coordination sites, e.g., either one- or threefold-coordinated Se atoms or both, appear to be present in chalcogenide glasses with characteristic densities of 10^{16}–10^{18} cm^{-3}. These centers are electronically active and give rise to interesting photoelectronic and spin resonance properties [8.5, 6, 25, 26] which are reviewed in Chap. 3. For the present, we assume idealized local coordinations, 8-N, with all bonds satisfied. We discuss coordination defects in a later section.

b) Bond Statistics

There are then two distinctly different ways to specify the distribution of the three bond types, A–A, A–B, and B–B, that are consistent with the assumed 8-N coordination. The first is a random covalent network model [8.27] in which the word random is used to indicate that the distribution of bond-types is

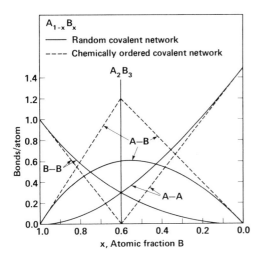

Fig. 8.1. Bond-counting statistics for a 3–2 network [8.21]

statistical and hence completely determined by (I) the respective local coordinations, y_a and y_b, and (II) the fractional concentrations of A and B atoms, $1-x$ and x, respectively. This approach to the bonding neglects factors such as the relative bond energies. The random covalent bonding description includes A–A, A–B, and B–B bonds at all compositions other than $x=0$ and $x=1$. The alternative model for bond statistics emphasizes the relative bond energies and is based on an assumption that heteropolar or A–B bonds are favored at all compositions. We designate this alternative model as a chemically ordered network model [8.21]. This model then contains a chemically ordered compound phase at a composition $x_c = y_a/(y_a + y_b)$. The compound is $A_{y_b}B_{y_a}$ and contains only A–B bonds. For compositions defined by $1 > x > x_c$, the alloys contain A–B and B–B bonds; whereas for $x_c > x > 0$, the alloys contain A–B and A–A bonds. Figures 8.1, 2 display the bond-counting statistics for two types of alloy systems defined by the local coordinations of the constituent atoms: Fig. 8.1: $y_a = 3$, $y_b = 2$, as in As–S and As–Se; Fig. 8.2: $y_a = 4$, $y_b = 2$, as in Ge–S and Ge–Se. Each figure contains the bond counting for the two models, random covalent bonding and chemical ordering. The existence of compound phases at a-As_2S_3, a-As_2Se_3, a-GeS_2, and a-$GeSe_2$, suggests that chemical ordering prevails in the four alloy systems, As–S, As–Se, Ge–S, and Ge–Se, at least over the range of bulk-glass formation. This is supported by other structural measurements, for example RDFs obtained by X-ray diffraction [8.28–31], and by structural interpretations of ir and Raman vibrational spectroscopy [8.20–22].

   It should be noted that the compound phase defined above is special in the sense that it contains only A–B bonds. Additional compound phases may also be present at other compositions without changing the bond-counting statistics. The criteria for existence of such compound compositions relates to the existence of a molecular entity from which the network can be generated. The

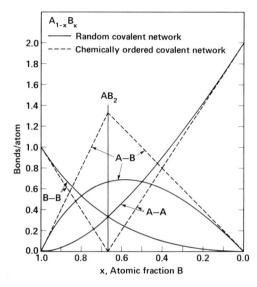

Fig. 8.2. Bond-counting statistics for a 4–2 network [8.21]

molecular building block can contain both A–B and B–B or A–B and A–A bonds in such a way that the distribution of bond types within and between these units is consistent with the bond-counting statistics. Examples of such additional compound phases have been identified in the Ge–S and Ge–Se alloy systems at compositions corresponding to 60% S or Se, Ge_2S_3 and Ge_2Se_3 [8.21, 24, 32, 33].

c) Local Molecular Clusters

Consider next the specification of local molecular clusters. This concept is somewhat arbitrary and has its origin in some of the early models for the atomic structure of fused silica, a-SiO_2 [8.34, 35]. Structural models for that system emphasized a molecular building block with tetrahedral geometry, $SiO_{4/2}$: a central Si atom with a tetrahedral distribution of four O atoms, each of which is bonded to a Si-atom in another tetrahedron, hence the 4/2 notation. Similar procedures have been applied in describing the molecular structure in other compound chalcogenide glasses [8.36, 37]. The "central atom" in the molecular cluster is chosen to be the one of higher coordination; see Fig. 8.3. The fractional concentration of the lower-coordination species is then the ratio of the two atomic coordinations and always greater than one. We use a convention in defining alloys $A_{1-x}B_x$ in which the coordination of the A atom is always greater than that of the B atom, so that the characteristic molecular building block at the special compound composition is then given by AB_{y_a/y_b}. The molecular cluster for As_2S_3 is $AsS_{3/2}$, and for GeS_2, $GeS_{4/2}$.

It is also necessary to specify the three-dimensional structure of these molecular building blocks. In general this reflects the normal covalent bond-

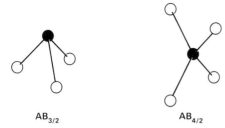

Fig. 8.3. Local clusters for A_2B_3 and AB_2 compound glasses

ing geometry of the central atomic species. $AB_{4/2}$ then has a tetrahedral geometry, whereas the $AB_{3/2}$ unit is a pyramidal structure with a face-apex angle close to $90°$. The tetrahedral geometry is determined by symmetry; however, as in the *Polk* model [8.10], as well as the real solid, the angles will generally display a statistical distribution with an average deviation of about $\pm 10°$. The actual pyramidal apex angle of the $AB_{3/2}$ unit may be estimated by a comparison with crystalline phases in which similar covalent bonding prevails [8.38] or obtained by an analysis of the X-ray RDFs. For AsS_3 and $AsSe_3$, this angle is approximately $96°$.

It is not meaningful to specify molecular units for elemental amorphous solids due to the equivalence of the constituent atoms. For elemental materials, the more important aspects of the SRO are then the coordination, the bond length and bond angle. The bond angle and coordination at the atomic sites define the local molecular symmetry.

So far we have considered two aspects of the SRO in a generalized way: (I) the local coordination which is assumed for the present to satisfy the 8-N rule, and (II) the nature of the bond-counting statistics. We have also described the local molecular order for compound phases. We now consider the local molecular order as it applies to binary alloys in a more general way. We assume that the bond-type distribution is determined by chemical driving-forces so that the alloy is divided into two different regimes at the compound composition, x_c. Consider first the B-rich regime, $1 > x > x_c$, and recall that B is the atom with the lower coordination. In this alloy regime there are A–B and B–B bonds, but not A–A bonds. Each A atom then has only B-type neighbors arranged in a local atomic cluster of the form AB_{y_a/y_b}. This is the same cluster that provides the molecular basis for the compound phase at x_c. In the B-rich alloys, these molecular clusters are connected through additional B atoms. In chalcogenide alloys $A_{1-x}B_x$ (B = S, Se, Te), this type of model for the B-rich alloy regime has been designated as a chain-crossing model [8.20, 21, 27, 30]. Pyramidal ($AB_{3/2}$) or tetrahedral ($AB_{4/2}$) molecular units are interconnected by chain segments of B atoms; hence, the threefold- or fourfold-coordinated A-type atoms provide points at which chains cross or branch. The model for the B-rich alloy regimes also allows for discrete molecular species. These have been identified as S_8 ring molecules [8.39, 40] in S-rich As–S [8.21, 22] and Ge–S [8.20] glasses. In both S-based alloy systems, bulk glass formation is restricted to values of $x < 0.9$. For higher S concentrations, crystalline orthorhombic S occurs as a second phase.

The local molecular order is not as simply specified in the A-rich alloy regimes, since some of the more highly coordinated A-type atoms must have A-type as well as B-type immediate neighbors. The most general molecular structures can be described in terms of a small set of basic network formers [8.21]. For the 4–2 alloys, these are tetrahedral units of the form AB_4, AB_3A, AB_2A_2, ABA_3, and AA_4 [8.21, 41] and are pictured schematically in Fig. 8.4. For 3–2 alloys, they are the corresponding pyramidal structures AB_3, AB_2A, ABA_2, and AA_3, also pictured in Fig. 8.4. For any given alloy composition the fraction of A-type atoms in any one of these local environments is not uniquely defined. One type of distribution is purely statistical: this is illustrated in Fig. 8.4. An alternative way of defining local molecular clusters is through a chemically (or physically) determined driving force that favors some set of specific molecular configurations. This latter type of consideration applies in the Ge–S and Ge–Se alloy systems. In the alloy regime between the compound phase, $x_c = 2/3$, and $x = 3/5$, there are two molecular species in the alloys, $GeX_{4/2}$ tetrahedral unit and a larger unit containing Ge–Ge bonds, $X_{3/2}\,Ge$–$Ge\,X_{3/2}$ [8.21, 24, 32, 33]. A similar situation prevails in the As-rich regime of $As_{1-x}S_x$ where the two molecular species are $AsS_{3/2}$ and $S_{2/2}\,As$–$As\,S_{2/2}$. An alternative possibility for the second local cluster is that it is part of a true molecular species, an As_4S_4 molecule which is equivalent to two interconnected $S_{2/2}\,As$–$As\,S_{2/2}$ units.

Binary alloys can also be formed from atoms with the same coordination, $Si_{1-x}Ge_x$, $Se_{1-x}Te_x$, etc. In this case, the SRO is generally the same as that of either member; however, the atomic species are randomly distributed on the sites. In the alloy systems mentioned above there are no chemically ordered compound crystalline phases [8.42]. However, there is a compound phase, SiC, in the $C_{1-x}Si_x$ binary alloy system. *Gorman* and *Solin* [8.43] have shown that amorphous alloys contain both C–C and Si–Si bonds for Si-rich and C-rich alloys relative to the SiC composition, implying no chemical ordering at the compound composition.

8.1.3 Defect Configurations

So far we have discussed SRO in idealized network structures in which there are no dangling bonds, and in which each atom satisfies its chemical valence bonding requirements. Departures from these idealizations are known to occur in real systems. There are significant differences in the defect structures of the tetrahedrally bonded and chalcogenide amorphous semiconductors [8.44]. a-As displays behavior with respect to defects that is closer to that of the chalcogenides. Therefore in order to provide a basis for understanding doping, or chemical modification, where the defects clearly play a role, it is convenient to consider the two classes of materials separately.

a) Tetrahedrally Bonded Materials

Consider first the tetrahedrally bonded materials a-Si and a-Ge. These materials cannot be prepared as glasses by quenching from a melt, but rather

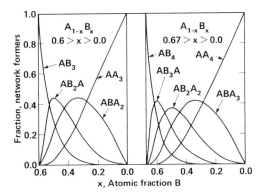

Fig. 8.4. Statistically determined local clusters for A-rich alloy regimes in 3–2 and 4–2 networks [8.21,41]

must be prepared by other means: quenching from the vapor phase as in sputtering or evaporation, or chemically, including chemical vapor deposition (CVD) of Si or Ge from an organometallic compound, or glow discharge decomposition of either silane (SiH_4) or germane (GeH_4). The films produced by evaporation or sputtering generally contain a large number of defects which give rise to electronically active states in the energy gap. These films also display densities lower than those of the corresponding crystalline phases, and also below those estimated from idealized random network models [8.10–12]. The defect states contribute to relatively large spin resonance signals indicative of a large number of defect sites ($\sim 10^{20}$ cm^{-3}) with unpaired spins [8.45]. The density deficiency implies that a substantial fraction of these defect states may reside on internal void surfaces [8.45, 46]. Material produced by evaporation or sputtering without any other network modifying atomic species generally cannot be doped due to these high defect densities.

The main thrust in the development of a-Si as a material for potential technological applications, has then been toward an elimination or compensation of the defect states. This activity is the subject of several other chapters in this book (see Chaps. 7, 9, and 10). Here we note that compensation of defect states can be accomplished by incorporation of H [8.47–50]. We have not presented a general discussion of 4–1 alloy systems; however, in the solid state, it is possible to have Si atoms bonded to either one, two, or three H atoms [8.51–54]. In the last section of this paper we discuss the local bonding arrangements of H in a-Si.

Once a sufficient number of the defects have been neutralized by H compensation, the material can be doped with either group III or V atoms just as in crystalline Si [8.7]. The dopants P or As added to a-Si then yield *n*-type material, whereas the introduction of B yields *p*-type material. The chemical nature of these dopants implies that for some range of concentrations, a substantial fraction of the doping atoms reside at fourfold-coordinated sites. Their activity in the doping process is essentially the same as it is in c-Si. At higher concentrations, their incorporation can be different. There is evidence that above some critical concentration, As is incorporated in a-Si in a threefold-coordinated local bonding environment yielding an alloy with intrinsic conductivity rather than a doped semiconductor [8.55].

b) Chalcogenides and Pnictides

The situation in the chalcogenide and pnictide amorphous semiconductors is in many ways more complex, principally due to the lower coordination of some of the atomic species. Rather than trace the historical evolution of the identification of the defect structure, we shall summarize the important results. One important difference in the nature of defect states relates to the observation of an electron spin resonance (ESR) signal. As-deposited evaporated and sputtered films of a-Ge and a-Si have very large ESR signals, corresponding to unpaired spin densities of 10^{20} cm^{-3} or more [8.45, 46]. In contrast, in chalcogenide glasses and films deposited near the glass-transition temperatures, there is virtually no ESR signal [8.25, 26]. In 1975, Anderson [8.56] noted that this behavior could be explained if two electrons with opposite spins close to the same atom are actually attracted to one another. The lowest energy state would then have no unpaired spin density. Mott et al. [8.5] proposed a model in which these spin-paired defect states were charged states that influenced the photoelectronic properties, principally the luminescence. In the Anderson model [8.56], the spin-paired states were assumed to arise in a general way from the disorder; however, Mott et al. [8.5] suggested that the states might be physical defects such as dangling bonds. The model proposed by Mott et al. [8.5] also explained the induced ESR and optical absorption first reported by Bishop et al. [8.25, 26]. Kastner et al. [8.6] suggested a different structural model based on pairs of sites with local coordinations that did not satisfy the 8-N rule. They called these sites valence-alternation pairs (VAPs). For a-Se, these defect states are singly and triply coordinated sites, previously described by Mott et al. [8.5] as D$^-$ and D$^+$, and designated by Kastner et al. [8.6] as C$_1^-$ and C$_3^+$ (see Fig. 8.5). In a-As, the corresponding pairs of defects would be labeled P$_2^-$ and P$_4^+$, where P species is a pnictide atom. The existence of these spin-paired, charged defect states is related in an intimate way to the nonbonding or lone-pair electrons of group V and VI atoms. This is discussed in more detail in Chap. 3 and elsewhere [8.6, 44, 53]. In a glass with both group V and VI atoms present, a-As$_2$Se$_3$, a larger number of coordination-type defect pairs are possible [8.57, 58].

It has been argued that these pairs of coordination defect states are (I) inherently more stable than isolated dangling bonds, and (II) that they can be characterized by an effective correlation energy that is negative, as originally required by Anderson [8.56]. Since the maximum coordination for normal covalent bonding is four, such pairs are not possible in the tetrahedrally coordinated amorphous semiconductors. This appears to be the origin of the very different defect-state behavior in the two classes of amorphous semiconductors. The characteristic defect states in tetrahedrally bonded amorphous semiconductors are then dangling bonds which are electrically neutral and have

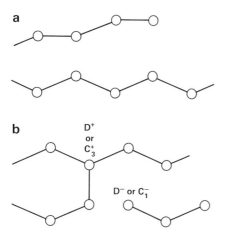

a

b   D⁺
or
C₃⁺

D⁻ or C₁⁻

Fig. 8.5. Valence-alternation defects in a network based on twofold-coordinated atoms: (**a**) idealized bonding; (**b**) valence-alternation pair

an unpaired spin. In contrast, the characteristic defect states in chalcogenide and pnictide semiconductors are valence-alternation pairs which are charged, but have no unpaired spin density.

There is yet another type of defect that can occur in chalcogenide glasses. It is associated with "wrong bonds" rather than "wrong coordination". This type of defect only has meaning in the context of an idealized model of the bonding. For example, if chemical ordering is assumed to be complete, then at the alloy composition $A_{y_b}B_{y_a}$, there are only A–B bonds. However, in any real system, the driving force for chemical ordering is finite, so that any set of preparative conditions will generally yield some fraction of A–A and B–B bonds. In bulk-quenched glasses, these deviations from idealized bond order appear to be small; however, in films prepared by vapor deposition onto cooled substrates, these effects are large and therefore readily observable [8.59, 60]. We discuss an example of this in Sect. 8.3.

We now consider the doping or chemical modification of chalcogenide glasses. It has been observed that small quantities of an alkali metal or halogen can have a significant effect on the electronic properties of chalcogenide glass films [8.8, 44, 61]. This is attributed to the compensation of a native defect, presumably one member of a valence-alternation pair, by the strongly electropositive or electronegative dopant atom. In order to be active the dopant atoms cannot be added to the melt or quenched in a bulk glass, but rather must be incorporated by some low temperature process, e.g., diffusion or ion implantation.

8.2 Probes of the Short-Range Order

This section treats two types of probes of the SRO: (I) RDFs derived by the traditional techniques of X-ray, neutron, and electron diffraction, and by a more recently developed technique based on EXAFS (Extended X-ray

Absorption Fine Structure); and (II) analysis of vibrational spectra obtained by infrared (ir) absorption and reflection, and Raman scattering. The RDF techniques yield direct structural information, bond lengths, coordination numbers, etc., whereas the analysis of the ir and Raman spectroscopy yields information about bond types, and the symmetries of the local atomic environments.

8.2.1 Measurements of Radial Distribution

a) X-Ray, Neutron, and Electron Diffraction

X-ray, neutron, and electron diffraction are the structural probes of choice for most solids. In the ideal case of a single-crystal sample, they yield not only the equilibrium positions of the atoms but also the averaged distributions about those sites. Unfortunately, this great wealth of precise information is reduced substantially in a liquid or amorphous solid. The disorder eliminates all of the angular information, leaving only the radial distribution function, or RDF. There have been numerous reviews of diffraction studies of the structure of amorphous solids. Those by *Wright* [8.3] and by *Wright* and *Leadbetter* [8.4] are of particular interest. Given this extensive literature, the present effort is limited to the most qualitative considerations.

The basic quantity determined in a diffraction study is the diffracted intensity as a function of momentum transfer, $i(q)$, where $q = 4\pi \sin\theta/\lambda$. θ is the angle of the diffracted beam, and λ is the wavelength in Å. A quantity related to the RDF may be obtained by Fourier transforming $qi(q)$ into real space. For the typical case of a two-component system, let the pair density of atom species A and B separated by a distance r be denoted by $\varrho_{AB}(r)$. The Fourier transform of $qi(q)$ can then be expressed as [8.3, 4]

$$\varphi^D(r) = \int_0^\infty du\, u[\varrho_{AA}(u)\, P_{AA}^D(r-u) + \varrho_{BB}(u)\, P_{BB}^D(r-u) + 2\varrho_{AB}(u)\, P_{AB}^D(r-u)],$$

(8.1)

where the superscript D refers to a diffraction study. φ^D is a sum of three terms, each the convolution of a pair density and a distinct peak function, P^D. For X-rays, neutrons, or electrons, $P_{AB}^D(r)$ is the Fourier transform of the product of the individual scattering factors for atoms A and B multiplied by a window function representing the range of q space transformed. More important than the precise form of P_{AB}^D, however, is the fact that it is completely known. The extraction of structural information from φ^D is then reduced to determining the three independent structure factors – $\varrho_{AA}(r)$, $\varrho_{AB}(r)$, and $\varrho_{BB}(r)$. In principle, this requires three independent measurements of φ^D in which the peak functions are varied (typically one X-ray and two neutron diffraction studies), and is only rarely done (for an application to Cu–Sn liquids, see [8.62]). It is much more common to analyze a single diffraction measurement with some simple structural model (for an application to a-SiO$_2$, see [8.63]). The uniqueness of the result of such a procedure is difficult to establish given the similarity of the various peak functions. The situation worsens as the number of atom species increases, since φ^D is a sum of $N(N+1)/2$ terms for an N-component system. For these reasons, the amount of local structural information which has been extracted from diffraction studies declines drastically as the chemical complexity of the system increases. The maximum amount of information is thus obtained for elemental systems.

There is another sort of structural information which can be obtained from diffraction studies, but which is not evident in the Fourier transform. By way of example, consider the case of glassy As_2S_3 and As_2Se_3 [8.64, 65]. A small but very sharp peak occurs in $qi(q)$ at low values of $q(\sim 1.2\,\text{Å}^{-1})$. This peak suggests the presence of a structural unit with at least one dimension $\sim 5\,\text{Å}$. Experiment has been unable to decide between two alternative explanations for this peak: the existence of layers in the amorphous solid, in analogy with crystalline phases [8.64], or the existence of large molecular units, such as As_4S_6 [8.65]. As will be discussed in a later section, this peak changes interestingly on annealing [8.66].

b) EXAFS (X-Ray Absorption)

The situation is somewhat different in the analysis of the extended fine structure on the cross section for X-ray absorption, or EXAFS. The recent availability of intense sources of continuous radiation in the X-ray region has stimulated the development of this technique (see, e.g. [8.67], and numerous papers by P. A. Lee and collaborators, and by T. M. Hayes et al.). The EXAFS are manifest as oscillations in the absorption cross section for the photoexcitation of a K-shell electron to a continuum state, measured as a function of incident X-ray energy. These oscillations are a final-state electron effect, arising from the interference between the outgoing wave function and that small fraction of itself which is scattered back from the near-neighbor atoms. The primary objective of studies of EXAFS is to determine the local environment of the excited atom species by analyzing the measured oscillations. The wide separation in energy of the atomic K edges allows the excitation of each atom species individually. As a consequence, the number of pair density functions observed in a given measurement is smaller than in a diffraction study. The quantity analogous to $i(q)$ is $\Delta(k)$ – the observed oscillations normalized to the atomic K-shell absorption and expressed as a function of final-state electron momentum, k. For the example of a two-component system, the Fourier transform of $k\Delta(k)$ measured on the K edge of element A can be expressed as [8.68].

$$\varphi_A^E(r) = c_A^{-1} \int_0^\infty du \left[\varrho_{AA}(u) P_{AA}^E(r-u) + \varrho_{AB}(u) P_{AB}^E(r-u) \right], \tag{8.2}$$

where c_A is the concentration of element A and the superscript E refers to an EXAFS study. The peak function involved, P^E, is not as well understood as in the diffraction case, and is often extracted from experiments on known systems [8.68]. The most obvious difference between (8.1) and (8.2), however, is in the presence of only two independent pair densities – $\varrho_{AA}(r)$ and $\varrho_{AB}(r)$. On a more subtle level, the EXAFS peak functions for different scattering atoms tend to be more distinct than in the diffraction studies. Taken in conjunction, these two differences simplify greatly the structural analysis of EXAFS data from multicomponent systems. This will be illustrated in later sections.

There are two further points to be made in the comparison between EXAFS and diffraction studies. Firstly, the absence of low-k components in the Fourier transform of the EXAFS data results in an extreme sensitivity to the broadening of peaks in $\varrho(r)$ [8.68]. As a consequence, unlike diffraction studies, EXAFS seldom yields structural information about atoms beyond the most sharply defined one or two shells of near neighbors in an amorphous solid. Secondly, the EXAFS is substantially more sensitive than a diffraction study to rare constituents. This attribute is evident from (8.1) and (8.2) given that $\varrho_{AB}(r)$ is proportional to $c_A c_B$. The enhanced sensitivity of EXAFS is essential in the study of arsenic doped a-Si:H alloys discussed in a later section. Using special techniques, it has even been possible to examine the environment of species present only in ppm [8.69] or of surface atoms preferentially [8.70].

Except in unusual circumstances, these probes of the radial distribution can be expected to yield definitive information about the distribution of bond lengths and the type and number of nearest neighbors in a multicomponent system. Insight into the higher orders of SRO can be gained only by inference from model fits to the pair densities. The situation is somewhat better in the analysis of the vibrational spectra, to which we now turn.

8.2.2 Vibrational Spectroscopy

The ir and Raman activity of the vibrational modes in a crystalline solid are determined by group-theoretical selection rules with the general result that only a very small number of long-wavelength or zone-center modes contribute discrete features to the first-order spectra. In contrast, all of the vibrational modes of an amorphous solid can become optically active due to a disorder-induced breakdown of selection rules. The resultant ir and Raman spectra are then expected to be continua which exhibit the general features of the one-phonon density of vibrational states, but modulated by matrix-element-dependent terms which reflect the symmetry of the local atomic environments [8.71–74]. This dependence on a one-phonon density of states, $g(v)$, and matrix element effects can be included in generalized expressions for the ir absorption, $\alpha(v)$, or the Raman scattering. Consider the ir absorption; this can be approximated by an expression of the form

$$n\alpha(v) = \pi^2 (4\pi N e^2 / \bar{M}) f(v) g(v), \tag{8.3}$$

where $f(v)$ is an oscillator strength that includes the frequency dependence of the matrix elements, N is the density of atomic oscillators, $\bar{M}$ is an oscillator mass, n is the index of refraction, and v is the frequency in cm^{-1}. If $f(v)$ is slowly varying, then $\alpha(v)$ reflects the general features in the density of states. This can be established through a comparison of the ir absorption and reduced Raman scattering. If the two spectra are very different, then it can be assumed that $f(v)$ and its Raman counterpart are strongly frequency dependent and that matrix-element effects dominate.

The vibrations that contribute to $f(v)$ can be qualitatively separated into different types of modes; for example, into modes characterized by three frequency regimes: low-frequency acoustic modes, intermediate-frequency bond-bending modes, and high-frequency bond-stretching modes. The bond-stretching modes generally yield the most direct structural information. The frequencies of the modes are determined primarily by nearest-neighbor interactions and their relative ir or Raman activity, by the local molecular symmetry.

A number of different models have been used to provide a structural basis for the interpretation of the ir and Raman results. These generally fall into four categories: (I) models based on comparisons between the vibrational spectra of crystalline and amorphous solids of the same composition; (II) calculations based on very large clusters containing hundreds of atoms; (III) calculations based on smaller clusters but "properly" terminated as, for example, in the cluster-Bethe-lattice method; and, finally, (IV) calculations based on local molecular clusters that are assumed to be vibrationally decoupled from other molecular clusters in the solid.

Vibrational spectra have proven to be a particularly useful tool in studying the local atomic structure in binary alloy systems. For example in the Ge–S system, analysis of the vibrational spectra allows one to discriminate between the different models for the bond-counting statistics, random covalent bonding versus chemical ordering, and provides additional information about the structure of the local network forming molecular clusters [8.20, 21, 24]. A second example concerns the application of vibrational spectroscopy of H in a-Si and a-Ge [8.51–54, 75]. Analysis of the ir and Raman spectra provides two types of information: the number of H atoms bonded to each Si or Ge site, and the local bonding topology.

Another interesting aspect of vibrational spectroscopy relates to the application of local cluster models to compound systems. The local molecular clusters that characterize the network structures of a-GeO$_2$, a-GeS$_2$, and a-GeSe$_2$ have the same internal structure: GeX$_{4/2}$ tetrahedra; however, the bond angles at the twofold-coordinated O and S and Se sites are markedly different. For a-GeS$_2$, bond angles of approximately 95–100° promote weak intercluster coupling, with the result that the major spectral features are well described by the internal modes of the GeX$_4$ tetrahedron; whereas for a-GeO$_2$, the bond angle is approximately 125° with a much stronger intermolecular coupling [8.76]. The coupling is so strong that the ir and Raman spectra are qualitatively different, and cannot be explained using the same local cluster modes. Therefore, even though two amorphous solids may have very similar local atomic structure, aspects of the network connectivity can play an important role in determining the details of the ir and Raman response. A second example of this relates to comparisons between a-As_2O_3 and a-As_2S_3 and a-As_2Se_3 [8.77].

8.2.3 Brief Survey

Having discussed in some detail the type of SRO information to be obtained from a measurement of the radial distribution function or the vibrational spectra, it is appropriate to give a very brief survey of interesting results.

a) Elemental Materials

a-Si and Ge. The strong preference of these atoms for fourfold-coordination gives rise to fully three-dimensional structures in both crystalline and amorphous forms. Early evidence for the amorphous phases being actually microcrystalline (see, for a discussion of a-Si and Ge, [8.78]) has been shown to be inconclusive, so that the random network model is now widely accepted. The remaining issue within the random network model concerns the proportion of five-, six-, and seven-membered rings. It has not been decisively resolved using RDF studies.

The ir and Raman spectra are qualitatively very similar to each other and have been demonstrated to reflect the major features in a one-phonon density of vibrational states [8.79–81]. Differences between the ir and Raman spectra and the density of states of the respective crystalline phases are also well understood.

a-C. There have also been studies of a-C [8.79]. The major issues revolve around the SRO; in most samples that have been studied, it is clearly threefold coordinated as in graphite.

a-As. The preference of As for threefold-coordination opens up the possibility of a layerlike structure in the amorphous phase similar to that found in crystalline phases. It is interesting to note in this regard that the crystalline form is semimetallic while the amorphous phase is a covalent semiconductor. The RDF of amorphous As is equally well described by either a quasi-crystalline layered structure [8.82] or a fully three-dimensional random network [8.14].

The ir [8.83] and Raman [8.19, 84, 85] spectra are also in good agreement with a calculation of the vibrational modes of a random network based on threefold-coordinated atoms [8.86, 87]. The one-phonon density of states is simply related to the electronic density of states for the valence and conduction bands through an isomorphism that is applicable to a-As and other group V amorphous solids. Additional weak features in the ir and Raman spectra have been assigned to defect states with coordinations other than three [8.19, 84, 85]. The Raman spectrum also contains a weak feature assigned to a molecular species, As_4 pyramidal molecules [8.19].

a-Se and Te. As might be expected for twofold-coordinated atoms, the SRO of the solid amorphous phases of these chalcogens are based on rings or chains of atoms in varying proportions. Diffraction studies of a-Se suggest that the bulk glass is comprised largely of bent helical chains [8.88], but that evaporated or sputtered thin films have varying proportions of rings and chains [8.89, 90]. An EXAFS comparison between trigonal and evaporated a-Se [8.91] has confirmed the number and bond length distribution of the nearest neighbors but contained no information on the network topology. The structure of vapor-deposited Te films has been shown to consist of short chains of ~ 10 atoms [8.92].

There are two important results from ir and Raman spectroscopy for a-Se: first the identification of a very weak feature in the Raman spectrum at about $112\,\text{cm}^{-1}$, that is assigned to a molecular component of the glass, Se_8 ring molecules [8.16–18]; and second, the establishment of a theoretical and quantitative basis for understanding differences in the vibrational frequencies of the optical modes of trigonal and a-Se. The differences are related to directional interchain forces present in the crystalline trigonal phase, but absent in the amorphous phase [8.38, 93–96]. These forces serve to decrease the frequency of the symmetric bond-stretching mode of the trigonal crystal relative to that of a similar stretching mode in the amorphous phase.

Similar considerations apply in comparisons between the vibrational modes of a-Te and trigonal Te; however, the interchain forces in trigonal Te are considerably greater than in trigonal Se, so that the differences in frequency are even more pronounced. The symmetrical Raman-active stretching mode is at $158\,\text{cm}^{-1}$ in a-Te [8.94], and at $122\,\text{cm}^{-1}$ in trigonal Te [8.97].

b) Compounds

a-Ge(S, Se)$_2$. In the absence of the requisite three independent measurements, diffraction studies have been unable to yield definitive information about the SRO in these materials. There is no evidence for larger structural units such as is observed in the arsenic chalcogenides. An EXAFS comparison between bulk and vapor-deposited a-GeSe$_2$ [8.98] suggests the existence of $\sim 5\%$ like neighbor bonds in the latter. This point is discussed further in Sect. 8.3.3.

The major features in the ir and Raman spectra have been interpreted in terms of the vibrations of a tetrahedral unit $GeX_{4/2}$ (X=S, Se) [8.20, 37]. The validity of this interpretation rests on a vibrational decoupling at the twofold-chalcogen-atom sites [8.76]. The Raman spectrum of a-GeSe$_2$ also displays an additional sharp polarized feature that cannot be explained in terms of the vibrational modes of the GeSe$_{4/2}$ molecular units; this mode then arises from a coupling of these molecular clusters [8.99, 100]. Two explanations have been put forth: (I) the mode, which occurs at about $220\,\text{cm}^{-1}$, is simply due to the Ge–Se–Ge units [8.99]; or (II) the mode is an indicator of larger molecular entities, specifically 12-membered rings of alternating Ge and Se atoms [8.100].

a-As$_2$(S, Se, Te)$_3$. The evidence for larger structural dimensions in these materials is discussed in Sects. 8.2.1a, 8.3.3a. On a local scale, diffraction studies on both bulk [8.65] and vapor-deposited [8.101] a-As$_2$S$_3$ have been interpreted as indicating the existence of like neighbor bonds in excess of the requirements of stoichiometry. This interpretation is supported by an EXAFS study [8.98], and will be discussed in Sect. 8.3.3a. Finally, an EXAFS study of vapor-deposited a-As$_2$Te$_3$ by *Pettifer* et al. [8.102] has yielded surprising evidence for two distinct and well-defined As–Te bond lengths. They interpret this as evidence for a metallic-to-covalent transition on moving from crystal to amorphous film.

The ir and Raman response are interpreted in terms of the local symmetry at the As sites, thereby relating the ir and Raman modes of the glass to the vibrations of the local molecular clusters, AsB$_{3/2}$ (B = S, Se) [8.36, 103–106]. The basis for this interpretation rests on a very weak vibrational coupling at the twofold-coordinated chalcogen-atom sites [8.36, 103, 107]. Similarities between the vibrational spectra of the amorphous and crystalline phases of both materials have been taken to imply some degree of a two-dimensional local topology in the glass phase [8.108, 109]. Studies of the ir absorption in these glasses have also established that the vibrational line shapes are Gaussian [8.110, 111] rather than Lorentzian as in crystalline solids.

a-Oxides. The initial controversy in these solids over random network versus microcrystalline models has developed into an active discussion concerning the range over which order exists in the amorphous solids. These issues are discussed in detail in *Wright* and *Leadbetter* [8.4], including the results of both diffraction and EXAFS studies. In addition, an EXAFS study of a-As$_2$O$_3$ by *Pettifer* and *McMillan* [8.112] has found little evidence for the existence of As$_4$O$_6$ molecular units in the glass, unlike the situation in the other arsenic chalcogenides (see the discussion earlier in this section).

Even though we have chosen not to emphasize insulating amorphous solids such as a-SiO$_2$, a-GeO$_2$, a-As$_2$O$_3$, and a-BeF$_2$, it is informative to comment on two aspects of their ir and Raman response. The first relates to the identification of features in the Raman spectrum that can be attributed to longitudinal optical phonons. This is established through a comparison of the Raman spectrum with the energy-loss spectrum, as obtained from a Kramers–Kronig analysis of the ir reflectance [8.113–115]. A second aspect relates to the structural interpretation of the features in the vibrational spectra. First, the ir and Raman spectra for a-SiO$_2$, a-GeO$_2$, and a-BeF$_2$ are well described by the model of *Bell* and *Dean* [8.23, 113–115], i.e., as the vibrational modes calculated by application of a simplified force field calculation to the "ball and stick" model of *Bell* and *Dean*. In making these comparisons, it is important to note that the *Bell–Dean* force field does not include the long-range electrostatic forces which give rise to the frequency separation between the longitudinal and transverse components of ir active modes. The longitudinal modes should not then be included in comparisons between the ir and Raman response and the

Bell–Dean one-phonon density-of-states calculation. A second point relates to the fact that the spectra for the SiO_2-like glasses cannot be explained in terms of the vibrations of a tetrahedral cluster, $AB_{4/2}$. This is a result of relatively strong coupling at the twofold-coordinated O and F sites [8.76, 116]. For a-SiO_2, it has been shown that a unit Si–O–Si provides a better molecular basis for describing the vibrational spectrum [8.116]. A similar situation prevails for a-As_2O_3. For this material, an interpretation of the spectra requires the consideration of two molecular groups, the $AsO_{3/2}$ pyramidal molecule, and the coupling through the O atoms as described by the vibrations of the water-molecule-like unit, As–O–As [8.38, 77].

c) Pseudobinary Alloys

a-As_2S_3–As_2Se_3, a-GeS_2–$GeSe_2$. The spectra of these alloy glasses have been shown to reflect their pseudobinary nature [8.117–119]. The alloy substitution can be understood in terms of a random distribution of S and Se atoms on the sites of the twofold-coordinated atoms of an otherwise idealized 3–2 or 4–2 network.

a-GeB_2–As_2B_3   **(B=S, Se).** These glasses also have vibrational spectra that are characteristic of a pseudobinary alloy. The vibrational modes in the ir and Raman spectra indicate Ge–B and As–B vibrations, and an absence of Ge–As modes [8.118–120]. This means that the local structure can be described as containing two local environments, $GeB_{4/2}$ and $AsB_{3/2}$, that are interconnected through the twofold-coordinated B-atom sites. As in the case of the end-member compound glasses, the vibrational coupling at these sites is very weak.

d) Binary Alloys

a-$Ge_{1-x}B_x$ **(B=S, Se).** Analysis of these spectra clearly demonstrate that the bond-counting statistics are dominated by chemical ordering [8.20–22], with Ge–B bonds favored at all alloy compositions within the glass-forming alloy regimes, $0.90 > x > 0.57$ for Ge–S, and $1.00 > x > 0.57$ for Ge–Se. The vibrational spectra in the B-rich alloy regimes, $x > x_c = 2/3$, are well described by a chain-crossing model, in which the two local molecular environments are $GeB_{4/2}$ and $B_{2/2}$ [8.20, 21]. In addition, the Ge–S spectra also have vibrational modes due to S_8 rings for high values of x, $0.90 > x > 0.80$ [8.20]. The spectra in the Ge-rich alloy regime yield an interesting new feature, a strongly polarized Raman mode that has been identified as characteristic of a new local environment $B_{3/2}Ge$–$GeB_{3/2}$ [8.20, 24]. Evidence for a chemically ordered compound phase at the composition Ge_2B_3 is also derived from other studies [8.32, 33].

a-$As_{1-x}(S, Se)_x$. In a diffraction study of a-Se as arsenic is added, *Renninger* and *Averbach* [8.121] found that the Se_8 rings tend to disappear and the low-q peak found in a-As_2Se_3 develops, suggesting a larger structural unit as discussed earlier.

The behavior in the (S, Se)-rich alloy regimes parallels that of the corresponding Ge glasses, with S_8 ring signatures readily evident in S-rich glasses of $As_{1-x}S_x$. The spectra in the As-rich regime are more difficult to interpret. Glass formation extends to rather high As-concentrations in $As_{1-x}Se_x$, to $x=0.4$, or 60% As. However, there are no sharp features evident in either the ir or Raman spectra which can be assigned to a specific molecular environment. On the other hand, there are features which are clearly attributable to As–As bonds [8.21, 22]. In contrast, the ir and Raman spectra of As-rich alloys in $As_{1-x}S_x$ clearly indicate a new molecular environment. It is either due to a network-forming unit of the form $S_{2/2}As$–$As\,S_{2/2}$, or to a discrete molecule As_4S_4 [8.21].

e) Other Systems

A number of other amorphous semiconductors have also been studied by ir and Raman spectroscopy with an emphasis on understanding the SRO. These systems include a-SiC [8.43], a-GaAs and other III–V systems [8.122, 123], a-$Ge_{1-x}Te_x$ [8.124], a-Sb [8.84, 85], and a-Ge–Sn [8.123, 125] alloys. A number of other ternary and quaternary alloys have also been investigated.

8.3 Recent Results

This section includes the results of some very recent experimental studies which give new information about the SRO in amorphous semiconductors. The emphasis in this section is on the nature of the local atomic environments at alloy atom sites which modify the electronic structure: H in a-Si; dopants from column III or V in a-Si; and finally structural and bonding defects in chalcogenide and pnictide materials.

8.3.1 Local Environments in a-Si

A recurrent theme in many of the chapters in this text is the technological breakthrough achieved by *Spear* and co-workers [8.6, 126] in the substitutional doping of a-Si. This in turn has lead to the fabrication of *p–n* junction devices [8.127] with a particular emphasis on solar energy conversion [8.128]. A number of very interesting questions relate to the substitutional doping of a-Si. Clearly H can play an important role in reducing the number of electronically active defects [8.47–50]; however, it is not completely certain that the H incorporation is necessary for all conditions of preparation [8.126]. Indeed one of the most interesting aspects of current research on a-Si and a-$Si_{1-x}H_x$ alloys concerns the relationships between the preparative technique and the resultant local atomic structure, in particular the complex interaction between defect states, H, and the dopants. This section addresses two of these issues: the local environments at the H sites, and at a characteristic *n* type dopant, As. The first study is based on ir and Raman spectroscopy, whereas the second is based on EXAFS.

a) Vibration Spectroscopy of a-Si$_{1-x}$H$_x$ Alloys

A number of groups [8.51–54] have studied the ir absorption and Raman scattering of a-Si containing substantial amounts of bonded H, from about 8 to 50 at. %. For this concentration range of atomic percent H, it is appropriate to describe the material as a Si–H binary alloy, a-Si$_{1-x}$H$_x$. This binary alloy can then be characterized as a 4–1 system. Since the singly coordinated H atoms act as terminators, it is convenient to describe the local atomic structure by first describing the multiplicity of H atoms at the Si sites, and then the resultant network topology. H atoms can be included in a-Si films in several different ways: by sputtering in a H ambient [8.47] and by glow discharge decomposition of SiH$_4$ [8.128–130]. Although the details of the ir absorption and Raman scattering differ quantitatively between samples produced in various ways, it is nevertheless possible to identify four qualitatively different types of local atomic configurations. However, before discussing these local environments in Si–H, it is informative to review some of the results obtained by *Connell* and *Pawlik* [8.75] for the local atomic structure of H in Ge. They obtained spectroscopic evidence for two different local environments, which they interpreted as: single H atoms at isolated defect sites (in effect, compensated dangling bonds) and single H attachment to Ge defect sites on the surface of an internal void. This interpretation was based on the lack of ir absorption in the spectral range from 700 to 900 cm^{-1}. Multiple attachment of H to Ge, as for example in GeH$_2$ sites, would be expected to yield a bond-bending mode in this region.

In contrast, spectroscopic studies of the local atomic structure in Si$_{1-x}$H$_x$ alloys yield evidence for multiple as well as single-H-atom attachment [8.51–54, 131]. Before discussing the vibrational spectra, it is useful to establish the nature of the vibrational modes expected at sites containing one, two, or three H atoms: SiH, SiH$_2$, and SiH$_3$, respectively. Figure 8.6 illustrates the atomic motions of these modes [8.51, 52, 131]. There are basically two types: those involving changes in the Si–H bond length (bond stretching) or H–Si–H bond angle (bond bending), and those involving the rotation of these groups as a rigid unit (rocking, wagging, or twisting modes). H can be incorporated into a-Si in a number of different ways: by the glow-discharge decomposition of SiH$_4$, or by sputtering in a H ambient. Figure 8.7 illustrates the ir absorption spectra for films of a-Si produced from the glow discharge decomposition of SiH$_4$ and deposited onto c-Si substrates held at 230 °C. The absorption spectra are different according to the rf power coupled into the system [8.51–53, 131] and to the bias (A is the anode, and C the cathode) of the surface on which the a-Si:H alloy is condensed; the hydrogen concentrations in these films range from 7 to 15 atomic percent [8.131].

There are two ir absorption bands for the SiH environment: a bond-stretching mode at 2000 cm^{-1}, and a bond-bending (rocking or wagging) mode at 630 cm^{-1}. These bands dominate in the 1-W (A) sample. There are three new vibrations that are assigned to the SiH$_2$ group: a bond-stretching mode at

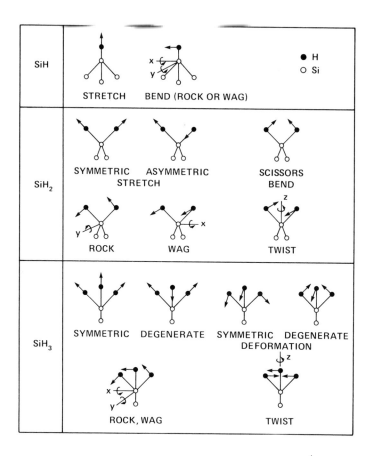

▲
Fig. 8.6. Local Si–H vibrations for SiH, SiH$_2$, and SiH$_3$ groups [8.51,52,131]

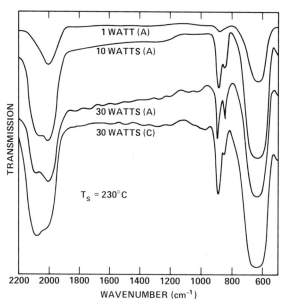

Fig. 8.7. The ir transmission of a-Si:H for $T_s = 230\,°C$ [8.131]

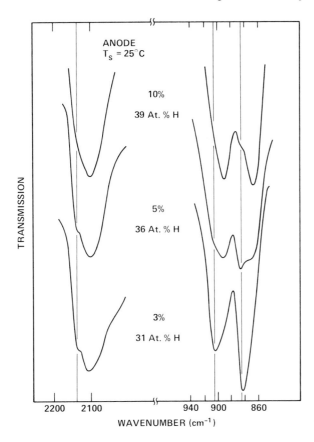

Fig. 8.8. The ir transmission of a-Si:H for $T_S = 25$ °C. The relative dilution of SiH_4 in Ar is given, as is the atomic percent of hydrogen [8.131]

2090 cm^{-1}, a scissors-type bond-bending mode at 890 cm^{-1}, and a weakly ir-active bond-wagging mode at 845 cm^{-1}. In addition there is a broad bond-rocking absorption at 630 cm^{-1}. Since this mode is at the frequency of the bond-bending mode of SiH, it does not aid in establishing the nature of the Si–H site. These SiH_2 features are evident in the 30-W (C) sample. In addition that sample contains isolated SiH groups, as evidenced by the absorption at 2000 cm^{-1}. The $(SiH_2)_n$ environment is characterized by an increase in the relative ir absorption of the 845 cm^{-1} bond-wagging mode. The behavior of the corresponding CH_2 bond-wagging mode is very similar in plasma-deposited polyethylene [8.132–134]. Films produced at substrate temperatures in excess of 200 °C do not exhibit SiH_3 vibrations [8.51, 52, 133]; however SiH_3 modes are clearly evident at samples produced on room-temperature substrates (see Fig. 8.8). These samples typically contain at least 30 atomic percent H. The bond-stretching vibration is at 2140 cm^{-1} and the two bond-bending modes are at 862 cm^{-1} (symmetric deformation) and 907 cm^{-1} (degenerate defor-mation). The frequency difference between the two bending modes of the SiH_3 group, 45 cm^{-1}, is very nearly equal to the frequency difference between the

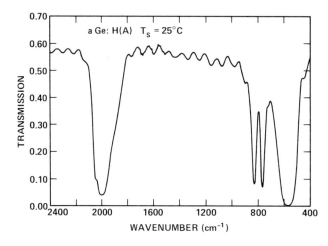

Fig. 8.9. The ir transmission of a-Ge:H for $T_S = 25\,^\circ$C [8.131]

SiH$_2$ scissors-type-bond-bending mode and bond-wagging mode. Note further that the Si–H bond-stretching frequencies in a-Si:H alloys are smaller than those found in silane molecules [8.135], but fall within the range of frequencies observed for H vibrations in c-Si [8.136]. The values for the bond-stretching frequencies in a-Si have been explained by an extension of the induction-effect mechanism which accounts for the shifts in frequency of SiH vibrations in substituted silane molecules, SiHR$_1$R$_2$R$_3$ [8.135]. The frequency of the SiH vibration increases as the sum of the electronegativities of the substituting R$_j$ atoms or groups is increased; it is 2314.5 cm^{-1} in SiHF$_3$ and decreases to 2118 cm^{-1} in SiH(CH$_3$)$_3$ [8.135].

H has also been incorporated in a-Ge via the glow discharge decomposition of GeH$_4$. The absorption spectra obtained in this way are qualitatively similar to those of a-SiH produced under the same set of deposition conditions [8.133]. Figure 8.9 displays the absorption spectrum for a sample deposited onto a room-temperature substrate. The absorption at 1976 cm^{-1} is assigned to GeH$_2$ and that at 2050 cm^{-1} to GeH$_3$. The two modes at 770 cm^{-1} and 830 cm^{-1} contain contributions from both GeH$_2$ and GeH$_3$ groups. The analysis of this spectrum then yields a different explanation for the absorption at 1976 cm^{-1} than that discussed above. *Connell* and *Pawlik* [8.75] attributed the absorption at 1976 cm^{-1} to a GeH group, whereas the comparisons with a-Si:H suggest it is due to GeH$_2$. The difference between the films studied in [8.75] and [8.131] is in the H content – less than 8 atomic percent in [8.75], but at least 30 atomic percent in [8.131]. This then accounts for the relatively weaker bond-bending mode not being observed in the earlier study.

b) Coordination of Arsenic Dopants in a-Si:H Alloys

It was believed for many years that amorphous semiconductors could not exhibit effects similar to those found in the substitutional doping of crystalline semiconductors [8.137]. Recently, however, large conductivity changes and

Table 8.1 Composition of samples deposited from silane-arsine mixtures

Percent AsH$_3$ in mixture	Substrate temperature T_s [°C]	Atomic percent arsenic	Atomic percent hydrogen	Atomic percent silicon
0.5	235	0.9	13.2	85.9
1.0	235	1.7	13.6	84.7
5.0	235	6.0	15.6	78.4
10.0	235	11.7	17.6	70.7
1.0	25	2.6	34.6	62.8

other effects characteristic of such doping have been observed [8.7, 126, 138] when phosphorous, boron, and arsenic are incorporated in thin films of amorphous "silicon" prepared by plasma decomposition of silane. In addition, changes in electronic properties indicative of a transition from doping to alloying behavior are observed as a function of increasing impurity concentration in the case of arsenic [8.138]. A study of the local environment about the arsenic atoms might well reveal the presence of fourfold-coordinated, or substitutional, arsenic in these materials, and might further reveal the active fraction of As. The low concentrations of As in these alloys, of the order of 1 %, precludes the use of standard diffraction techniques. Accordingly, *Knights* et al. [8.55] undertook an EXAFS study. This work is summarized below in sufficient detail to illustrate the capabilities of this technique.

EXAFS and hydrogen-evolution measurements were performed [8.55] on five As-doped a-Si:H alloys. The first information extracted from these measurements was the precise chemical compositions, shown in Table 8.1. The hydrogen content was determined by an evolution measurement and the arsenic content from the As K-shell absorption step height in the EXAFS measurement. Not only is the hydrogen content of these films large, but, as is discussed later, the hydrogen is bonded. Thus these films are truly silicon–hydrogen alloys.

The Fourier transform of the EXAFS, $\varphi^E(r)$, from the As K-shell absorption is shown in Fig. 8.10 for the a-Si:H alloy doped with 1.7 atomic percent As, crystalline SiAs, and a-As:H. The latter two materials were measured to obtain the peak functions necessary to extract the structural information from EXAFS, as discussed in conjunction with (8.2). φ^E is a complex function, and both its real part and magnitude are shown. The first shell of nearest neighbors gives rise to the prominent feature near 2 Å in each case. Note the complete absence of structure beyond 3 Å in the two amorphous samples. This is strong evidence for the lack of any order extending beyond the first nearest neighbors about the As sites – specifically, the second-nearest-neighbor distribution must be at least 0.2 Å broader than the first. This rules out an explanation of the electronic behavior based on long-range order, or some quasicrystallinity.

Examining the nearest neighbors of As more closely, bear in mind that As in c-SiAs has only Si nearest neighbors. The specific shape of the peak at ∼2 Å in

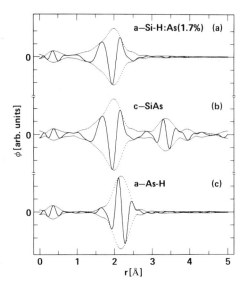

Fig. 8.10a–c. The real part (solid line) and the magnitude of the Fourier transform $\phi(r)$ of the EXAFS on the As K-shell absorption in (**a**) 1.7 at % As in amorphous Si–H; (**b**) c-SiAs, and (**c**) amorphous As–H. The units of all three vertical scales are arbitrary but identical

Fig. 8.10b, both real part and magnitude, is the peak function which characterizes a Si nearest neighbor. Similarly, the peak at ∼2.2 Å in Fig. 8.10c characterizes an As nearest neighbor. Note, for example, that the peak at ∼3.3 Å in Fig. 8.10b is clearly As, as expected. Considering the EXAFS from the doped sample, in Fig. 8.10a, it is clear that the nearest neighbors of As are Si. More precisely, using a fitting technique described in detail elsewhere [8.55], it has been shown that, in the 1.7 % As sample, As has 2.80 ± 0.04 Si nearest neighbors at a mean distance of 2.38 ± 0.01 Å.

The detailed analysis was repeated for each sample, and the resulting coordination numbers are shown in Fig. 8.11 as open and closed circles. The coordination distances are identical to within 0.002 Å. Note that only Si nearest neighbors are found in each case. For the 12 % As sample, an estimate was made of the upper bound on the number of As–As nearest neighbors. It was found to be less than half that expected for randomly selected nearest neighbors, strong evidence for unlike neighbor bonds being energetically preferred. There are two important features evident in Fig. 8.11: (I) the number of Si nearest neighbors of As decreases monotonically with increasing As concentration; (II) the Si coordination of As at high concentrations is less than three, the lowest valence of As.

At this juncture, the effect of hydrogen on the EXAFS results must be considered. Since hydrogen does not scatter electrons effectively [8.139], it will not yield an EXAFS signal. Each H bonded to As will reduce the apparent number of As neighbors by one. In order to quantify this reduction, assumptions must be made concerning the bonding to and the distribution of hydrogen between As and Si. The observation of strong infrared absorption bands in these samples corresponding to bonded hydrogen (the As–H and Si–H

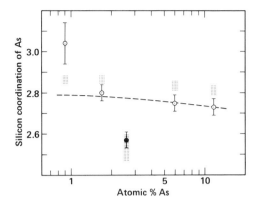

Fig. 8.11. Silicon coordination of As as a function of As concentration in amorphous Si–H:As compounds. Points represent values determined from EXAFS; shaded bars represent range of values predicted from hydrogen content (see text)

frequencies are too close to differentiate) suggests that essentially all the H is bonded. The H is assumed to be divided between Si and As equally on either a per atom or a per bond basis. The range of reduced coordinations predicted by these models is represented by the shaded bars in Fig. 8.11, the lowest value being that predicted by the "per atom" model.

Note that the model values parallel the systematic decrease in Si coordination of As with increasing As concentration, and account quite well for the reduced coordination found in the sample deposited at 25 °C. In addition, at the lowest concentration, the measured Si coordination is 0.2 ± 0.1 atom higher than predicted by these trends. Since there is no evidence to suggest a change in the manner in which hydrogen is bonded at low arsenic concentrations, this difference is attributed to an increase in the total coordination of As. Specifically, 20 % of the As atoms are fourfold coordinated with Si. The onset of fourfold coordination with decreasing As coordination coincides with the enhancement in electrical conductivity [8.138]. This is analogous to the substitutional doping found in crystalline semiconductors – except for the pronounced absence of long-range order.

In summary, this EXAFS study revealed the degree and nature of H incorporation in these materials, properly described as silicon–hydrogen alloys. The SRO was shown to reflect a preference for unlike atom bonds, and to be radially disordered at the second nearest neighbor. Finally, the enhancement of electronic conductivity was shown to correlate with the onset of fourfold coordination, providing the first direct evidence for substitutional doping in an amorphous semiconductor.

8.3.2 Bonding Coordination Defects in a-As

Infrared absorption [8.83] and Raman [8.19, 84, 85] spectra for bulk a-As are shown in Fig. 8.12. The dominant features of the spectra, indicated by the solid arrows and occurring at 160, 145, 200, and 230 cm^{-1}, have been assigned to the vibrational modes of threefold-coordinated As atoms. This assignment is based

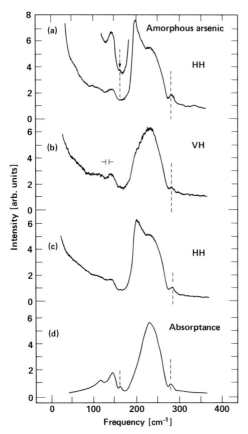

Fig. 8.12a-d. Polarized Raman spectra and ir-absorption of amorphous As. The spectra in (a) and (b) and the absorptance (d) were obtained from bulk amorphous As while (c) is the HH Raman spectra of a glow discharge produced As film

on theoretical calculations of the one-phonon density of states, ir absorption and Raman scattering [8.86, 87]. None of the calculations can account for the weak sharp spectral features at 165 and 280 cm^{-1}. These features are not due to vibrational modes of discrete molecular components of the bulk a-As, e.g., As$_4$ pyramidal molecules, or to chemical impurities. It has been proposed that the vibrational modes are due to bond coordination defects [8.19]. Calculations based on the cluster-Bethe-lattice method [8.19, 140], indicate that neither a one- nor twofold-coordinated defect could yield the observed spectral features. However, the same calculations, as shown in Fig. 8.13 indicate that a fourfold-coordinated center can give rise to both spectral features. Estimates of the defect density range from 10^{18} to 10^{20} cm^{-3}. The major uncertainty is in the magnitude of matrix-element differences between threefold- and fourfold-coordinated As sites. The study [8.19] by ir and Raman spectroscopy then represents the first direct spectroscopic evidence for bond-coordination defects in an amorphous semiconductor. One expects bond-coordination defects to come in pairs [8.6], so that it is natural to expect the positively charged fourfold-coordinated sites in As to be "compensated" by equal numbers of negatively

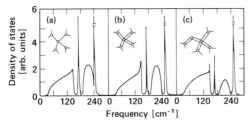

Fig. 8.13a-c. The vibrational DOS for a tetrahedrally coordinated defect (obtained by averaging over the local DOS of the fourfold coordinated atom and its nearest neighbors) before (**a**) and after (**b**) relaxation. (**c**) The DOS for a relaxed nearest-neighbor twofold and fourfold coordinated defect (obtained by averaging over the local DOS of the twofold and fourfold coordinated atoms). The Bethe lattice is again used [in (**a**), (**b**) and (**c**)] to simulate the amorphous structure

charged twofold-coordinated sites. Included in Fig. 8.13 is the calculated one-phonon density of states for a near-neighbor pair of fourfold- and twofold-coordinated defect sites. The calculated spectrum is qualitatively similar to that of an isolated fourfold-coordinated site.

8.3.3 Wrong Bonds in Vapor Deposited Chalcogenide Amorphous Semiconductors

We have already noted that compound amorphous semiconductors such as a-As_2S_3 contain primarily As–S bonds when prepared as bulk-quenched glasses. Very small numbers of wrong bonds relative to a chemically ordered network may also be present, as evinced by very weak Raman features at 490 and 230 cm^{-1}. The 490 cm^{-1} feature is attributed to S–S bonds and the 230 cm^{-1} to As–As bonds [8.104, 105]. Films of nominally the same compound composition, but prepared by vapor deposition onto substrates held at temperatures below the glass transition temperature, display a very different bond-type distribution. This is evident in studies of the SRO by X-ray diffraction [8.31] and by Raman scattering [8.59, 60]. In this section we discuss each of these studies of the SRO.

a) Annealing Effects on SRO

Irreversible changes upon annealing have been observed in both the structure and the optical properties of as-evaporated a-As_2S_3 and a-As_2Se_3 thin films [8.66, 141]. The $qi(q)$ diffraction measured on as-deposited samples exhibits a sharp peak at $q \sim 1.2$ Å^{-1} and a broad peak at $q \sim 2.2$ Å^{-1}. On annealing, the first peak decreases substantially while both peaks broaden and move to larger q. This change has been attributed to the polymerization of either As_4S_6 or As_4Se_6 molecules, as appropriate [8.66]. With the aim of examining nearest-neighbor changes on annealing, an EXAFS study was undertaken on evaporated amorphous As_2S_3, As_2Se_3, and $GeSe_2$ [8.60]. The EXAFS structural

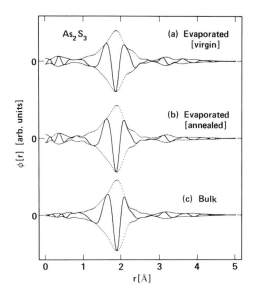

Fig. 8.14a–c. The real part (solid line) and the magnitude of the Fourier transform $\phi(r)$ of the EXAFS on the As K-shell absorption in amorphous thin film As_2S_3 (a) as-evaporated and (b) annaled, and (c) in bulk glassy As_2S_3. The units of all three vertical scales are arbitrary but identical

analysis is simplified by the influence of only two pair densities on each measured spectrum, instead of three as in a diffraction study. In addition, the peak functions of As and S are so radically different that the As environment in a-As_2S_3 is easily determined.

The Fourier transforms of the EXAFS from the As K-shell absorption are shown in Fig. 8.14 for a vapor-deposited As_2S_3 thin film, both as-deposited and after annealing, and for the corresponding bulk glass. Two effects are obvious: the annealed film and bulk glass are very similar in nearest-neighbor environment about As; the as-deposited film has a substantial additional feature on the large-r side of the first neighbor peak. A detailed analysis of these peaks [8.60] has led to the following picture. The first neighbor shell about As in the as-deposited film contains 2.4 S atoms at 2.26 Å and 0.6 As atoms at 2.54 Å. The presence of the As atoms is clearly indicated by the As peak function, which is quite distinct from the S peak function. Upon annealing, the first neighbor shell changes to 2.8 S atoms at 2.27 Å and 0.2 As atoms at 2.56 Å. These remaining As atoms correspond to the reduced bump on the high-r side of the first peak in Fig. 8.14b. The as-deposited films have a substantial number of wrong bonds (As–As), most of which are eliminated on annealing. The residual amount after annealing may be due to a slight departure from the nominal composition of As_2S_3. These results are consistent with the findings of *Leadbetter* and *Apling* [8.65] that there are like-neighbor bonds in evaporated a-As_2S_3 in excess of the requirements of stoichiometry.

The EXAFS measurements on As_2Se_2 and $GeSe_2$ are much more difficult to interpret definitively because the scattering signals of the constituents are nearly identical in shape and position. Consequently, the analysis was limited to a simple question: does adding $n\%$ wrong bonds improve the comparison

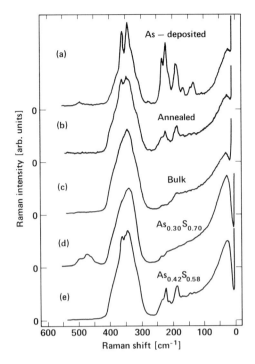

Fig. 8.15a–e. Raman spectra for (**a**) as-deposited As_2S_3 film, (**b**) an annealed film, and for bulk quenched glasses (**c**) As_2S_3 (**d**) $As_{0.3}S_{0.7}$ and (**e**) $As_{0.42}Se_{0.58}$

between the bulk glass and the as-deposited film? In the case of As_2Se_3, the comparison between the bulk glass and the as-deposited film suggests that there may be as many as 10–15% like nearest neighbors or wrong bonds, compared with the strongly indicated 20% in As_2S_3. In the case of $GeSe_2$, the EXAFS gave no indication of wrong bonds in the as-deposited film. These results have been combined with the results of ir and Raman studies in a model for the change in optical properties on annealing [8.60, 142].

In an earlier EXAFS study, *Sayers* et al. [8.98] found evidence for $\sim 5\%$ wrong bonds in $GeSe_2$ and $\sim 35\%$ wrong bonds in As_2Se_3. These results for a-As_2Se_3 must be considered as different from those already mentioned [8.60], even given the difficulty in interpreting this experiment. It is possible, of course, that this reflects a real difference in the samples.

b) Raman Scattering from a-As_2S_3 Vapor-Deposited Thin Films

Figure 8.15 indicates a comparison between the Raman spectra of a bulk sample of a-As_2S_3 and an as-deposited thin film [8.59, 60]. The thin film was prepared by an open-boat evaporation of a-As_2S_3 onto an unheated substrate. The spectrum of the as-deposited film clearly contains features not evident in a bulk sample. These additional features occur in three frequency regimes: as a weak feature at about $490 \, cm^{-1}$; as two sharp features superposed on the main a-As_2S_3 Raman mode (at about $340\,cm^{-1}$); and finally as a number of sharp modes, approximately six, between 100 and $250\,cm^{-1}$. Also shown in the figure is the Raman spectrum of the thin film subsequent to annealing. The annealed film more closely resembles the bulk glass.

The nature of the local order in the as-deposited film has been established by comparison with Raman spectra of glasses in the $As_{1-x}S_x$ alloy system from the S-rich ($x=0.70$) and As-rich ($x=0.58$) alloy regimes. The feature at $490\,cm^{-1}$ in the as-deposited film is then identified as being due to S–S vibrations, whereas the other features are associated with molecular clusters having As–As bonds. It was originally suggested [8.59] that the sharp features in the as-deposited glass were due to a molecular species As_4S_6 which does not include As–As bonds. Further comparisons with the Raman spectra of As-rich glasses in the alloy system $As_{1-x}S_x$ [8.60] suggest that the sharp features in the thin films are either due to a network forming cluster $S_{2/2}\,As$–$As_{2/2}$ or the molecule As_4S_4 [8.21]. Comparisons with the Raman spectra of realgar, a molecular crystal based on As_4S_4, suggest that an explanation based on the molecule is more probable [8.119].

8.4 Summary

We have attempted to describe the SRO in amorphous semiconductors with an eye toward the future development of the field. The emphasis in the past has been on defining the local structure of representative classes of amorphous solids, primarily elements and compounds. One aspect of current and most probably future interest is in the control of the electronic properties of amorphous semiconductors by chemical modification [8.8] and by doping [8.7]. We have focused on this emerging area of interest through the emphasis in our general discussion of probes of SRO, and through the specific examples of Sect. 8.3. There are other techniques that give information about defects and impurities. We have however emphasized those techniques which relate directly to the nature of the local atomic environments rather than to purely electronic or photoelectronic properties, which although they depend on the local atomic environments, do not yield specific structural information.

References

8.1   D.Turnbull, D.E.Polk: J. Non-Cryst. Solids **8–10**, 19 (1972)
8.2   G.S.Cargill III: Solid State Phys. **30**, 227 (1975)
8.3   A.C.Wright: In *Advances in Structure Research by Diffraction Methods*, Vol. 5, ed. by W.Hoppe, R.Mason (Pergamon Press, Oxford 1974) p. 1
8.4   A.C.Wright, A.J.Leadbetter: Phys. Chem. Glasses **17**, 122 (1976)
8.5   N.F.Mott, E.A.Davis, R.A.Street: Philos. Mag. **32**, 961 (1975)
       R.A.Street, N.F.Mott: Phys. Rev. Lett. **35**, 1293 (1975)
8.6   M.Kastner, D.Adler, H.Fritzsche: Phys. Rev. Lett. **37**, 1504 (1976)
8.7   W.E.Spear, P.G.LeComber: Solid State Commun. **17**, 1193 (1975)
8.8   S.R.Ovshinsky: In *Amorphous and Liquid Semiconductors*, ed. by W.E.Spear (G.G.Stevenson, Dundee 1977) p. 519

8.9   N.F.Mott: Adv. Phys. **16**, 49 (1967)

8.10   D.E.Polk: J. Non-Cryst. Solids **5**, 365 (1971)

8.11   N.J.Shevchik, W.Paul: J. Non-Cryst. Solids **8–10**, 381 (1972)

8.12   D.Henderson, F.Herman: J. Non-Cryst. Solids **8–10**, 359 (1972)

8.13   D.E.Polk, D.S.Boudreaux: Phys. Rev. Lett. **31**, 92 (1973)

8.14   G.N.Greaves, E.A.Davis: Philos. Mag. **29**, 1201 (1974)

8.15   M.Long, P.Galison, R.Alben, G.A.N.Connell: Phys. Rev. B**13**, 1821 (1976)

8.16   G.Lucovsky, A.Mooradian, W.Taylor, G.B.Wright, R.C.Keezer: Solid State Commun. **5**, 113 (1967)

8.17   A.Mooradian, G.B.Wright: In *Physics of Selenium and Tellurium*, ed. by W.C.Cooper (Pergamon Press, Oxford 1967) p. 269

8.18   M.Gorman, S.A.Solin: Solid State Commun. **18**, 1401 (1976)

8.19   R.J.Nemanich, G.Lucovsky, W.Pollard, J.D.Joannopoulos: Solid State Commun. **26**, 137 (1978)

8.20   G.Lucovsky, F.L.Galeener, R.C.Keezer, R.H.Geils, H.A.Six: Phys. Rev. B**10**, 5134 (1974)

8.21   G.Lucovsky, F.L.Galeener, R.H.Geils, R.C.Keezer: In *The Structure of Non-Crystalline Materials*, ed. by P.H.Gaskell (Taylor and Francis, London 1977) p. 127

8.22   P.S.E.Ewen, M.Sik, A.E.Owen: In *The Structure of Non-Crystalline Materials*, ed. by P.Gaskell (Taylor and Francis, London 1977) p. 231

8.23   R.J.Bell, P.Dean: Discuss. Faraday Soc. **50**, 55 (1970)

8.24   G.Lucovsky, R.J.Nemanich, F.L.Galeener: In *Amorphous and Liquid Semiconductors*, ed. by W.E.Spear (G.G.Stevenson, Dundee 1977) p. 130

8.25   S.G.Bishop, U.Strom, P.C.Taylor: Phys. Rev. Lett. **34**, 1346 (1975); Phys. Rev. Lett. **36**, 543 (1976)

8.26   S.G.Bishop, U.Strom: In *Optical Properties of Highly Transparent Solids*, ed. by S.S.Mitra, B.Bendow (Plenum, New York 1975) p. 317

8.27   F.Betts, A.Bienenstock, S.R.Ovshinsky: J. Non-Cryst. Solids **4**, 554 (1970)

8.28   S.C.Rowland, S.C.Narasimhan, A.Bienenstock: J. Appl. Phys. **43**, 2741 (1972)

8.29   L.Cervinka, A.Hruby: In *Amorphous and Liquid Semiconductors*, ed. by J.Stuke, W.Brenig (Taylor and Francis, London 1974) p. 431

8.30   R.W.Fawcett, C.N.J.Wagner, C.S.Cargill III: J. Non-Cryst. Solids **8–10**, 369 (1972)

8.31   G.Yu.Poltavtsev, V.M.Pozdnyadova: Ukr. Fiz. Zh. **18**, 679 (1973)

8.32   A.Feltz, B.Voight, E.Schlenzig: In *Amorphous and Liquid Semiconductors*, ed. by J.Stuke, W.Brenig (Taylor and Francis, London 1974) p. 261

8.33   A.Feltz, B.Voight, W.Burckhardt, L.Senf, G.Leonhardt: In *Proc. 6th Int. Conf. on Amorphous and Liquid Semiconductors*, ed. by B.T.Kolomiets (Nauka, Leningrad 1976) p. 88

8.34   F.Ordway: Science **143**, 800 (1964)

8.35   D.L.Evans, S.R.King: Nature (London) **212**, 1353 (1966)

8.36   G.Lucovsky, J.P.deNeufville, F.L.Galeener: Phys. Rev. B**8**, 5947 (1973)

8.37   G.Lucovsky, R.M.Martin: J. Non-Cryst. Solids **8–10**, 185 (1972)

8.38   G.Lucovsky, R.M.White: Phys. Rev. B**8**, 660 (1973)

8.39   A.T.Ward: J. Phys. Chem. **72**, 744, 4133 (1968)

8.40   D.W.Scott, J.P.McCullough, F.H.Kruse: J. Molec. Spectrosc. **13**, 313 (1964)

8.41   H.R.Philipp: J. Non-Cryst. Solids **8–10**, 627 (1972)

8.42   N.J.Shevchik: Phys. Rev. Lett. **31**, 1245 (1973)

8.43   M.Gorman, S.A.Solin: Solid State Commun. **15**, 761 (1974)

8.44   H.Fritzsche: In *Amorphous and Liquid Semiconductors*, ed. by W.E.Spear (G.G.Stevenson, Dundee 1977) p. 3

8.45   M.H.Brodsky, R.S.Title: Phys. Rev. Lett. **23**, 581 (1969)

8.46   M.H.Brodsky, R.S.Title: In *Structure and Excitations in Amorphous Solids*, ed. by G.Lucovsky, F.L.Galeener (American Institute of Physics, New York 1976) p. 97

8.47   W.Paul, A.J.Lewis, G.A.N.Connell, T.Moustakas: Solid State Commun. **20**, 969 (1976)

8.48   M.Araki, H.Ozaki: Solid State Commun. **18**, 1603 (1976)

8.49   P.Nath, S.K.Barthwal, K.L.Chopra: Solid State Commun. **16**, 301, 723 (1975)

8.50 P.G.LeComber, R.J.Loveland, W.E.Spear, R.A.Vaughan: In *Amorphous and Liquid Semiconductors*, ed. by J.Stuke, W.Brenig (Taylor and Francis, London 1974) p. 245

8.51 M.H.Brodsky, M.Cardona, J.J.Cuomo: Phys. Rev. B**16**, 3556 (1977)

8.52 J.C.Knights, G.Lucovsky, R.J.Nemanich: Philos. Mag. B**37**, 467 (1978)

8.53 C.C.Tsai, H. Fritzsche, M.H.Tanielian, P.J.Gaczi, P.D.Persans, M.A.Vesaghi: In *Amorphous and Liquid Semiconductors*, ed. by J.Stuke, W.Brenig (Taylor and Francis, London 1974) p. 339

8.54 P.J.Zanucchi, C.R.Wronski, D.E.Carlson: J. Appl. Phys. **48**, 5227 (1977)

8.55 J.C.Knights, T.M.Hayes, J.C.Mikkelsen,Jr.: Phys. Rev. Lett. **39**, 712 (1977)

8.56 P.W.Anderson: Phys. Rev. Lett. **34**, 953 (1976)

8.57 M.Kastner: In *Amorphous and Liquid Semiconductors*, ed. by J.Stuke, W.Brenig (Taylor and Francis, London 1974) p. 504

8.58 M.Kastner, H.Fritzsche: Philos. Mag. B**37**, 199 (1978)

8.59 S.A.Solin, G.N.Papatheodorou: Phys. Rev. B**15**, 2084 (1977)

8.60 R.J.Nemanich, G.A.N.Connell, T.M.Hayes, R.A.Street: Phys. Rev. B **18**, 6900 (1978)

8.61 S.R.Ovshinsky: In *Structure and Excitations in Amorphous Solids*, ed. by G.Lucovsky, F.L.Galeener (American Institute of Physics, New York 1976) p. 31

8.62 J.E.Enderby, D.M.North, P.A.Egelstaff: Philos. Mag. **14**, 961 (1966)

8.63 R.L.Mozzi, B.E.Warren: J. Appl. Crystallogr. **2**, 164 (1969)

8.64 S.G.Bishop, N.J.Shevchik: Solid State Commun. **15**, 629 (1974)

8.65 A.J.Leadbetter, A.J.Apling: J. Non-Cryst. Solids **15**, 250 (1974)

8.66 J.D.deNeufville, S.C.Moss, S.R.Ovshinsky: J. Non-Cryst. Solids **13**, 191 (1974)

8.67 F.W.Lytle, D.E.Sayers, E.A.Stern: Phys. Rev. B**11**, 4825 (1975)
      E.A.Stern, D.E.Sayers, F.W.Lytle: Phys. Rev. B**11**, 4836 (1975)

8.68 T.M.Hayes, P.N.Sen, S.H.Hunter: J. Phys. C**9**, 4357 (1976)

8.69 J.Jaklevic, J.A.Kirby, M.P.Klein, A.S.Robertson, G.S.Brown, P.Eisenberger: Solid State Commun. **23**, 679 (1977)

8.70 P.H.Citrin, P.Eisenberger, R.C.Hewitt: Phys. Rev. Lett. **41**, 309 (1978)

8.71 G.Lucovsky: In *Amorphous and Liquid Semiconductors*, ed. by J.Stuke, W.Brenig (Taylor and Francis, London 1974) p. 1009

8.72 M.H.Brodsky: In "*Light Scattering in Solids*," Top. Appl. Phys. Vol. 8, ed. by M.Cardona (Springer, Berlin, Heidelberg, New York 1975) p. 208

8.73 S.A.Solin: In *Structure and Excitations in Amorphous Solids*, ed. by G.Lucovsky, F.L. Galeener (American Institute of Physics, New York 1976) p. 205

8.74 G.Lucovsky, F.L.Galeener: In *Proc. 6th Int. Conf. on Amorphous and Liquid Semiconductors*, ed. by B.T.Kolomiets (Nauka, Leningrad 1976) p. 207

8.75 G.A.N.Connell, J.R.Pawlik: Phys. Rev. B**13**, 787 (1976)

8.76 P.N.Sen, M.F.Thorpe: Phys. Rev. B**15**, 4030 (1977)

8.77 G.N.Papatheodorou, S.A.Solin: Phys. Rev. B**13**, 1741 (1976)

8.78 S.C.Moss: In *Amorphous and Liquid Semiconductors*, ed. by J.Stuke, W.Brenig (Taylor and Francis, London 1974) p. 17

8.79 J.E.Smith,Jr., M.H.Brodsky, B.L.Crowder, M.I.Nathan, A.Pinczuk: Phys. Rev. Lett. **26**, 642 (1972)
      J.E.Smith,Jr., M.H.Brodsky, B.L.Crowder, M.I.Nathan: J. Non-Cryst. Solids **8–10**, 179 (1972)

8.80 M.Wihl, M.Cardona, J.Tauc: J. Non-Cryst. Solids **8–10**, 172 (1972)

8.81 R.Alben, D.Weaire, J.E.Smith,Jr., M.H.Brodsky: Phys. Rev. B**11**, 2271 (1975)

8.82 See, for example, P.M.Smith, A.J.Leadbetter, A.J.Apling: Philos. Mag. **31**, 57 (1975)

8.83 G.Lucovsky, J.C.Knights: Phys. Rev. B**10**, 4324 (1974)

8.84 J.S.Lanin: In *Structure and Excitations in Amorphous Solids*, ed. by G.Lucovsky, F.L. Galeener (American Institute of Physics, New York 1976) p. 123

8.85 J.S.Lanin: Phys. Rev. B**15**, 3863 (1977)

8.86 J.D.Joannopoulos: In *Structure and Excitations in Amorphous Solids*, ed. by G.Lucovsky, F.L.Galeener (American Institute of Physics, New York 1976) p. 108

8.87 J.D.Joannopoulos, W.Pollard: Solid State Commun. **20**, 947 (1976)

8.88 J.C.Malaurent, J.Dixmier: In *The Structure of Non-Crystalline Materials*, ed. by P.H. Gaskell (Taylor and Francis, London 1977) p. 49

8.89 A.I.Andrievskii, I.D.Nabitovich, Ya.V.Voloshchuk: Sov. Phys. Crystallog. **5**, 349 (1960)

8.90 M.D.Rechtin, B.L.Averbach: J. Non-Cryst. Solids **12**, 391 (1973); Solid State Commun. **13**, 491 (1973)

8.91 T.M.Hayes, S.H.Hunter: In *The Structure of Non-Crystalline Materials*, ed. by P.H.Gaskell (Taylor and Francis, London 1977) p. 69

8.92 T.Ichikawa: J. Phys. Soc. Jpn. **33**, 1729 (1972); Phys. Status Solidi B**56**, 707 (1973)
T.Ichikawa, S.Ogawa: Acta Crystallogr. A**28**, S129 (1972)

8.93 G.Lucovsky: Phys. Status Solidi **49**, 633 (1972)

8.94 M.H.Brodsky, R.J.Gambino, J.E.Smith, Jr., Y.Yacoby: Phys. Status Solidi B**52**, 609 (1972)

8.95 G.Lucovsky, R.M.Martin: J. Phys. Soc. Jpn. **40**, 161 (1976)

8.96 R.M.Martin, G.Lucovsky, K.Helliwell: Phys. Rev. B**13**, 1383 (1976)

8.97 A.S.Pine, G.Dresselhaus: Phys. Rev. B**4**, 356 (1971)

8.98 D.E.Sayers, F.W.Lytle, E.A.Stern: In *Amorphous and Liquid Semiconductors*, ed. by J.Stuke, W.Brenig (Taylor and Francis, London 1974) p. 403

8.99 P.Tronc, M.Bensoussan, A.Brenac: Phys. Rev. B**8**, 5947 (1973)

8.100 R.J.Nemanich, S.A.Solin, G.Lucovsky: Solid State Commun. **21**, 273 (1977)

8.101 A.J.Apling, A.J.Leadbetter, A.C.Wright: J. Non-Cryst. Solids **23**, 369 (1977)

8.102 R.F.Pettifer, P.W.McMillan, S.J.Gurman: In *The Structure of Non-Crystalline Materials*, ed. by P.H.Gaskell (Taylor and Francis, London 1977) p. 63

8.103 G.Lucovsky: Phys. Rev. B**6**, 1480 (1972)

8.104 R.J.Kobliska, S.A.Solin: J. Non-Cryst. Solids **8–10**, 191 (1972)

8.105 R.J.Kobliska, S.A.Solin: Phys. Rev. B**8**, 756 (1973)

8.106 I.G.Austin, E.S.Garbett: Philos. Mag. **23**, 17 (1971)

8.107 A.P.DeFonzo, J.Tauc: Solid State Commun. **18**, 937 (1976)

8.108 P.C.Taylor, S.G.Bishop, D.L.Mitchell, D.Treacy: In Ref. 8.29, p. 1267

8.109 M.Rubinstein, P.C.Taylor: Phys. Rev. Lett. **29**, 119 (1972)

8.110 P.C.Taylor, S.G.Bishop, D.L.Mitchell: Solid State Commun. **8**, 1783 (1970)

8.111 D.L.Mitchell, S.G.Bishop, P.C.Taylor: J. Non-Cryst. Solids **8–10**, 231 (1972)

8.112 R.F.Pettifer, P.W.McMillan: Philos. Mag. **35**, 871 (1977)

8.113 F.L.Galeener, G.Lucovsky: In *Structure and Excitations in Amorphous Solids*, ed. by G. Lucovsky, F.L.Galeener (American Institute of Physics, New York 1976) p. 223

8.114 F.L.Galeener, G.Lucovsky: Phys. Rev. Lett. **37**, 1474 (1976)

8.115 F.L.Galeener, G.Lucovsky: Solid State Commun. (in press)

8.116 R.B.Laughlin, J.D.Joannopoulos: Phys. Rev. B**16**, 2942 (1977)

8.117 E.J.Felty, G.Lucovsky, M.B.Myers: Solid State Commun. **5**, 555 (1967)

8.118 G.Lucovsky, R.C.Keezer, H.A.Six, R.H.Geils: In *Proc. 6th Int. Conf. on Amorphous and Liquid Semiconductors*, ed. by B.T.Kolomiets (Nauka, Leningrad 1976) p. 296

8.119 R.J.Nemanich, S.A.Solin, G.Lucovsky: In *Proc. 6th Int. Conf. on Amorphous and Liquid Semiconductors*, ed. by B.T.Kolomiets (Nauka, Leningrad 1976) p. 518

8.120 G.Lucovsky, R.J.Nemanich, S.A.Solin, R.C.Keezer: Solid State Commun. **17**, 1567 (1975)

8.121 A.L.Renninger, B.L.Averbach: Phys. Rev. B**8**, 1507 (1973)

8.122 W.Prettl, N.J.Shevchik, M.Cardona: Phys. Status Solidi B**59**, 241 (1973)

8.123 R.W.Stimets, J.Waldman, J.Lin, F.Chang, R.J.Temkin, G.A.N.Connell, H.R.Fetterman: In *Amorphous and Liquid Semiconductors*, ed. by J.Stuke, W.Brenig (Taylor and Francis, London 1974) p. 1239

8.124 G.B.Fisher, J.Tauc, Y.Verhelle: In *Amorphous and Liquid Semiconductors*, ed. by J.Stuke, W.Brenig (Taylor and Francis, London 1974) p. 1259

8.125 J.S.Lanin: Solid State Commun. **11**, 1523 (1972)

8.126 P.G.LeComber, W.E.Spear: In *Structure and Excitations in Amorphous Solids*, ed. by G.Lucovsky, F.L.Galeener (American Institute of Physics, New York 1976) p. 284

8.127 W.E.Spear, P.G.LeComber, S.Kinond, M.H.Brodsky: Appl. Phys. Lett. **28**, 105 (1976)

8.128 D.E.Carlson, C.R.Wronski: Appl. Phys. Lett. **28**, 671 (1976)

8.129 P.G.LeComber, A.Madan, W.E.Spear: J. Non-Cryst. Solids **11**, 219 (1972)

8.130 J.C.Knights: In *Structure and Excitations in Amorphous Solids*, ed. by G.Lucovsky, F.L. Galeener (American Institute of Physics, New York 1976) p. 296

8.131 G.Lucovsky, R.J.Nemanich, J.C.Knights: Phys. Rev. B **19**, 2064 (1979)

8.132 R. Zbinden: *Infrared Spectroscopy of High Polymers* (Academic Press, New York 1964)

8.133 L.F.Thompson, K.G.Mayhan: J. Appl. Polym. Sci. **16**, 2291 (1972)

8.134 H.Kobayashi, A.T.Bell, M.Shen: J. Appl. Polym. Sci. **17**, 885 (1973)

8.135 L.J.Bellamy: *The Infra-Red Spectra of Complex Molecules* (Chapman and Hall, London 1975)

8.136 H.J.Stein: J. Elec. Mater. **4**, 159 (1975)

8.137 N.F.Mott, E.A.Davis: *Electronic Processes in Non-Crystalline Materials* (Clarendon Press, Oxford 1971) p. 280

8.138 J.C.Knights: Philos. Mag. **34**, 663 (1976)

8.139 P.H.Citrin, P.Eisenberger, B.M.Kincaid: Phys. Rev. Lett. **36**, 1346 (1976)

8.140 J.D.Joannopolous, F.Yndurian: Phys. Rev. B**10**, 5164 (1974)

8.141 A.J.Apling, A.J.Leadbetter: In *Amorphous and Liquid Semiconductors*, ed. by J.Stuke, W.Brenig (Taylor and Francis, London 1974) p. 457

8.142 R.A.Street, R.J.Nemanich, G.A.N.Connell: Phys. Rev. B (in press)

9. Doped Amorphous Semiconductors

By P. G. LeComber and W. E. Spear

With 26 Figures

The possibility of substitutional doping in crystalline semiconductors has been one of the most important factors in the development of semiconductor physics and solid-state electronics. The lack of similar sensitive control in the case of amorphous (a-) semiconductors had been a serious limitation in this field. However, in 1975 the authors showed [9.1] that the electronic properties of a-Si and a-Ge prepared by glow discharge decomposition could be controlled over a remarkably wide range by substitutional doping in the gas phase. The work has opened up new possibilities for the study and application of these materials. Many research groups have since entered this field and, as the proceedings of the last International Conference show [9.2], interesting developments have taken place in the subject.

In the following sections an attempt has been made to provide an overall view of our present understanding of doped amorphous semiconductors. Most of the discussion will be concerned with a-Si – it is the material which has so far been studied in greatest detail and also appears to be of greatest applied interest. We shall begin with an introductory section on the density-of-states distribution in a-Si, a topic of considerable importance to the subject of this chapter. This is followed by a discussion of the preparation of doped amorphous semiconductors and the role of hydrogen. We shall then consider the effects of doping on the electronic properties of a-Si and a-Ge and discuss the interpretation of drift mobility, Hall mobility, and thermoelectric power data which have provided new information on donor band hopping in a-Si. The next section is devoted to photoconductivity and a discussion of the remarkable increase of photosensitivity that has been obtained by slight phosphorus doping. Finally, we shall take a look at our recent work on the formation and electronic properties of the amorphous surface barrier and p–n junction which are proving to be of increasing applied interest.

9.1 The Density of States Distribution and the Possibility of Doping

The work during the last few years in the field of amorphous semiconductors has brought out the close dependence of the electronic properties of these materials on the density and distribution of localized states in the mobility gap.

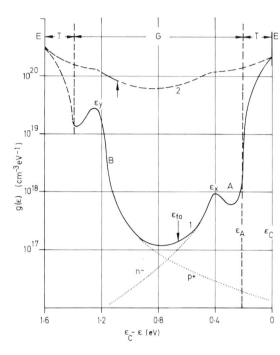

Fig. 9.1. Density of states distributions for a-Si specimens. The full lines are the total density of states determined from field-effect experiments and the arrow on each curve shows the position of the Fermi level. Curve *1*: glow discharge specimen, $T_d \simeq 520$ K. *A* and *B* are the assumed division of this data into acceptor-like and donor-like states respectively. Curve *2*: evaporated specimen. *E*, extended states; *T*, tail states; *G*, gap states. [9.4, 6, 11]

The possiblity of efficient doping also depends on the overall state density and we shall begin by discussing this important topic.

Curve *1* in Fig. 9.1 represents the experimentally determined density of states distribution $g(\varepsilon)$ for a-Si specimens deposited by the glow discharge technique (see Sect. 9.2.1) on a substrate held at about 520 K. The function $g(\varepsilon)$ denotes the number of states per unit volume in unit energy interval and is plotted here against the electron energy ε, measured from ε_c, the energy at which localization sets in.

The solid line represents the results of extensive field effect measurements on undoped and doped a-Si specimens [9.3–6]. It has been necessary to combine the data from a considerable number of specimens with different Fermi level positions to cover the experimental energy range from about 0.1 to 1.3 eV. The arrow marks the position of the Fermi level ε_{f0} normally observed in undoped a-Si specimens ($\simeq 0.65$ eV below ε_c). As indicated, the energy spectrum can be divided according to the electronic character of the states into extended states *E*, band-tail states *T*, and gap states *G*.

The field effect experiment and its interpretation presents a number of difficult problems which have been discussed in the above-mentioned papers. The validity of assumptions on which the $g(\varepsilon)$ curve is based can ultimately be justified only if predictions derived from the distribution are verified experimentally. We are encouraged by the fact that this appears to be the case in a growing number of instances such as the interpretation of the trap-controlled drift mobility [9.7, 8], the spectral dependence of photoconductivity and optical absorption [9.9] (experimental evidence for the 1.2 eV peak), the observed movement of ε_f with doping [9.5], the interpretation of photoconductivity and recombination in doped specimens [9.10] and the predicted behavior of barrier capacitance [9.11]. We shall therefore regard the $g(\varepsilon)$ curve in Fig. 9.1 as a reasonable approximation to the density of states distribution in glow discharge a-Si.

Field effect measurements lead to the total density of states irrespective of their charge state. If the latter is important, such as in the discussion of recombination [9.10] or in the calculation of the barrier space charge [9.11], the following simple model has been used with reasonable success. It is suggested that the $g(\varepsilon)$ minimum arises from overlapping tails of two distributions of defect centers which differ by their charge state. The dotted curves in Fig. 9.1 show a likely division of $g(\varepsilon)$ into its two components, which is also consistent with the observed Fermi level position. Distribution A, extending from the ε_c side into the gap contains acceptor-like states which are neutral when empty, so that below ε_f they will carry a negative charge denoted by n^-. Curve B contains states from the opposité side of the mobility gap which are neutral when full and therefore provide positively charged donor-like states above ε_f, density p^+. The charge states described are identical to those envisaged in the CFO model [9.12] but the important difference is that here we are dealing with the overlap of defect distributions, whereas the original CFO model considered the tail state overlap in alloy glasses.

For comparison, curve 2 in Fig. 9.1 is a representation of the density of states distribution for an a-Si specimen produced by thermal evaporation. Because of the high level of gap states, it was only possible to investigate by the field effect technique a small energy range (full line) close to ε_f; the rest of the curve is speculative. Nevertheless, the high values of $g(\varepsilon) \sim 10^{20}\,\mathrm{cm}^{-3}\,\mathrm{eV}^{-1}$ in evaporated Si have been confirmed by other workers [9.13].

We shall now discuss the possibility of substitutional doping on an a-Si specimen with a density of states distribution such as that shown by curve 1. Suppose for example that a band of N_D substitutional phosphorus atoms has been incorporated leading to donor states between ε_c and ε_A in the tail states. Practically all the N_D excess electrons will condense into empty gap states above ε_{f0} displacing the Fermi level towards ε_c by an amount $\Delta\varepsilon_f$. The new Fermi level position is determined by the condition for charge neutrality:

$$N_D \simeq N_D^+ \simeq \int_{\varepsilon_{f0}}^{\varepsilon_c} \frac{g(\varepsilon)d\varepsilon}{1+\exp\left(\dfrac{\varepsilon-\varepsilon_f}{kT}\right)} + \Delta n(\varepsilon_c). \tag{9.1}$$

In the amorphous semiconductor, the increase in the extended state electron density $\Delta n(\varepsilon_c)$ will normally be quite negligible in comparison to the first term of (9.1). This is basically different from the crystalline case, where in the exhaustion range the density of ionized donors $N_D^+ \simeq \Delta n(\varepsilon_c)$. It is important to realize that in the amorphous material, changes in electrical properties are brought about primarily by changes in the gap-state occupation. For sensitive doping $\Delta \varepsilon_f$, which is of the order of $N_D/g(\varepsilon_f)$, should be as large as possible; a low level of gap states is therefore an essential condition. Here curve *1* in Fig. 9.1 suggests that preparation by the glow discharge technique should be the most promising approach. It was in fact on the basis of such field-effect results that we started the first doping experiments on a-Si [9.1].

Clearly, curve *2* in Fig. 9.1 explains the reason for the insensitive behavior to doping of evaporated or sputtered amorphous semiconductors; $g(\varepsilon_f)$ is far too high to allow much change in the Fermi level position and consequently in the electrical properties.

In the beginning of this discussion we assumed that N_D impurity atoms can be incorporated substitutionally into the random network. To what extent is this justified? It has for instance been argued [9.14] that in the amorphous phase, few phosphorus atoms would retain their donor-like character because the additional bond is most likely taken up in the random network during the growth of the film. Although this mechanism obviously plays a role, the experimental evidence in the following sections shows that in doping from the gas phase quite an appreciable fraction of the incorporated phosphorus atoms survive as donors in the network; the same applies to the boron acceptors. However, it is possible that this fraction will depend, perhaps critically, on the method of preparation and of doping.

9.2 Preparation of Doped Amorphous Semiconductors

In this section we shall discuss the methods by which doped amorphous semiconductors have been prepared: glow discharge deposition, rf sputtering, ion implantation, and diffusion. There has been a considerable amount of interest in the role of hydrogen in materials prepared by the first two techniques and a discussion of this aspect has been included.

9.2.1 Deposition and Doping by the Glow Discharge Technique

This approach is likely to play an increasingly important role in the field of amorphous materials and some of its basic aspects will now be reviewed.

It has long been recognized that a plasma provides a useful means of transferring energy to gaseous molecules. Generally a variety of new species are produced: atoms, free radicals, and ions, both stable and unstable. These are chemically active and often represent an intermediate stage in the formation of new stable compounds. It is interesting to note that one of the first applications of the plasma technique was to the preparation of ozone, described by Siemens in 1857. During the last few decades, plasma chemistry has developed rapidly [9.15, 16]. Different types of electrical discharges have been used and described in the literature, but of these the glow discharge has proved to be of particular interest. It contains electrons with energies between 1 and 10 eV and their density, approximately equal to the positive ion density, amounts to about 10^{10} cm^{-3}. The electron temperature may be 10 or even 100 times that of the

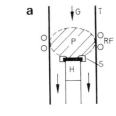

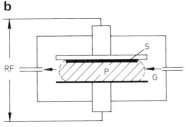

Fig. 9.2a, b. Diagrams illustrating experimental methods for gas phase deposition of an amorphous semiconductor; (**a**) inductive coupling of the rf, (**b**) capacitance coupling

gas so that electrons possess sufficient energy to break molecular bonds. It is this property which makes the glow discharge plasma so useful in promoting chemical reactions at comparatively low ambient temperatures.

Sterling and his collaborators [9.17, 18] were the first to apply the glow discharge technique to the preparation of the amorphous semiconductors Si and Ge.

Thin film specimens were formed by decomposing the corresponding hydrides, silane or germane. Figure 9.2a illustrates the experimental method. The gas *G* flows through the quartz reaction tube *T* past the substrate *S*, which is held on a heated pedestal *H*. The plasma *P* is maintained by predominantly inductive coupling of rf power into the gas. The power level is small, typically 10–20 W, and frequencies between 1 and 100 MHz have been tried. The simplicity of the arrangement is deceptive. The electronic properties of the deposited specimens are critically dependent on a number of variables such as the substrate temperature [9.8, 19], flow rate, pressure, rf power level, the floating potentials on specimen and other surfaces, the tube diameter, and also the position of the coil with respect to the substrate. The above system is difficult to scale up and the capacitatively coupled arrangement of Fig. 9.2b is now being used by a number of laboratories for depositing larger area specimens.

The significant point of both the arrangements in Fig. 9.2 is that the plasma remains in close contact with the specimen surface. This is a most important feature of glow discharge deposition. For instance, in the preparation of a-Si from silane, complex surface reactions will take place during growth involving electrons and positive ion fragments such as SiH, SiH_2, and SiH_3. The experimental control of these surface reactions to obtain well defined and reproducible electronic specimen properties remains a major problem of the technique. It is perhaps not surprising that there are pronounced differences in the properties of glow discharge specimens from different laboratories.

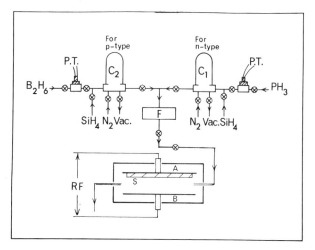

Fig. 9.3. Schematic diagram of the preparation unit for n- and p-type a-Si and a-Ge specimens developed by the Dundee group. C_1, C_2, glass cylinders; P.T., pressure transducer; F, flow meter; S, substrate [9.5]

Doping from the gas phase can be achieved by adding small but accurately determined amounts of phosphine or diborane, the hydrides of a pentavalent or trivalent impurity, to the silane. A schematic diagram of the preparation unit developed at Dundee [9.5] is shown in Fig. 9.3. The silane is mixed with the doping gases in the glass cylinders C_1 and C_2 respectively. For the preparation of an n-type mixture, phosphine is first admitted into a small known volume and its pressure measured with the pressure transducer PT. It is then expanded into the evacuated cylinder C_1 and silane added to attain a standard pressure. In this way it is possible to add with a reasonable degree of accuracy a few volume parts per million of phosphine – that is about 10^{17} molecules per cm^3 to the silane.

Similarly, diborane can be premixed in cylinder C_2 for doping with a p-type impurity. Therefore in principle any desired sequence of n- and p-type layers can be deposited simply by opening and closing the appropriate taps. The flow rate is measured by the electronic flow meter F, and decomposition takes place in the rf glow discharge between plates A and B of the chamber. Because of the toxic nature of the gases used, independent evacuation and nitrogen flushing facilities are incorporated in the apparatus. Doped a-Ge films have been prepared in the same way from germane.

9.2.2 Hydrogen and the Properties of Glow Discharge Amorphous Semiconductors

We have seen that in the deposition of a-Si the plasma contains silicon–hydrogen fragments and also H$^+$ ions. It is therefore relevant to ask to what extent hydrogen is incorporated into the random network and whether this influences the structural and electronic properties of the deposited material.

It has often been argued that the presence of hydrogen tends to saturate dangling bonds on internal surfaces of microvoids and on point defects in the structure. This is undoubtedly a likely interaction and may well be responsible in glow discharge Si for the absence of the "hard" paramagnetic states (see, Chap. 7). However, it has by no means been established that bond saturation by hydrogen is the only reason for the very much lower overall density of states in glow discharge a-Si as compared with the levels found in evaporated or sputtered material. An equally important reason may be the glow discharge method itself in which the film grows under basically different conditions than those achieved by other experimental techniques [9.4]. It is very likely that different deposition methods lead to different defect structures. Comparatively large microvoids, containing numerous dangling bonds, appear to dominate the properties of evaporated or sputtered a-Si [9.20] whereas in the glow discharge material point defects, possibly of the divacancy type, predominate.

Little is known about the nature of the remaining "soft" states (see Chap. 7), which are expected to make only a very small contribution to the ESR signal. The field effect experiments on glow discharge specimens (Sect. 9.1) show that these states form a quasi continuum in the gap and lead to the conclusion that this distribution is very sensitive to plasma conditions and of course also to the method of preparation.

The hydrogen problem has aroused much current interest. *Brodsky* and his collaborators [9.21, 22] as well as others (see Chap. 8) have investigated the infrared and Raman spectra from a-Si specimens prepared by glow discharge decomposition at $T_d \simeq 300\,\mathrm{K}$ and $520\,\mathrm{K}$. The spectral features show that the structure contains hydrogen in the form of SiH, SiH_2, and SiH_3 groups. Quantitative estimates based on the absorption measurements suggest that the amount of bonded hydrogen is typically 35–52 atomic percent for the 300 K specimens and 14–25 atomic percent for those deposited at 520 K. The same specimens were subsequently investigated by mass-spectrographic analysis and by a nuclear resonance reaction method, both of which substantially confirmed the above results. *Fritzsche* [9.23] and *Knights* [9.24] prepared a-Si specimens by the method illustrated in Fig. 9.3b, but employed a highly dilute argon–silane mixture (60:1) and a large flow rate. They also found a hydrogen content comparable to that in Brodsky's experiments.

These results are interesting, but raise several questions. Earlier infrared measurements on high-T_d silicon specimens by *Chittick* [9.25] gave no evidence of the Si–H mode; similar measurements in the authors' laboratory led to the same negative result. It is surprising that the very large hydrogen concentrations found in the recent work should have been missed in earlier experiments. A likely explanation is that the hydrogen incorporated during deposition may depend, perhaps critically, on the detailed plasma conditions at the specimen surface, as has been established for some of the other properties mentioned in Sect. 9.2.1. A conclusive answer to the hydrogen problem requires a systematic investigation of hydrogen content, density of states, transport properties, etc., as a function of preparation conditions. Work along these lines is in progress in our laboratory.

The recent work by *Moustakas* et al. [9.26] on the photoconductivity of a-Si sputtered in different partial pressures of hydrogen gas has led to an interesting suggestion. It was found that after an initial sharp rise in photoconductivity with hydrogen content, which can be associated with saturation of dangling bonds, the photoresponse dropped rapidly in specimens containing more than about 6 atomic percent of hydrogen. *Moustakas* et al. attribute this effect to an increase in the density of gap states around ε_c and tentatively suggest that the increase is associated with the growing number of Si–H antibonding states in the system.

If this is correct, then the electronic properties of the above glow discharge specimens, containing about 25 atomic percent of hydrogen, may well be adversely affected by the large excess of hydrogen which amounts to something like 10^3 times the density of dangling bonds. It is the authors' view that in glow discharge Si, hydrogen is only beneficial at relatively low concentrations ($\gtrsim 1$ atomic percent) to reduce the density of the "hard" states; plasma conditions and control will have to be improved to achieve a minimum hydrogen level as well as a low overall density of "soft" states.

9.2.3 Hydrogenated Sputtered Amorphous Semiconductors

The work of the Harvard group on hydrogenated sputtered a-Ge and Si has made an important contribution to the subject. It is generally accepted that specimens prepared by cathodic sputtering contain a high overall level of gap states associated with dangling bonds in probably quite an extensive microvoid structure. *Paul* and his collaborators [9.27, 28] showed that by adding hydrogen to the argon used in the sputtering process, a few atomic percent of hydrogen can be incorporated into the films. This has a marked effect on the electronic properties of the material, greatly reduces the ESR signal [9.29], and is consistent with the suggestion that dangling bonds have been saturated by hydrogen atoms. In the present context, the significant result is that fairly sensitive doping from the gas phase using phosphine and diborane now becomes possible [9.30].

In the absence of field effect measurements it is difficult to say how far $g(\varepsilon)$ has been decreased by hydrogenation during sputtering. However, comparing the temperature dependence of the conductivity and the photoconductivity in undoped glow discharge Si specimens with corresponding results in the hydrogenated sputtered material [9.26, 31], the latter has undoubtedly a higher $g(\varepsilon)$. The same seems to apply to a-Ge prepared by the two methods as can be seen from the comparison of photoconductive properties at the end of Sect. 9.4.

9.2.4 Doping by Ion Implantation and Diffusion

Doping by ion implantation has achieved considerable importance in crystalline device technology. Attempts to apply this method to evaporated Ge specimens have not been encouraging, probably for the reasons discussed already in Sect. 9.1. However, recent work on substitutional doping of glow discharge Si by ion implantation [9.32] has led to more promising results (Sect. 9.3.1) mainly because of the low overall density of gap states. The implantations were carried out on originally undoped glow discharge a-Si at 520 K, the deposition temperature of the specimens. In this way, the damage produced by the ion bombardment is annealed during irradiation, which seems to prevent the clustering of smaller defects into larger and more stable microvoid structures.

Beyer and *Fischer* [9.33] have recently doped glow discharge a-Si interstitially with Li both by ion implantation and by diffusion. The Li implantations were performed with 30 keV ions giving a penetration depth of about 0.2 μm, smaller than the sample thickness. The authors suggested, however, that at annealing temperatures above 50 °C, the implanted Li should be homogeneously distributed by diffusion. Samples were also doped by evaporating a Li layer onto the undoped material followed by heating to 400 °C for one hour to allow the Li to diffuse into the sample. Under these conditions the diffusion length of Li is likely to be greater than the sample thickness so that a homogeneous distribution is obtained. The results of these experiments are described in Sect. 9.3.1.

9.3 Electronic Transport Properties of Doped Amorphous Semiconductors

In this section we shall discuss the effect of doping on the transport properties of amorphous semiconductors, mainly a-Si. Conductivity measurements show the remarkable range of control that can be achieved and the results of drift mobility, Hall effect, and thermoelectric power measurements lead to the conclusion that with increasing donor concentration, a new current path is established in which carriers hop between donor sites. The section ends with a brief account of recent crystallization experiments on a-Si, which are useful for determining the doping efficiency in the amorphous material.

9.3.1 Control of Conductivity by Doping

Figure 9.4 summarizes some of the results [9.5] that have been obtained in our laboratory for n- and p-type a-Si prepared by the technique described in Sect. 9.2.1. The room-temperature conductivity σ_{RT} of the specimens is plotted against the gaseous impurity ratio. On the right-hand side this is the ratio of the number of phosphine to the number of silane molecules in the gaseous mixture, N_{PH_3}/N_{SiH_4}, whereas on the left the corresponding diborane-to-silane ratio is shown. In the centre of the graph, the conductivities of 10^{-8}–$10^{-9}\,\Omega^{-1}\,cm^{-1}$ are representative of undoped glow discharge specimens, the properties of which are largely determined by structural defect states in the mobility gap.

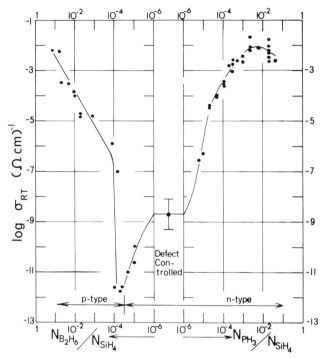

Fig. 9.4. Room-temperature conductivity σ_{RT} of n- and p-type amorphous Si specimens, plotted as a function of the gaseous impurity ratio. For the right-hand curve, this is the number of phosphine to silane molecules in the gas mixture used for specimen preparation. On the left it is the corresponding diborane to silane ratio. The centre refers to undoped specimens [9.5]

Consider first the n side. The encouraging feature is that even a minute quantity of phosphine, 6×10^{-6} ppv, increases σ_{RT} by over two orders of magnitude. This clearly supports deductions from field effect measurements concerning the low value of $g(\varepsilon_f)$. With increasing phosphorus doping, curve *1* in Fig. 9.1 shows that ε_f moves into regions of higher state density which slows down the rate of rise of conductivity. With a phosphine-to-silane ratio of 10^{-3}, one approaches $\sigma_{RT} \simeq 10^{-2} \, \Omega^{-1} \, cm^{-1}$; that is, σ_{RT} has been increased by seven orders of magnitude by phosphine doping. Further increase in the impurity level does not seem to lead to higher conductivities.

Turning now to the effect of boron doping, we see that initially σ_{RT} decreases to about $10^{-12} \, \Omega^{-1} \, cm^{-1}$. This feature can be understood from the density of state distribution (Fig. 9.1). In an undoped specimen, ε_f lies at about 0.6 eV below ε_c on the n side of the minimum. We know from our previous work that hole conduction sets in when $\varepsilon_c - \varepsilon_f$ exceeds 0.85 eV. Thus the initial boron doping pulls ε_f to the p side of the density of states minimum and the decreasing σ reflects the rapid reduction in the number of electrons at ε_c. At a diborane-to-silane ratio of 10^{-4} there occurs an almost vertical rise in σ_{RT} by six orders of magnitude, indicating that hole conduction has taken over.

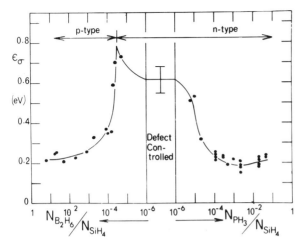

Fig. 9.5. Activation energy ε_σ of the electrical conductivity plotted as a function of the gaseous impurity ratio for n- and p-type amorphous Si specimens. For n-type specimens, $(\varepsilon_\sigma)_n = (\varepsilon_c - \varepsilon_f)_0$. The centre of the diagram refers to undoped specimens [9.5]

The temperature dependence of the conductivity has also been measured for the specimens shown in Fig. 9.4. Practically all of these showed a reasonably well defined activation energy ε_σ, which is plotted against the impurity ratio in Fig. 9.5. On the n side, where at room temperature extended state conduction at ε_c predominates (except at high doping levels, see Sects. 9.3.3–5), the observed activation energy $(\varepsilon_\sigma)_n$ can be identified with $(\varepsilon_c - \varepsilon_f)_0$. This quantity denotes the energy difference between extended states and Fermi level extrapolated to $T = 0$. Figure 9.5 shows that with increasing phosphorus doping, ε_f moves from 0.6 eV to about 0.2 eV below ε_c. When maximum conductivity is reached, ε_f just lies in the rapidly rising part of the tail-state distribution (Fig. 9.1).

When diborane is added to the silane, electron transport persists until compensation of filled defect states has moved the Fermi level approximately to the centre of the density of states minimum at $\varepsilon_c - \varepsilon_f = 0.85$ eV. The specimen then turns p type, and the observed $(\varepsilon_\sigma)_p$ mirrors the movement of $\varepsilon_c - \varepsilon_f$ on the n side. At the higher doping levels, $(\varepsilon_\sigma)_p$ approaches 0.2 eV which would represent $(\varepsilon_f - \varepsilon_v)_0$ with predominant extended state conduction. However, the nature of hole transport in a-Si is still somewhat uncertain (Sect. 9.3.3), so that $(\varepsilon_\sigma)_p$ can only be regarded as an approximate measure of the Fermi level position above ε_v.

The information summarized in Figs. 9.4, 5 illustrates the systematic and wide-ranging control of the electronic properties that can be achieved in glow discharge Si by gas phase doping. The "intrinsic" conductivity of about $10^{-12}\,\Omega^{-1}\,cm^{-1}$ can be increased to $10^{-2}\,\Omega^{-1}\,cm^{-1}$ on both the n and p sides of the density of states minimum. This corresponds to a movement in ε_f of 1.2 eV, essentially between the onset of the electron and hole tail states. It would be difficult to explain these results other than by a model of substitutional

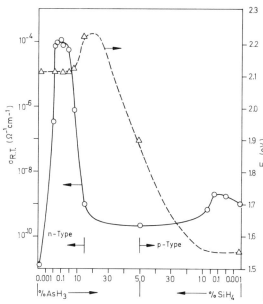

Fig. 9.6. Room-temperature conductivity σ_{RT} and E_α, a measure of optical absorption edge, as a function of the gas composition used to prepare amorphous As–Si specimens [9.34]

impurities, which in most respects is similar to that used for crystalline materials.

The work by *Knights* [9.34] on the As–Si system, using mixtures of arsine and silane, shows clearly that glow discharge a-Si can equally well be doped with As donors. In Fig. 9.6, the room-temperature conductivity is plotted against the percentage composition of the gas used for the sample preparation. The broken curve represents E_α, the photon energy at which the absorption coefficient α is equal to $10^{4.5}$; although this is an arbitrary measure of the optical absorption edge, it indicates the changes in this quantity. The initial sharp rise in σ_{RT} as the arsine level is increased from 0 to 0.1 % does not lead to any measurable shift in the absorption edge, and is almost certainly caused by a shift in ε_f as a result of the substitutional As donors. However, the right-hand side of the graph shows that no donor states are formed when Si is incorporated in a-As.

Some detailed experiments on the properties of doped a-Ge prepared by the glow discharge technique have recently been carried out in our laboratory [9.35]. In the undoped material deposited at 500 K, ε_f lies between 0.46 and 0.41 eV [9.36]. Phosphorus doping shifts ε_f towards ε_c and, with an impurity ratio of 1.2×10^{-2} ppv, ε_f lies about 0.2 eV below ε_c. In this way σ_{RT} has been increased from 10^{-5} to $3 \times 10^{-2} \, \Omega^{-1} \, cm^{-1}$. It is likely that the overall level of the state density is larger in glow discharge Ge [9.36] than in Si, and this makes Ge a less attractive material in doping experiments.

As mentioned in Sect. 9.2.3, *Paul* and his colleagues showed that it is possible to dope rf sputtered a-Ge and a-Si by the addition of either phosphine or diborane, to the argon–hydrogen sputtering gas. In this way, the room

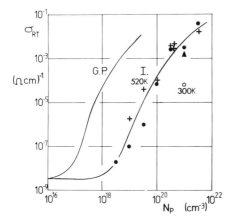

Fig. 9.7. Doping by ion implantation. Curve *I*: room-temperature conductivity in irradiated volume as a function of implanted P atom density for three batches of undoped specimens. All implantations at 520 K. Curve G.P.: doping from gas phase, $N_P \simeq 3N_D^+$. [9.32]

temperature conductivity of films containing approximately 6 atomic percent hydrogen could be increased from 2×10^{-10} to $4 \times 10^{-6}\,\Omega^{-1}\,\mathrm{cm}^{-1}$ by adding 2180 ppm of phosphine [9.30]. Although the range of control is not as large as that obtainable with the glow discharge process, p–n junctions and Schottky barrier diodes have been produced by this technique, which suggests that it may become of applied interest.

The third successful approach has been the doping of glow discharge a-Si by ion implantation at elevated temperatures [9.32]. When the specimen is implanted at about 520 K the radiation damage is largely removed during the experiment, leading to the same control of conductivity as that achievable by doping from the gas phase. For example, Fig. 9.7 shows the room-temperature conductivity σ_{RT} in the irradiated volume (extending to about 1000 Å in recent experiments) plotted against the density of implanted phosphorus atoms. The conductivity can be increased from its initial value of $\sim 10^{-9}\,\Omega^{-1}\,\mathrm{cm}^{-1}$ to more than $10^{-2}\,\Omega^{-1}\,\mathrm{cm}^{-1}$. Similar effects can be obtained with B implantations although σ_{RT} initially decreases to about $2 \times 10^{-12}\,\Omega^{-1}\,\mathrm{cm}^{-1}$ and then increases over nine orders of magnitude with increasing boron concentration (cf. Fig. 9.4). However, the doping efficiency appears significantly smaller than that obtainable from gas phase doping, as is evident from the curve marked GP in Fig. 9.7, representing typical results for the latter. Nevertheless, the range of control of the electrical properties is practically identical for the two methods and it is therefore likely that doping of amorphous semiconductors by ion implantation may prove to be of interest in future device developments.

Figure 9.8a and b show the results of conductivity and thermoelectric power measurements [9.33] on a-Si specimens implanted with approximately 1 % of Li. Curve *1* in each figure refers to an undoped reference specimen, whereas curves *2a* to *2d* show the effect of progressive annealing after implantation. In a specimen annealed at 20 °C (*2a*), the defects introduced by the ion bombardments still dominate the transport. But curves *2d*, after annealing at 300 °C, indicate that the Fermi level has been moved by the Li doping to within 0.2 eV

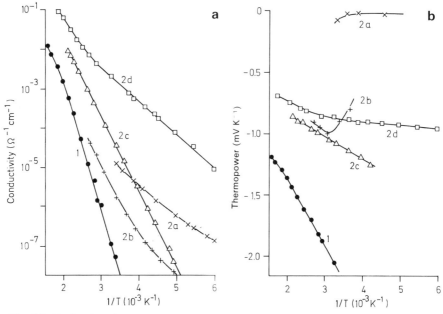

Fig. 9.8 (a) Conductivity and (b) thermoelectric power plotted as a function of reciprocal temperature for a-Si doped with Li by ion implantation: (*1*) undoped film; (*2a–d*) implanted with 1 at % Li. Annealing temperatures: (*2a*) 20 °C; (*2b*) 100 °C; (*2c*) 200 °C; (*2d*) 300 °C [9.33]

of the extended electron states, just as in phosphorus doped specimens. Similar results have been obtained with Li diffusion. For example, after doping with 4 atomic percent of Li, the conductivity increases from its undoped value of 8×10^{-10} to $6 \times 10^{-4} \Omega^{-1} cm^{-1}$ and the conductivity activation energy decreases from 0.76 to 0.18 eV. The negative sign of the thermoelectric power confirms that Li acts as a donor in a-Si.

9.3.2 Efficiency of Gas Phase Doping in a-Si

In Sect. 9.1 we asked what fraction of the incorporated phosphorus atoms might act as donors in the amorphous phase. We shall now attempt to give an approximate answer to this important problem for a-Si. Two quantities will have to be determined: the density of ionized donors N_D^+, which in our case is closely equal to the donor density N_D, and N_P, the density of phosphorus atoms incorporated into the specimen. The first has been obtained by fitting (9.1) to the experimental $(\varepsilon_\sigma)_n$ curve (Fig. 9.5), using the known density of states distribution. This gives N_D^+ as a function of the impurity ratio PH_3/SiH_4. The second quantity N_P was determined by ion probe analysis for a series of phosphorus doped specimens with known impurity ratio. Comparison of N_D^+ and N_P then led to the conclusion [9.5] that

$$N_P \simeq 3N_D,$$

which means that about one-third of the incorporated phosphorus atoms act as donors; additional valence bonds of the remaining phosphorus atoms are most probably accommodated into the random network. Phosphorus doping from the gas phase is therefore a relatively efficient process. The same seems to apply to doping with boron.

9.3.3 Drift Mobility Studies in a-Si

The application of drift mobility techniques [9.37] to highly resistive amorphous layers and thin crystalline specimens has greatly contributed to the understanding of their transport properties. Earlier drift mobility and conductivity measurements on undoped glow discharge Si [9.7, 8] had shown fairly conclusively that in a high T_d specimen, the transport above about 250 K is predominantly by electrons in the extended states. Below 250 K, hopping conduction through the tail states takes over and with further reduction in T, the predominant conduction path will sink to perhaps the ε_x maximum (Fig. 9.1) and eventually to within a few kT of the Fermi level.

Recent drift mobility experiments on doped a-Si specimens have led to interesting results [9.38], some of which will now be discussed. A series of compensated a-Si specimens was prepared containing donor densities in the range 5×10^{17}–6×10^{18} cm^{-3}. The average distance between donor sites, R_D, ranged from 90 to 39 Å. Compensation with boron acceptors was essential to maintain the high specimen resistance needed for drift mobility measurements. Figure 9.9 shows the room temperature drift mobility μ plotted against R_D^{-1}. The point U refers to an undoped specimen. It must be stressed that the value of about 10^{-1} cm^2 V^{-1} s^{-1} is not the extended state mobility, but is representative of a trap-controlled transport which includes the interaction of the drifting carriers with tail states near ε_A through trapping and thermal release [9.7]; extended state drift mobilities between 1 and 10 cm^{2}V^{-1}s^{-1} have been deduced from such measurements. In lightly doped specimens, two mobilities, μ_a and μ_b, can be derived from the transit pulse. Points 1b and 2b for μ_b are probably still associated with a trap controlled mobility, but points *1a* and *2a* refer to a new transport path. The mobility μ_a depends strongly on the doping level (and therefore R_D^{-1}) and its magnitude suggests that we are observing hopping transport through the band formed by the donor states.

If this interpretation of donor band hopping is correct, then the mobility should be given by

$$\mu_a \simeq \frac{1}{6} \frac{eR_D^2}{kT} v_{ph} \exp(-2\alpha R_D) \exp(-W/kT). \qquad (9.2)$$

The term $\exp(-2\alpha R_D)$ describes the overlap of the wavefunctions on neighboring hopping sites, with the parameter α^{-1} denoting the spatial extension of a localized donor wave function; $v_{ph} \exp(-W/kT)$ represents the probability per unit time that the localized electron hops to a new site, at an energy W above

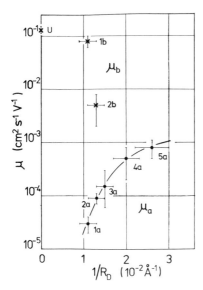
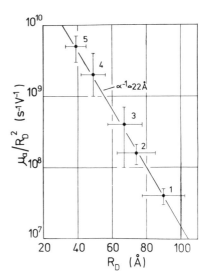

Fig. 9.9. Measured drift mobilities μ_a and μ_b at 295 K in n-type a-Si specimens as a function of the inverse donor separation. The numbers denote increasing phosphorus content; U refers to an undoped specimen [9.38]

Fig. 9.10. Plot to test the applicability of (9.2). It leads to an experimental value of $\alpha^{-1} \simeq 22$ Å [9.38]

the original one. To explore the applicability of (9.2) in this case, μ_a/R_D^2 is plotted semilogarithmically versus R_D in Fig. 9.10. The plot illustrates the changes in $\exp(-2\alpha R_D)$ with donor separation and leads to a value of $\alpha^{-1} \simeq 22$ Å. This is closely similar to the 20 Å obtained for phosphorus donors in crystalline Si [9.39] and the 21 Å predicted by the well-known hydrogenic model and suggests that the environments of the donor atoms in the two phases are probably quite similar. From the temperature dependence of μ_a, an average value of 0.05 eV for the hopping energy has been deduced.

Hole drift mobility measurements on a-Si have recently been carried out by *Moore* [9.40] and in our laboratory [9.41]. The results lead to a hole drift mobility of about 10^{-4} cm^2 V^{-1} s^{-1} at room temperature, and μ versus $1/T$ curves show an activation energy of 0.4 eV. A likely interpretation is that excess holes drift in the extended states below ε_v with a mobility similar to that of electrons above ε_c (1–10 cm^{2}V^{-1}s^{-1}). During transit, the holes interact through trapping and thermal release with states about 0.4 eV above ε_v, that is in the region of the ε_y maximum (Fig. 9.1). However, on the basis of the drift mobility measurements alone it is not possible to exclude an interpretation on the trap-controlled model in which the predominant transport path lies in the tail states, just above ε_v, instead of the extended hole states.

The measurements in our laboratory have also been extended to specimens with different boron concentrations. In contrast to the electron drift mobility results, no evidence for impurity hopping was found in this case. It appears that the larger density of gap states in the lower half of the mobility gap favors the trap-controlled mechanism (as well as hole hopping transport through the distribution at lower T) which masks any clear observation of hopping between acceptor sites.

9.3.4 The Hall Effect in a-Si

Hall effect measurements have been a most important tool in the study of wide-band crystalline semiconductors. They provide a method for determining the carrier concentration and, together with the conductivity, lead to a mobility value. The sign of the Hall constant is unambiguously determined by the sign of the charge carriers. This useful interpretation breaks down completely when strong disorder scattering reduces the mean free path of the carriers to a few atomic spacings, as in the random phase region of an amorphous semiconductor. The theories that have been developed for this case conclude that the magnitude and sign of the Hall constant will depend on the local geometry of the sites involved in the tunneling processes [9.42] and also on the nature of the local electronic states [9.43]. (See Chap. 5.)

An intriguing aspect of Hall effect measurements in amorphous materials is the anomaly in the sign of the Hall mobility. This has been observed in measurements on chalcogenide glasses [9.44–47], in evaporated a-Ge [9.48] and recently also in a-As [9.49]. In most of these materials (with the exception of a-As) the sign of μ_H was negative, regardless of the sign of the thermoelectric power. The experiments on substitutionally doped n- and p-type a-Si [9.50] have led to an interesting and clear-cut extension of these results. They show a double reversal in sign: in all n-type specimens, where thermoelectric power and field effect verified electron conduction, μ_H had the sign normally expected for positive carriers; p-type specimens, on the other hand, gave a Hall effect corresponding to free electrons. These results have also been confirmed by the *Marburg* group [9.51]. A possible explanation of the double reversal has recently been put forward by *Emin* [9.43], who took into account the nature of the electronic states in the elementary tunneling paths.

The dependence of the conductivity and of the magnitude of the Hall mobility $|\mu_H|$ on temperature and doping provide important information on the conduction processes in a-Si. In Fig. 9.11, $|\mu_H|$ is shown as a function of the inverse temperature for a number of n-type a-Si specimens, labelled 1 to 7, with donor densities from 8.5×10^{17} to $9 \times 10^{18} \, \text{cm}^{-3}$ respectively [9.50]. The doping range is similar to that used in the drift mobility experiments (Sect. 9.3.3). Results for three p-type specimens have also been included. The experimental difficulties in determining the small Hall potentials ($\simeq 200 \, \mu\text{V}$) restricted the low-temperature range that could be investigated, particularly with the more resistive specimens. Like the drift mobility results, the Hall mobility curves show a systematic dependence on the donor density. Curves *1* and also *2* for the smallest doping levels show little temperature dependence

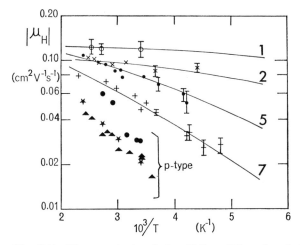

Fig. 9.11. The magnitude of the Hall mobility plotted logarithmically against inverse temperature for four *n*-type samples and three *p*-type specimens. The lines through the data have been calculated on the basis of the model discussed in the text [9.50]

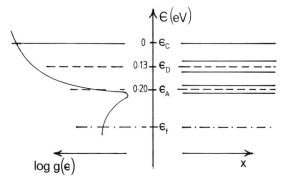

Fig. 9.12. Model used in the analysis of Hall effect and thermoelectric power data. It shows the density of states distribution $g(\varepsilon)$ near ε_c and, on the right, the three current paths in the extended states, the donor band, and the tail states

with $|\mu_H|$ values between 0.1 and $0.2\,\mathrm{cm^2\,V^{-1}\,s^{-1}}$. Both these observations agree with the predictions of the *Friedman* theory [9.42].

The main features of the Hall effect and the associated conductivity results can be explained by the model in Fig. 9.12 which we shall also use for the analysis of the thermoelectric power data presented in the following section. It is similar to that used previously by *Nagels* et al. [9.44] for the interpretation of the Hall effect data in chalcogenide glasses. Three possible conduction paths are considered: in the extended states above ε_c, in the donor band around an energy ε_D, and in the tail states near ε_A. For the last two the transport is by phonon-assisted hopping and an identical hopping energy, W, will be assumed. It should perhaps be emphasized that although we associate conduction through the donor states with a single energy, ε_D, it would appear more likely that the donors are distributed in energy, perhaps throughout most of the tail states [9.5]. ε_D would then be equivalent to the centroid of the distribution.

The drift mobility results in Fig. 9.9 suggested that once the donor density exceeds about $10^{18}\,\mathrm{cm}^{-3}$ (i.e., $R_{\mathrm{D}}^{-1} \simeq 1.4 \times 10^{-2}\,\mathrm{\AA}^{-1}$), the predominant hopping path through the system will have moved to the donor states, so that with the possible exception of specimen 1 we can neglect σ_{A}. The total conductivity σ can then be written as

$$\sigma = \sigma_{\mathrm{c}} + \sigma_{\mathrm{D}} \tag{9.3}$$

with

$$\sigma_{\mathrm{c}} = \sigma_{0\mathrm{c}} \exp\left[-\frac{(\varepsilon_{\mathrm{c}} - \varepsilon_{\mathrm{F}})_0}{kT} \right] \tag{9.4}$$

and

$$\sigma_{\mathrm{D}} = \sigma_{0\mathrm{D}} \exp\left[-\frac{(\varepsilon_{\mathrm{D}} - \varepsilon_{\mathrm{F}})_0 + W}{kT} \right]. \tag{9.5}$$

The observed Hall mobility μ_{H} is the weighted mean of the Hall mobility μ_{c} at ε_{c} and μ_{D} at ε_{D}:

$$\mu_{\mathrm{H}} = \frac{\sigma_{\mathrm{c}}\mu_{\mathrm{c}} + \sigma_{\mathrm{D}}\mu_{\mathrm{D}}}{\sigma_{\mathrm{c}} + \sigma_{\mathrm{D}}}. \tag{9.6}$$

The experimental data were fitted to (9.3–6) on the basis of two assumptions: (I) μ_{c} is independent of temperature; (II) μ_{D} is sufficiently small that $\mu_{\mathrm{D}}\sigma_{\mathrm{D}} \ll \mu_{\mathrm{c}}\sigma_{\mathrm{c}}$ throughout the entire temperature range. The solid lines in Fig. 9.11 show that a very satisfactory fit to the experimental data can be obtained. The dependence of the deduced parameters on the gaseous impurity ratio (and the donor concentration) is shown in Fig. 9.13. In the model proposed here, we would expect the magnitude of $\sigma_{0\mathrm{c}}$ to be independent of doping and, within the experimental error, the data confirms this. The pre-exponent for the hopping conduction, however, is a strong function of the donor concentration and supports the suggestion from drift mobility experiments that the hopping path is associated directly with the impurities. A further result of the analysis is that the energy ε_{D}, the centroid of the donor band, lies about 0.13 eV below ε_{c}. Both Hall and drift mobility experiments therefore lead to the conclusion that the donor band can play an important part in doped amorphous semiconductors. For example, using $\sigma_{0\mathrm{c}}$ and $\sigma_{0\mathrm{D}}$ values from Fig. 9.13 in (9.4) and (9.5), it can easily be seen that even at room temperature with an impurity ratio of 10^3 vppm ($N_{\mathrm{D}} \simeq 5 \times 10^{18}\,\mathrm{cm}^{-3}$), the current through the donor band equals that through the extended states.

9.3.5 Thermoelectric Power

The temperature dependence of the thermoelectric power S can yield useful information on the transport properties of amorphous semiconductors; it is a.

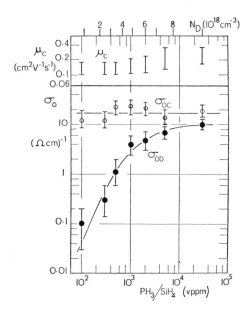

Fig. 9.13. The values of μ_c, σ_{0c}, and σ_{0D} (9.6, 4, and 5) as a function of the gaseous impurity ratio, deduced from the fit of the experimental data to the model. The estimated values of the donor concentration N_D are shown on the top scale [9.50]

far more sensitive experimental approach than the measurement of conductivity. For a current path at an energy ε', S is given by

$$S(\varepsilon) = - \frac{k}{|e|} \left[\frac{(\varepsilon' - \varepsilon_f)_0}{kT} \right] + S_0 , \qquad (9.7)$$

irrespective of the conduction mechanism. Thus a plot of S versus $1/T$ yields the energy difference between the Fermi level and the predominant current path. The significance of the intercept S_0 will be discussed later. Unlike the Hall effect, the sign of S should always reflect the sign of the charge carrier.

The substitutional doping of a-Si has made it possible to extend the study of thermoelectric power in this material over a much wider range of temperatures and Fermi level positions. Results on phosphorus-doped a-Si have recently been reported both by the Marburg [9.52] and Dundee [9.53] groups. It is reassuring that, unlike the Hall effect, S has the expected sign of the charge carrier. Although experimental results from the two laboratories are in reasonable agreement, there are a number of significant differences which have not yet been resolved. Figure 9.14 shows the temperature dependence of S obtained for n-type specimens by *Jones* et al. [9.53]. The solid lines were calculated from the model of Fig. 9.12. Curve *2* for $N_D \simeq 2 \times 10^{17}\,\mathrm{cm}^{-3}$ it typical of the results for very lightly doped specimens; the linear portion leads according to (9.7) to an energy $\varepsilon_s = (\varepsilon' - \varepsilon_f)_0$.

It was found that within the experimental accuracy $\varepsilon_s = \varepsilon_\sigma$, the conductivity activation energy. This suggests that for lightly doped specimens, $\varepsilon' = \varepsilon_c$, so that predominant transport takes place in the extended electron states [see (9.4)]. *Beyer* et al. [9.52], on the other hand, deduced that $\varepsilon_\sigma - \varepsilon_s = 0.07\,\mathrm{eV}$, which

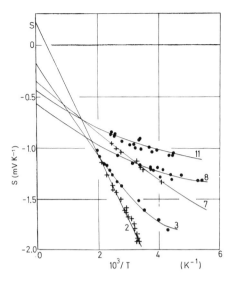

Fig. 9.14. Observed thermoelectric power S plotted as a function of reciprocal temperature for five n-type specimens. The sample numbers increase with increasing phosphorus content. The lines through the experimental points were calculated on the basis of the model in Fig. 9.12 [9.53]

would imply a predominant hopping path through localized states. With increasing doping both the magnitude of S and its temperature dependence decrease (e.g., specimens 7 and 11), and the curves could no longer be described by a single energy ε_s. In particular, these results suggested large negative values of S_0.

An interpretation of the thermopower and conductivity data has been attempted based on two main conclusions from earlier work: first the existence of the parallel conduction paths shown in Fig. 9.12 and secondly the contribution to S_0 from the statistical shift of ε_f with temperature [9.5]. Both these factors have also been considered in the work of the Marburg group [9.52].

The first part of the analysis considers the transport through parallel current paths using a similar approach to that for the Hall effect in the previous section. However, there is an interesting new result: in lightly doped specimens, containing between 5×10^{17} and $10^{18}\,\mathrm{cm}^{-3}$ donors, the analysis of the thermoelectric data showed that at the lower temperatures the predominant hopping contribution comes from a current path about 0.2 eV below ε_c. This corresponds closely to the values of $\varepsilon_c - \varepsilon_A$ obtained for tail-state hopping in the earlier work on drift mobility and conductivity in undoped a-Si [9.8]. However, in specimens containing more than about $10^{18}\,\mathrm{cm}^{-3}$ donors, it was found that the hopping path moves to 0.13 eV below ε_c. This is in agreement with the conclusions from the Hall and drift mobility experiments, and provides further evidence for a transition from tail state to donor hopping once the donor density exceeds $10^{18}\,\mathrm{cm}^{-3}$. The solid lines in Fig. 9.14 represent the calculated results; their general agreement with the experimental points is encouraging.

The intercept S_0 in Fig. 9.14 requires some further discussion. It can be seen that its value varies rapidly with doping from approximately $+0.2\,\mathrm{mV\,K}^{-1}$ in very lightly doped specimens to $-0.6\,\mathrm{mV\,K}^{-1}$ at the highest doping levels

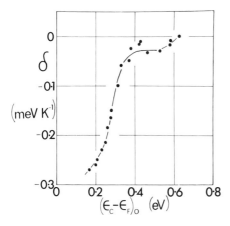

Fig. 9.15. The average temperature coefficient δ of the Fermi level position between 200 and 500 K plotted as a function of $(\varepsilon_c - \varepsilon_f)_0$ [9.53]

where $(\varepsilon_c - \varepsilon_f)_0 \simeq 0.20\,\text{eV}$. As suggested by *Beyer* et al. [9.52], the main reason for this variation is the temperature dependence of ε_f. In a linear approximation this can be expressed as

$$(\varepsilon_c - \varepsilon_f)_T = (\varepsilon_c - \varepsilon_f)_0 - \delta T. \tag{9.8}$$

S_0 is now given by

$$S_0 = -\frac{k}{|e|}\left(A - \frac{\delta}{k}\right). \tag{9.9}$$

A, the "heat of transport," is determined by the energy dependence of σ, whereas the coefficient δ in the more highly doped specimens of interest here arises from the statistical "overflow" of carriers to energies above ε_f in the rapidly rising density of state distribution near ε_A (Fig. 9.1). As discussed by *Spear* and *LeComber* [9.5], the effect causes a movement of ε_f away from ε_c with increasing $T (\delta < 0)$. Figure 9.15 shows the average value of δ between 200 and 500 K as a function of $(\varepsilon_c - \varepsilon_f)_0$, calculated from the known $g(\varepsilon)$. With $A \simeq 2$, which does not seem an unreasonable value, the changes in S_0 with doping can be accounted for by the energy dependence of δ, except in the most highly doped specimens. Once $(\varepsilon_c - \varepsilon_f)_0$ decreases below 0.25 eV, this interpretation breaks down, suggesting that some effect, as yet unidentified, causes A to increase appreciably.

Thermoelectric power experiments on *p*-type a-Si samples doped with B from the gas phase have been carried out by *Beyer* and *Mell* [9.54]. The results are analyzed in terms of two models: one involving two parallel hole conduction paths, the other suggesting transport by small polarons. According to the authors, the analysis indicates that the latter description might be more appropriate, although not entirely consistent with the results from Hall effect experiments.

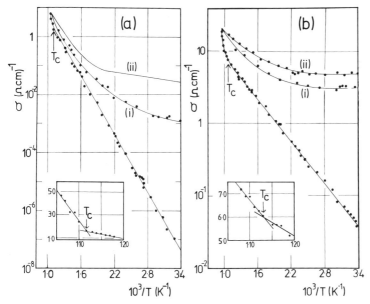

Fig. 9.16a, b. Temperature dependence of conductivity for two a-Si specimens: (**a**) undoped glow discharge specimen (*i*) after crystallization; (*ii*) results of *Pearson* and *Bardeen* [9.56]; (**b**) phosphorus doped a-Si specimen, (*i*) after crystallization (*ii*) after annealing for 30 min. at 700 °C. The inset shows the onset of crystallization at T_c [9.55]

9.3.6 Crystallization of Doped a-Si

The recent work on crystallization of doped a-Si [9.55] has shown that measurements after conversion to the polycrystalline state can provide useful information on some of the properties of the initial amorphous state.

Figures 9.16a and b show typical $\log \sigma$ against $10^3/T$ curves for two a-Si specimens; graph (a) refers to an undoped layer, (b) to a specimen containing about $5.5 \times 10^{18}\,\text{cm}^{-3}$ phosphorus donors in the amorphous phase. In these crystallization experiments, the temperature was raised at a rate of 1.5 °C per minute from room temperature to about 700 °C. The results show a well defined activation energy up to a temperature T_c indicated in the graphs. At T_c a discontinuity in gradient occurs, which is more clearly shown by the insets to these figures. The significant point is that heating to $T > T_c$ produces an irreversible change in the conductivity. For instance, curve (*i*) in Fig. 9.16a was obtained after heating the specimen to 700 °C at the above rate and then allowing it to cool. σ has increased by over four orders of magnitude at room temperature and the X-ray pattern shows without doubt that the specimen has been crystallized. For comparison, we show curve (*ii*), obtained by *Pearson* and *Bardeen* [9.56] on an undoped specimen of polycrystalline Si. In Fig. 9.16b for the doped specimen, curve (*i*) corresponds to the same crystallization procedure. After annealing the specimen for 30 min at 700 °C [curve (*ii*)], the larger

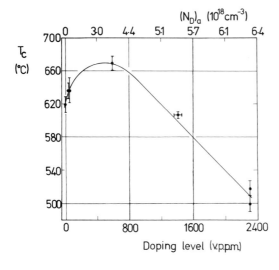

Fig. 9.17. Dependence of crystallization temperature T_c on gaseous doping level. The density $(N_D)_a$ of donors in the amorphous phase is given in the top scale [9.55]

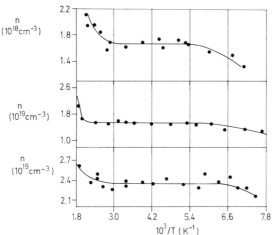

Fig. 9.18. Temperature dependence of the electron density in three crystallized phosphorus doped specimens [9.55]

σ suggests that the grain size in the polycrystalline structure has been increased by this treatment.

To compare the results from different specimens, we shall regard T_c as representing the onset of crystallization under the stated experimental conditions. The dependence of T_c on impurity content is brought out clearly in Fig. 9.17 where it is plotted against the gaseous doping level (in vppm). The approximate donor concentration $(N_D)_a$ originally present in the amorphous phase is also indicated. In undoped specimens, the onset of crystallization lies around 620 °C; with increasing phosphorus content, T_c goes through a maximum and then decreases to about 510 °C at the highest doping levels. This drastic reduction in T_c could in principle be associated with an increase in the growth rate and/or nucleation rate. A similar effect has been observed in the

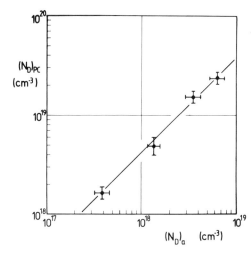

Fig. 9.19. Plot of the donor density $(N_D)_{pc}$ in the polycrystalline state against the donor density $(N_D)_a$ in the same specimens before (crystallization) [9.55]

epitaxial regrowth of (amorphized) Si after ion implantation with impurities such as P or B which have high solid solubilities in Si [9.57].

The sign anomaly in μ_H discussed in Sect. 9.3.4 is an intriguing aspect of Hall effect measurements on a-Si. It is therefore of interest to note that crystallization removes the anomaly in both n- and p-type specimens. In crystallized n-type specimens, μ_H increases up to $36\,\mathrm{cm^2\,V^{-1}\,s^{-1}}$. Figure 9.18 shows the temperature dependence of the electron density n deduced from Hall effect measurements on three crystallized phosphorus-doped specimens.

By analogy with the single-crystal results it is suggested that the flat regions represent the exhaustion range in which practically all the donors are ionized. This value will be used as an approximate measure of $(N_D)_{pc}$, the density of donors in the polycrystalline state.

As explained in Sect. 9.3.2, the density of donors in the amorphous phase $(N_D)_a$ was determined by comparing the measured Fermi level position with that calculated from (9.1) using the experimental density of states distribution (Fig. 9.1). It is of interest to compare $(N_D)_{pc}$ and $(N_D)_a$ for the same specimens, as a consistency check. This is done in Fig. 9.19, showing that

$$(N_D)_{pc} \simeq 3.8\,(N_D)_a\,.$$

On the other hand, the ion-probe analysis on phosphorus-doped specimens established that the density N_P of phosphorus atoms incorporated into the random network is

$$N_P \simeq 3\,(N_D)_a\,.$$

Comparison of the above two results leads to the not unexpected relation

$$(N_D)_{pc} \simeq N_P\,.$$

It means that phosphorus atoms whose additional bonds had been accommodated in the random network during deposition are restored as donors when the specimen is crystallized. Thus conversion to the polycrystalline state provides a useful method to determine N_P for assessing the doping efficiency (Sect. 9.3.2).

9.4 Photoconductivity and Recombination in Doped a-Si

The study of photoconductivity is a promising approach towards a closer understanding of recombination processes in solids. Glow discharge a-Si is a sensitive photoconductor [9.9] which attains its peak response in the red spectral region. In previous work on the undoped material, it has been attempted to relate the recombination of the photogenerated excess carriers to transitions in the known state distribution [9.58]. Substitutional doping has provided interesting new possibilities for the extension of this work in that the photoconductivity σ_{ph} can now be explored at a fixed temperature and light intensity with the Fermi energy in widely different parts of the state distribution.

Figure 9.20 summarises some of the results [9.10]. The room-temperature photoconductivity normalized to an incident flux of 7×10^{14} photons s^{-1} cm^{-2} ($\hbar\omega = 2\,\mathrm{eV}$) is plotted against the position of the dark Fermi level for a number of phosphorus-doped ($\bullet$), undoped ($+$), and boron-doped ($\circ$) specimens. The line P.G. $\simeq 1$ refers to unit photoconductive gain at the comparatively moderate applied field of 3×10^3 V cm^{-1} used throughout. The interesting result is that slight phosphorus doping (~ 10 vppm of phosphine) rapidly increases σ_{ph} and the recombination lifetime by one or two orders of magnitude. At this stage the recombination also changes from a predominantly monomolecular to a bimolecular process, as is indicated by curve v representing the exponent in the intensity dependence $\sigma_{ph} \propto I^v$. On the other hand, slight boron doping, which moves the dark Fermi level into and beyond the density of states minimum, leads to a considerable reduction in σ_{ph}. A qualitatively similar dependence on doping has been established by the Marburg group [9.59].

In trying to relate the observed changes in σ_{ph} and v to changes in gap-state occupation, it will be more meaningful to introduce the quasi-Fermi level position for electrons, $\varepsilon_c - \varepsilon_{fn}$, at the given temperature and intensity, as an independent variable. In Fig. 9.21 the results for 295 and 143 K have been replotted against the new variable. The graphs lead to the general conclusion that the recombination lifetime attains its optimum value when ε_{fn} has moved to a position between 0.35 and 0.30 eV below ε_c; at this stage bimolecular recombination also takes over.

What is the physical significance of this result? In Fig. 9.22 the solid line is a linear plot of $g(\varepsilon)$ between 0.7 and 0.25 eV below ε_c. The broken lines represent the distribution of occupied states $n(\varepsilon) = g(\varepsilon) f(\varepsilon)$ at 143 K for the $\varepsilon_c - \varepsilon_f$ values marked by the dots. Clearly when ε_{fn} reaches the critical energies suggested by

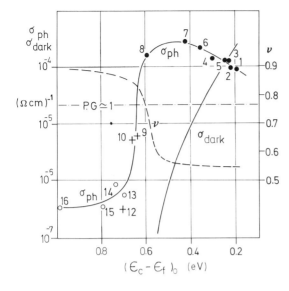

Fig. 9.20. Photoconductivity at 295 K plotted against $(\varepsilon_c - \varepsilon_f)_0$, the position of the dark Fermi level for phosphorus-doped (●), undoped (+), and boron-doped (○) specimens. The broken line represents the exponent ν (see right-hand ordinate) in the intensity dependence $\sigma_{ph} \alpha I^\nu$. σ_{dark} is a typical dark-conductivity curve. P.G.$\simeq 1$ refers to unit photoconductive gain at a field of 3×10^3 V cm^{-1} [9.10]

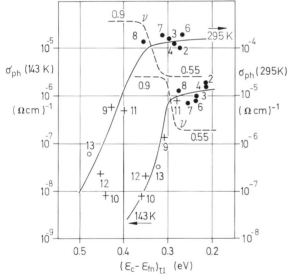

Fig. 9.21. Photoconductivity at 295 K (right-hand ordinate) and at 143 K (left-hand ordinate) plotted against the steady-state Fermi level position at a photon flux of 7×10^{14} s^{-1} cm^{-2}. The broken lines illustrate at each temperature the exponent in the intensity dependence $\sigma_{ph} \alpha I^\nu$ plotted on a linear scale [9.10]

Fig. 9.21, the ε_x peak in the density of states distribution (Fig. 9.1) will just be fully occupied and become the center of negative charge density in the system.

This suggests that changes in the charge distribution produced by doping have an important bearing on the detailed photoconductive behavior. In explanation, consider the charge states indicated in Fig. 9.1. In an undoped specimen at low light intensities, the lifetimes of the photogenerated electrons, which are the predominant charge carriers, are determined by capture into the p^+ distribution, leading to a monomolecular process. Suppose the specimen is

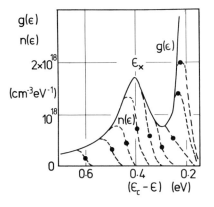

Fig. 9.22. Linear plot of the density of states $g(\varepsilon)$ against the energy measured from ε_c. The broken lines represent the thermal equilibrium electron occupation $n(\varepsilon)$ at 143 K, when the Fermi level lies at the energies marked by the dots [9.10]

now lightly doped, for example with $5 \times 10^{16}\,\mathrm{cm}^{-3}$ phosphorus donors, which will move ε_f to about 0.5 eV below ε_c. The significant point is that n^- has been increased and p^+ decreased by a net total amount N_D. Thus, doping has removed most of the charged p^+ centers which present comparatively high Coulomb cross sections for recombination. As a result, the electron lifetime and the photosensitivity increase rapidly, as is observed in Fig. 9.20. However, further doping (and/or increased intensity) will move both ε_{f_n} and ε_{f_p} into regions of higher state densities and the new charge distribution so produced will set up a competing recombination path. This happens when $(\varepsilon_c - \varepsilon_{f_n}) \simeq 0.33$ eV. Most of the excess charges in the system will now be concentrated in two well-defined regions of the mobility gap and the new recombination path prevents any further increase of the electron lifetime. It is likely that the magnitudes of these two charge distributions will remain proportional to one another with changes in intensity and thus give mainly bimolecular recombination.

The desensitization observed in Fig. 9.20 with light boron doping can be understood in a similar way. The acceptors will increase the extent and density of the p^+ distribution because in Fig. 9.1 the Fermi level will move towards the left on doping with a p-type impurity. Therefore, as long as the photocurrent is carried predominantly by the excess electrons, their lifetime is severely reduced by the increased density of positive Coulomb centers. σ_{ph} thus decreases and the recombination remains monomolecular.

It has generally been found that the photosensitivity of a-Ge is smaller than that of a-Si. In the following comparison it will be useful to consider the product $\mu\tau$ of carrier mobility and recombination lifetime, which can be obtained directly from photoconductivity measurements. For an undoped a-Si specimen this quantity lies between 10^{-6} and $10^{-5}\,\mathrm{cm}^2\,\mathrm{V}^{-1}$ at room temperature [9.9] and is considerably increased by phosphorus doping as shown by Fig. 9.20. Recent work on glow discharge a-Ge [9.60] has led to $\mu\tau \simeq 10^{-7}\,\mathrm{cm}^2\,\mathrm{V}^{-1}$ for the undoped films; appreciable sensitization by doping is observed as in the case of a-Si. In comparison with glow discharge a-Ge, evaporated [9.61] and sputtered [9.62] Ge specimens are much less photo-sensitive with $\mu\tau$ values between 10^{-11} and 10^{-10} cm^2 V^{-1}. It seems surprising that hydrogenation during sputtering [9.62] does not seem to improve the photosensitivity of a-Ge.

9.5 The Amorphous Barrier and p–n Junction

The recent developments in the field of doped amorphous semiconductors have opened up interesting possibilities for the application of these materials. In particular the thin-film a-Si p–n junction [9.63] and the Schottky barrier [9.64] hold considerable promise for large-area device applications, such as photo-voltaic energy conversion [9.65, 66]. It is therefore relevant to consider in this final section the formation and electronic properties of the amorphous barrier and p–n junction. As has already been pointed out [9.67] an amorphous barrier differs basically from its crystalline counterpart in that the net space charge in the barrier region is determined by both the ionized impurities and the localized states. Thus the barrier profile and properties may depend critically on the density of states distribution $g(\varepsilon)$ in the mobility gap of amorphous materials.

9.5.1 Charge Distribution and Barrier Profile

The information needed for calculating the charge density in the a-Si barrier is contained in the density of states curves A and B in Fig. 9.1. The inset to Fig. 9.23 illustrates the model and the notation that will be used in the following. The barrier profile $\varepsilon_b(x)$ is measured from ε_c, the onset of the extended states in the bulk of the specimen. It can represent an amorphous Si metal barrier or the n side of a p–n junction. $\varepsilon_c - \varepsilon_f$, the position of the Fermi level in the volume, is constant for a given specimen, but can of course be controlled during preparation by the gaseous doping level.

The solid curves in Fig. 9.23 represent $(p^+ - n^-)$ and $(n^- - p^+)$, the net positive and negative carrier densities in localized states at 300 K lying in a section through the barrier where the barrier height is ε_b. They were obtained from Fig. 9.1 by integrating over the A and B distributions as a function of the Fermi level position [9.11]. With $\varepsilon_b = 0$, Fig. 9.23 refers to bulk properties. In the undoped specimen, charge neutrality demands that $p^+ = n^-$ at $\varepsilon_c - \varepsilon_{f0}$. With phosphorus or boron doping, the conditions $N_D^+ = (n^- - p^+)$ and $N_A^- = (p^+ - n^-)$ must be satisfied in the specimen volume and the corresponding Fermi level positions can be read off directly.

In discussing the charge distribution ϱ within the barrier, consider as an example the barrier on an n-type a-Si specimen with $\varepsilon_c - \varepsilon_f = 0.45$ eV. According to Fig. 9.23 the positive charge of 10^{17} ionized donors per cm^3 is balanced in the volume by an equal negative charge $n^- - p^+$ in localized (and extended) states. In the barrier, $n^- - p^+$ will at first decrease so that the net space-charge density will be

$$\varrho^+(\varepsilon_b) = |e|\,N_D^+ - |e|\,(n^- - p^+)_{(\varepsilon_b + \varepsilon_c - \varepsilon_f)}, \tag{9.10}$$

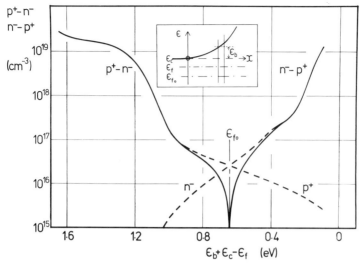

Fig. 9.23. Integrated charge densities in a-Si at room temperature as a function of the Fermi level position with respect to the electron conduction path in the barrier region. The full line shows the net charge $(p^+ - n^-)$ and $(n^- - p^+)$ and the dashed lines n^- and p^+ separately. The inset clarifies the notation used for the energy [9.11]

where the subscript to $n^- - p^+$ denotes the position on the energy axis of Fig. 9.23. When $\varepsilon_b = \varepsilon_{f0} - \varepsilon_f \simeq 0.2\,\mathrm{eV}$, the net charge in localized states vanishes; with larger ε_b it becomes positive and is then given by the $p^+ - n^-$ curve. Thus for $\varepsilon_b > \varepsilon_{f0} - \varepsilon_f$, we have

$$\varrho^+(\varepsilon_b) = |e|\,N_D^+ + |e|\,(p^+ - n^-)_{(\varepsilon_b + \varepsilon_c - \varepsilon_f)}. \tag{9.11}$$

These considerations show that the initial part of the a-Si barrier (9.10) is characterized by a lower space-charge density than the constant value $|e|\,N_D^+$ generally assumed for the crystalline case. However, once a critical value of ε_b is exceeded, the rapidly rising $p^+ - n^-$ contribution will cause a sharp increase in barrier space charge above the crystalline value (9.11).

The space-charge density $\varrho^+(\varepsilon_b)$ can be obtained directly from Fig. 9.23 as a function of ε_b for any given donor (or acceptor) volume density. To calculate the corresponding spatial profile, $\varepsilon_b(x)$, requires a solution of Poisson's equation. Figure 9.24 shows a number of $\varepsilon_b(x)$ curves for different values of $\varepsilon_c - \varepsilon_f$, obtained from $\varrho^+(\varepsilon_b)$ by a step-by-step method of solution [9.11]. The recent work on a-Si [9.64] indicates that metals such as Au or Pt with high work functions lead to barrier heights of about 1 eV. The corresponding $\varepsilon_b(W_0)$ is indicated by the cross on each profile, so that the barrier width $x = W_0$ can be read off directly.

It can be seen that the amorphous barrier in Fig. 9.24 for undoped or lightly doped specimens is characterized by a long, shallow initial part, going over into

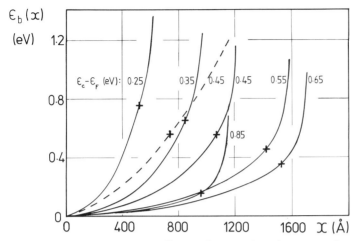

Fig. 9.24. Calculated barrier profile $\varepsilon_b(x)$ for a number of n-type specimens with $\varepsilon_c - \varepsilon_f$ values between 0.25 and 0.85 eV. The dashed curve has been calculated for $\varepsilon_c - \varepsilon_f = 0.45$ eV with a constant space-charge density [9.11]

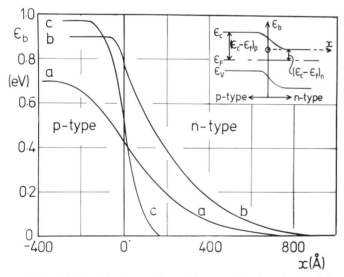

Fig. 9.25. Calculated barrier profiles of three p–n junctions for the following values of $(\varepsilon_c - \varepsilon_f)_n$ and $(\varepsilon_c - \varepsilon_f)_p$: (a) 0.35 and 1.05 eV; (b) 0.35 and 1.25 eV; (c) 0.18 and 1.15 eV. The inset illustrates the choice of origin and the positions of the bands with respect to this point [9.11]

a sharply rising profile. This is a direct consequence of the charge distribution discussed above [i.e., (9.10, 11)]. For comparison, the dashed curve represents $\varepsilon_b(x)$ for $\varepsilon_c - \varepsilon_f = 0.45$ eV with a constant $\varrho^+(\varepsilon_b) = |e| N_D$.

The barrier calculations have also been extended to the a-Si p–n junction and Fig. 9.25 shows three examples. The inset illustrates the choice of origin.

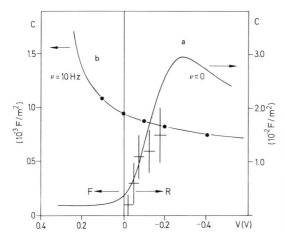

Fig. 9.26. Calculated capacity curves as a function of forward (F) and reverse (R) bias for an undoped a-Si–metal barrier. The curves are calculated for $\varepsilon_c - \varepsilon_f = 0.65$ eV in the a-Si: (a) for a very low frequency $v \simeq 0$; (b) for $v = 10$ Hz. The experimental points were obtained with a Au electrode on undoped a-Si and at $v = 10^{-3}$ Hz (+) and $v = 10$ Hz (●) [9.11]

Curve (a) is the profile for the junction between an n region with $(\varepsilon_c - \varepsilon_f)_n = 0.35$ eV and a p region with $(\varepsilon_c - \varepsilon_f)_p = 1.05$ eV, giving a barrier height of 0.7 eV. The very narrow p-type barrier in curve (b) arises when ε_f for the p material lies 1.25 eV below ε_c in a region of high state density. This curve may be considered as the amorphous analogue of the "one-sided" crystalline junction.

9.5.2 Barrier Capacitance and Its Dependence on Applied Potential and Frequency

The measurement of the barrier capacitance is a useful tool for the investigation of the impurity distribution and has been used extensively with crystalline semiconductors. As shown by the work of *Roberts* and *Crowell* [9.68] such measurements as a function of frequency can also give information on deep lying centers in the gap.

With the information on the space-charge distribution it is possible to calculate the differential capacitance C of the amorphous barrier as a function of the applied potential V. An interesting problem arises: the dependence of C on the measuring frequency v. So far we have assumed that all the states in the barrier attain their thermal equilibrium occupation, so that the net space-charge density at each value of ε_b is given by the curves in Fig. 9.23. However, in measuring the barrier capacitance (for instance on an ac bridge), we determine the effect on the charge distribution of a small change in the applied potential and it is then relevant to ask how quickly states in the barrier region near ε_f can respond to the change in potential. This can be estimated if we assume that the thermal release time τ' is a measure of the time required to approach thermal equilibrium in states at $\varepsilon_c - \varepsilon'$. Therefore if the measuring frequency exceeds a value $v' = 1/\tau'$, then these states will cease to contribute to the differential barrier capacity. For instance, states around ε_f near the metal–semiconductor interface $(\varepsilon_c - \varepsilon' \simeq 1\,\text{eV})$ require about 500 s at room temperature to approach the thermal equilibrium distribution; their contribution to C can therefore be obtained only in a dc experiment. On the other hand a measuring frequency of 10 Hz leads to $\varepsilon_c - \varepsilon' \simeq 0.79\,\text{eV}$. Referring to Fig. 9.23 this means that only the charge $(p^+ - n^-)$ in states for which $\varepsilon'_b + \varepsilon_c - \varepsilon_f \lesssim 0.79\,\text{eV}$ can follow the changes in applied field. This can be taken into account in the capacity calculations by a "cut-off" in the barrier charge distribution such that $\varrho(\varepsilon_b) \simeq$ constant for $\varepsilon_b \gtrsim \varepsilon'_b$.

Figure 9.26 shows the capacity versus applied voltage curves predicted for the two cases mentioned above. In (a) it is assumed that $v \simeq 0$, so that all the barrier states can contribute. This curve is diametrically opposite to that of a crystalline barrier; the unusual behavior in the reverse direction arises from the steep increase in $p^+ - n^-$ (Fig. 9.23) with reverse bias. The experimental points were obtained from a capacity measurement in which the current produced by a 20 mV potential step applied to the barrier was integrated over 10^3 s. In view of the fact that there are no adjustable parameters in the calculation, the agreement is very encouraging and provides a further independent check of the density of states distribution in Fig. 9.1.

Curve (b) shows calculated results and ac bridge measurements for a frequency of 10 Hz. Without the contribution of the deeper lying states, the behavior of C with bias is now similar to that in crystalline semiconductors. Nevertheless it has been shown [9.11] that the interpretation of barrier capacity and $1/C^2$ measurements in amorphous semiconductors require a different approach than that used in the crystalline case.

9.6 Concluding Remarks

The aim of this article has been to provide, as far as is possible at present, a reasonably consistent picture of substitutional doping in the amorphous phase and of the effect of the impurities on some of the electronic properties of a-Si and to lesser extent of a-Ge. There is little doubt that the developments described have greatly increased the scope for fundamental experiments on amorphous semiconductors. The rate of future developments in this field will probably depend to a large extent on the progress that is being made on the applied side. At present gas phase deposition and doping appear to be the most promising approach to the preparation of these materials and an increasing number of laboratories are turning towards these techniques. It may therefore be appropriate to conclude this article by emphasizing that the accurate control of the plasma, necessary for reproducible and optimized electronic specimen properties, is still a major problem. Its solution requires a careful and systematic experimental effort and it is likely that large scale applications of amorphous semiconductors will have to await further improvement and development in preparation techniques.

References

9.1   W. E. Spear, P. G. LeComber: Solid State Commun. **17**, 1193–1196 (1975)
9.2   W. E. Spear (ed.): *Proceedings of the 7th International Conference on Amorphous and Liquid Semiconductors* (CICL, University of Edinburgh 1977)
9.3   W. E. Spear, P. G. LeComber: J. Non-Cryst. Solids **8–10**, 727–738 (1972)
9.4   A. Madan, P. G. LeComber, W. E. Spear: J. Non-Cryst. Solids **20**, 239–257 (1976)
9.5   W. E. Spear, P. G. LeComber: Philos. Mag. **33**, 935–949 (1976)
9.6   A. Madan, P. G. LeComber: In Ref. 9.2, pp. 377–381
9.7   P. G. LeComber, W. E. Spear: Phys. Rev. Lett. **25**, 509–511 (1970)
9.8   P. G. LeComber, A. Madan, W. E. Spear: J. Non-Cryst. Solids **11**, 219–234 (1972)
9.9   R. J. Loveland, W. E. Spear, A. Al-Sharbaty: J. Non-Cryst. Solids **13**, 55–68 (1973/4)
9.10   D. A. Anderson, W. E. Spear: Philos. Mag. **36**, 695–712 (1977)
9.11   W. E. Spear, P. G. LeComber, A. J. Snell: Philos. Mag. B **38**, 303–317 (1978)
9.12   M. H. Cohen, H. Fritzsche, S. Ovshinsky: Phys. Rev. Lett. **22**, 1065–1068 (1969)
9.13   G. W. Neudeck, A. K. Malhotra: J. Appl. Phys. **46**, 239–246 (1975)
9.14   N. F. Mott: Philos. Mag. **19**, 835–852 (1969)
       N. F. Mott, E. A. Davis: *Electronic Processes in Non-Crystalline Materials* (Oxford University Press, Oxford 1971)
9.15   J. R. Hollahan, A. T. Bell (eds.): *Techniques and Application of Plasma Chemistry* (Wiley, New York 1974)
9.16   F. K. McTaggart: *Plasma Chemistry in Electrical Discharges* (Elsevier, New York 1967)
9.17   H. F. Sterling, R. C. G. Swann: Solid State Electron. **8**, 653–654 (1965)
9.18   R. C. Chittick, J. H. Alexander, H. F. Sterling: J. Electrochem. Soc. **116**, 77–81 (1969)
9.19   W. E. Spear: In *Amorphous and Liquid Semiconductors*, ed. by J. Stuke, W. Brenig (Taylor and Francis, London 1974) p. 1–16
9.20   G. A. N. Connell, J. R. Pawlik: Phys. Rev. B **13**, 787–804 (1976)
9.21   M. H. Brodsky, M. Cardona, J. J. Cuomo: Phys. Rev. B **16**, 3556–3571 (1977)
9.22   M. H. Brodsky, M. A. Frisch, J. F. Ziegler, W. A. Lanford: Appl. Phys. Lett. **30**, 561–563 (1977)
9.23   H. Fritzsche: In Ref. 9.2, pp. 3–15
9.24   J. C. Knights: In Ref. 9.2, pp. 433–436
9.25   R. C. Chittick: J. Non-Cryst. Solids **3**, 255–270 (1970)
9.26   T. D. Moustakas, D. A. Anderson, W. Paul: Solid State Commun. **23**, 155–158 (1977)
9.27   A. J. Lewis, G. A. N. Connell, W. Paul, J. R. Pawlik, R. J. Temkin: *Tetrahedrally Bonded Amorphous Semiconductors*, ed. by M. Brodsky, S. Kirkpatrick, D. Weaire (American Institute of Physics, New York 1974) pp. 27–32
9.28   A. J. Lewis: Phys. Rev. B **14**, 658 (1976)
9.29   J. R. Pawlik, W. Paul: In Ref. 9.2, pp. 437–441
9.30   W. Paul, A. J. Lewis, G. A. N. Connell, T. D. Moustakas: Solid State Commun. **20**, 969–972 (1976)
9.31   D. A. Anderson, T. D. Moustakas, W. Paul: In Ref. 9.2, pp. 334–338
9.32   G. Muller, S. Kalbitzer, W. E. Spear, P. G. LeComber: In Ref. 9.2, pp. 442–446
9.33   W. Beyer, R. Fischer: Appl. Phys. Lett. **31**, 850–852 (1977)
9.34   J. C. Knights: Philos. Mag. **34**, 663–667 (1976)
9.35   S. Li: M. Sc. Thesis (University of Dundee 1977)
9.36   D. I. Jones, W. E. Spear, P. G. LeComber: J. Non-Cryst. Solids **20**, 259–270 (1976)
9.37   W. E. Spear: J. Non-Cryst. Solids **1**, 197–214 (1969)
9.38   D. Allan, P. G. LeComber, W. E. Spear: In Ref. 9.2, pp. 323–327
9.39   M. Finetti, A. M. Mazzone, L. Passari, B. Ricco, E. Susi: Philos. Mag. **35**, 1141–1151 (1977)
9.40   A. R. Moore: Appl. Phys. Lett. **31**, 762–764 (1977)
9.41   D. Allan: Philos. Mag. B **38**, 381–392 (1978)
9.42   L. Friedman: J. Non-Cryst. Solids **6**, 329–341 (1971)
9.43   D. Emin: Philos. Mag. **35**, 1189–1198 (1977); in Ref. 9.2, pp. 249–260

9.44 P.Nagels, R.Callearts, M.Denayer: In *Amorphous and Liquid Semiconductors*, ed. by J.Stuke, W.Brenig (Taylor and Francis, London 1974) pp. 867–876

9.45 A.J.Grant, T.D.Moustakas, T.Penny, K.Weiser: In *Amorphous and Liquid Semiconductors*, ed. by J. Stuke, W.Brenig (Taylor and Francis, London 1974) pp. 325–333

9.46 M.Roilos, E. Mytilineou: In *Amorphous and Liquid Semiconductors*, ed. by J. Stuke, W. Brenig (Taylor and Francis, London 1974) pp. 319–324

9.47 C.H.Seager, D.Emin, R.K.Quinn: Phys. Rev. B**8**, 4746–4760 (1973)

9.48 C.H.Seager, M.L.Knotek, A.H.Clark: In *Amorphous and Liquid Semiconductors*, ed. by J. Stuke, W.Brenig (Taylor and Francis, London 1974) pp. 1133–1138

9.49 E.Mytilineou, E.A.Davis: In Ref. 9.2, pp. 632–636

9.50 P.G.LeComber, D.I.Jones, W.E.Spear: Philos. Mag. **35**, 1173–1187 (1977)

9.51 W.Beyer, H.Mell, H.Overhof: In Ref. 9.2, pp. 328–332

9.52 W.Beyer, A.Medeišis, H.Mell: Comm. Phys. **2**, 121 (1977)

9.53 D.I.Jones, P.G.LeComber, W.E.Spear: Philos. Mag. **36**, 541–551 (1977)

9.54 W.Beyer, H.Mell: In Ref. 9.2, p. 333

9.55 O.J.Reilly, W.E.Spear: Philos. Mag. B**38**, 295–302 (1978)

9.56 G.L.Pearson, J.Bardeen: Phys. Rev. **75**, 865–883 (1949)

9.57 L.Csepregi, E.F.Kennedy, T.J.Gallagher, J.W.Mayer, T.W.Sigmon: J. Appl. Phys. **48**, 4234–4240 (1977)

9.58 W.E.Spear, R.J.Loveland, A.Al-Sharbaty: J. Non-Cryst. Solids **15**, 410–422 (1974)

9.59 W.Rehm, R.Fischer, J.Stuke, H.Wagner: Phys. Status Solidi (b) **79**, 539–547 (1977)

9.60 D.I.Jones, W.E.Spear, P.G.LeComber, S.Li, R.Martins: Philos. Mag. B**39**, 147–158 (1979)

9.61 N.Croiteru, L.Vescan: Conference on Amorphous Semiconductors, Reinhardsbrunn (Ac. d. Wiss. der DDR 1974)
      T.M.Donovan, M.L.Knotek, J.E.Fischer: In *Amorphous and Liquid Semiconductors*, ed. by J. Stuke, W.Brenig (Taylor and Francis, London 1974) pp. 549–555

9.62 T.D.Moustakas, W.Paul: Phys. Rev. B**16**, 1564–1576 (1977)

9.63 W.E.Spear, P.G.LeComber, S.Kinmond, M.H.Brodsky: Appl. Phys. Lett. **28**, 105–107 (1976)

9.64 C.R.Wronski, D.E.Carlson, R.E.Daniel: Appl. Phys. Lett. **29**, 602–605 (1976)

9.65 D.E.Carlson, C.R.Wronski: Appl. Phys. Lett. **28**, 671–673 (1976)

9.66 D.E.Carlson: IEEE Trans. ED-**24**, 449–453 (1977)

9.67 W.E.Spear: In *Proc. 13th Int. Conf. Physics of Semiconductors*, Rome, 1976, ed. by F.G.Fumi (North-Holland, Amsterdam 1976), pp. 515–524

9.68 G.I.Roberts, C.R.Crowell: J. Appl. Phys. **41**, 1767–1776 (1970)

10. Amorphous Silicon Solar Cells

By D. E. Carlson and C. R. Wronski

With 33 Figures

The first solar cell was made in 1954 by *Chapin* et al. [10.1] when they demonstrated that sunlight could be converted directly into electrical power with a conversion efficiency of $\sim 6\%$ using a *p–n* junction in single-crystal silicon. Solar cell research thrived in the early 1960s mainly as a result of the utilization of solar cells in space. Recently, the possibility of terrestrial applications has generated new interest in this area.

Considerable progress has been made since 1954 in improving the conversion efficiency. Single-crystal silicon cells have exhibited conversion efficiencies as high as 19% [10.2] while GaAs cells have exhibited efficiencies as high as 23% [10.3]. At present, the cost of solar cells is at least ten dollars per peak watt, and a reduction in price by more than an order of magnitude is necessary before large-scale terrestrial applications become practical. The amorphous semiconductor solar cells discussed in this chapter represent a promising new approach toward the development of low-cost solar cells for terrestrial applications.

10.1 Amorphous Semiconductor Cells

10.1.1 Background

The first amorphous semiconductor solar cells were made at RCA Laboratories in 1974 [10.4, 5]. Photovoltaic energy conversion was observed in several types of devices such as *p–n*, *p–i–n*, and Schottky-barrier junctions as well as heterojunctions. The amorphous semiconductor films were deposited from a glow discharge in silane (SiH_4), and subsequently, these films have been shown to consist of a glassy alloy of silicon and hydrogen [10.6–10].

This hydrogenated form of amorphous silicon (a-Si:H) was first studied by *Chittick* et al. in 1969 [10.11]. They observed a strong dependence of the film resistivity on substrate temperature and also a photoconductive effect much larger than that observed in evaporated or sputtered a-Si films. They also reported some preliminary data indicating that the discharge-produced films could be substitutionally doped with phosphorus by adding small amounts of phosphine (PH_3) to the SiH_4 discharge atmosphere.

A much more detailed and definitive study of the substitutional doping of a-Si:H was pursued by *Spear* and *Le Comber* [10.12], and they independently developed the technology to make *p–n* junctions in a-Si:H [10.13]. These investigators also showed that a-Si:H possesses a relatively low density of gap states and that this density depends on the substrate temperature [10.14]. Chapter 9 reviews the subject of doped amorphous semiconductors.

Amorphous Si:H films have also been produced by sputtering in an atmosphere of Ar and H_2 [10.15], and devices made from these films have also exhibited photovoltaic effects [10.16]. In fact, this research effort was one of the first to recognize the role of hydrogenation in reducing gap states in tetrahedral amorphous semiconductors [10.17].

At present, efficient photovoltaic energy conversion has not been demonstrated in any amorphous material other than a-Si:H, but some scientists believe that amorphous chalcogenide materials may be used to make solar cells [10.18]. Amorphous organic semiconductor films have exhibited photovoltaic energy conversion and efficiencies of $\sim 1\%$ have been achieved [10.19]. Attempts were made at RCA Laboratories to make solar cells using a-Ge:H but the photovoltaic effect was negligible.

10.1.2 Conditions for Efficient Photovoltaic Energy Conversion

The efficiency of a solar cell is defined by

$$n = \frac{J_m V_m}{P_i} = \frac{(FF)J_{sc}V_{oc}}{P_i}, \tag{10.1}$$

where J_m and V_m are the output current density and voltage for a cell operating under maximum output power conditions, and P_i is the total power incident on the cell ($P_i \approx 100\,\text{mW cm}^{-2}$ for the sun directly overhead on a clear day; AM1 condition). J_{sc} is the current density of the cell under short-circuit conditions, and V_{oc} is the voltage under open-circuit conditions. The fill factor (FF) is defined by (10.1), i.e., $(FF) = \dfrac{J_m V_m}{J_{sc} V_{oc}}$, and can approach values in the range 0.85–0.90 for ideal diodes [10.20].

We now consider the conditions that must be satisfied for a thin-film semiconductor device to act as an efficient solar cell. First, the optical absorption coefficient must be sufficiently large to absorb a significant fraction of the solar energy in the thin film employed. For films of the order of 1 µm in thickness, the absorption coefficient α must be greater than $10^4\,\text{cm}^{-1}$ over at least the visible portion of the solar spectrum.

A second condition that must be satisfied is that the photogenerated electrons and holes be efficiently collected by contacting electrodes on both sides of the semiconductor film. This condition implies that the minority carrier diffusion length be comparable to the film thickness or that a built-in space-charge field be present in most of the film.

A large built-in potential is also necessary for efficient photovoltaic energy conversion since this potential determines the output voltage of the cell. A built-in potential is generated by the formation of a semiconductor junction such as a p–n junction, heterojunction, or a Schottky-barrier junction [10.21].

Finally, the total resistance in series with the solar cell (excluding load resistance) must be kept small so that the IR drop during operation is only a small percentage of the output voltage. Contributions to the series resistance can come from the contacts, the bulk resistivity of the semiconductor film, the sheet resistance of thin conductive coatings, the current collection grids, and the electrical wiring.

10.2 Deposition of Hydrogenated Amorphous Silicon

Since a–Si:H is the only amorphous semiconductor that has been used to make relatively efficient solar cells, the remainder of this chapter will concentrate on the deposition and properties of a–Si:H films as well as the electrical characteristics of a–Si:H solar cells. In this section we consider the glow discharge deposition process and the effect of various substrates on the a–Si:H films. As mentioned in Sect. 10.1.1, a–Si:H films can also be deposited by sputtering in an atmosphere of Ar and H_2 [10.15], but since efficient solar cells have not yet been produced by this technique, the following discussion will be limited to glow discharge deposition.

10.2.1 Glow Discharge Deposition Conditions

Several types of glow discharge systems have been used to deposit a–Si:H films. Most of the early work utilized an rf electrodeless glow discharge generated by a coil external to the discharge chamber [10.11]. Such systems usually operate within a frequency range of 0.5–13.5 MHz and at SiH_4 pressures of ~0.1–2.0 Torr; flow rates are typically in the range of ~0.2–5.0 standard cm^3 per min (sccm), and the deposition rates are usually in the range of 100–1000 Å/min. The major disadvantage of the rf electrodeless systems is that the uniformity of the a–Si:H film tends to be poor due to the relatively small discharge chamber (typically 5–7 cm inner diameter).

Much better uniformity is obtained with an rf capacitive glow-discharge system similar to those used for sputter deposition [10.22]. The capacitive discharge system uses parallel-plate electrodes inside the discharge chamber and generally operates at a frequency of 13.5 MHz [10.22, 23]. The SiH_4 pressure is usually in a range of 5–250 mTorr, flow rates are typically 10–30 sccm, and deposition rates are generally 500 Å/min.

A dc glow discharge in SiH_4 can also be used to deposit a–Si:H films [10.4]. If the substrate is made the cathode, then a–Si:H films can be deposited at rates ranging from 0.1 to 1.0 μm/min by varying the cathodic current density from 0.2 to 2.0 mA cm^{-2} in ~1.0 Torr of SiH_4. The deposition rate on anodic substrates is roughly an order of magnitude less than on cathodic substrates.

A dc discharge can also be used to deposit films on insulating or conductive substrates by placing a cathodic screen in the proximity of the substrate (see

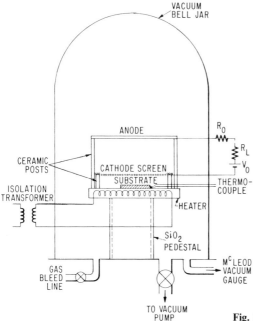

VACUUM
BELL JAR

ANODE

R_0

R_L

CERAMIC
POSTS

CATHODE SCREEN
SUBSTRATE

V_0

ISOLATION
TRANSFORMER

THERMO-
COUPLE

HEATER

SiO$_2$
PEDESTAL

M^{C}LEOD
VACUUM
GAUGE

GAS
BLEED
LINE

TO VACUUM
PUMP

Fig. 10.1. dc glow-discharge deposition system

Fig. 10.1). In this case, the deposition rate will be less than that for a cathodic substrate and will depend on the SiH$_4$ pressure, the current density to the screen, and the screen–substrate spacing. One disadvantage to this technique is that flakes of a–Si:H tend to pop off the cathodic screen, especially for long deposition times ($\gtrsim 10$ min).

Many of the systems described above have been built with substrate heaters capable of reaching temperatures of $\sim 600\,°C$. However, as discussed in Sect. 10.5.2, good electronic properties are only obtained if the a–Si:H films are deposited at substrate temperatures in the range of 200–400 °C.

Generally, gas-phase nucleation or glow-discharge polymetrization will occur in all discharge systems at high SiH$_4$ pressures [10.24]. In such cases, the a–Si:H films will appear hazy due to the precipitation of particles from the discharge. These films often exhibit good electronic properties except in extreme cases where the films appear matty and scratch easily. Gas phase nucleation can be suppressed by operating at low SiH$_4$ pressures [10.24] (this condition can also be met by using SiH$_4$ diluted in Ar [10.25]).

10.2.2 Substrate Effects

The properties of a–Si:H films are not strongly affected by the substrate material in most cases. Thus, for substrate temperatures $<400\,°C$, metals such as Cr, Ti, V, Nb, Ta, and Mo have all been used to make a–Si:H solar cells.

Auger electron spectroscopy has been used to profile several of these cells and diffusion effects do not appear to be significant for temperatures $<400\,°C$. For example, the diffusion coefficient of Mo in a–Si:H is estimated to be $\lesssim 10^{-18}$ $cm^2\,s^{-1}$ at $450\,°C$.

However, other substrates such as Fe and Al cannot be used at deposition temperatures near $400\,°C$ since the diffusion coefficients are ~ 2–3×10^{-15} $cm^2\,s^{-1}$. Good solar cells can be fabricated on Fe substrates at temperatures $\lesssim 330\,°C$ (the diffusion coefficient for Fe in a–Si:H is $<10^{-18}$ $cm^2\,s^{-1}$ at $300\,°C$). Some good solar cells have also been fabricated on Al substrates at temperatures $\lesssim 300\,°C$, but results have not been as reproducible as with other substrates. The presence of an oxide on the substrate surface can cause variations in the contact resistance (see Sect. 10.5.2), and may also influence the formation of silicides; Herd et al. [10.26] found that the eutectic temperatures for Al in contact with evaporated a–Si was $\sim 275\,°C$. Generally, all metal substrates should be chemically cleaned or sputter cleaned just before deposition of the a–Si:H.

Some metals appear to be unsuitable as substrates due to interdiffusion and silicide formation even at low temperatures (Au, Cu), and poor adhesion has been observed on Ag (perhaps due to an oxide layer). Poor adhesion is evident in the form of small blisters in the a–Si:H films, and these blisters often occur on crystalline substrates such as Si, Ge, and GaAs. Infrared absorption data indicate that some ordering of the a–Si:H structure may be occurring on crystalline Si substrates at deposition temperatures as low as $300\,°C$ [10.9].

Insulating substrates such as fused silica do not appear to influence the properties of a–Si:H, but alkali-containing glasses may dope the a–Si:H with alkali impurities that act as donors [10.27].

10.3 Properties of Hydrogenated Amorphous Silicon Films

In this section, we consider the properties of a–Si:H that make it attractive as a solar cell material.

10.3.1 The Role of Hydrogen

Recently, several investigators [10.6–10] have shown that films prepared by the glow discharge decomposition of SiH_4 contain considerable amounts of bonded hydrogen (~ 5–50 at. %). The hydrogen apparently compensates bonds that would otherwise be dangling (or reconstructed dangling) bonds, and thus the density of states in the gap is relatively small [10.14, 28, 29].

The hydrogen concentration of a–Si:H films depend on deposition conditions such as substrate temperature, type of discharge, discharge power, SiH_4 pressure, and SiH_4 flow rate. In general, the hydrogen content will decrease as the substrate temperature increases [10.9, 30, 31]. The hydrogen content has been shown to increase with power for an rf electrodeless discharge in ~ 0.8 Torr of SiH_4 with a substrate temperature of $\sim 310\,°C$ [10.31]. The same study also showed that the hydrogen content decreased as the SiH_4 pressure increased. However, since the hydrogen content depends on many deposition parameters, much more experimental work is needed before the discharge conditions can be correlated with the actual film compositions.

The a–Si:H films appear to be a glassy silicon–hydrogen alloy, and no specific concentration of hydrogen is required for good electronic properties. In fact, reasonably good solar cells have been made using a–Si:H films with hydrogen concentrations in the range of 10–50 at. % [10.31]. However, all these films were deposited at substrate temperatures in the range 200–400 °C.

Films deposited at substrate temperatures below $\sim 200\,°C$ do not exhibit good electronic properties. These films contain dihydride and possibly trihydride groups [10.9, 30, 32], and these groups may act as recombination centers. Films deposited near room temperature may be more accurately classified as polysilanes [10.30].

Above $\sim 200\,°C$, the hydrogen appears to exist only in the monohydride form, i.e., SiH groups [10.9, 30, 32]. There is no evidence for interstitial atomic or molecular hydrogen [10.29, 30], and it is likely that such interstitial species would rapidly diffuse out of the films. Recent electron diffraction studies of an a–Si:H film deposited at $\sim 205\,°C$ indicate that the film possesses a greater degree of short-range tetrahedral ordering than evaporated a–Si films [10.33].

One possible model for such films is that the hydrogen is incorporated in the form of compensated microvoids such as monovacancies containing 4 hydrogen atoms, or divacancies containing 6 hydrogen atoms. (The compensated monovacancy should be relatively stable since a silicon atom and four hydrogen atoms must simultaneously jump for it to move.) These compensated microvoids should not give rise to localized states in the gap [10.34]. The size and concentration of these microvoids would probably depend strongly on the deposition conditions.

Tanielian et al. [10.35] found that the density of a–Si:H films increased from 1.92 to $2.17\,g\,cm^{-3}$ as the H to Si ratio decreased from 0.35 to 0.13. If we assume that the density of void-free a–Si is close to that of crystalline Si, then the density of an a–Si:H film containing 30 at. % of hydrogen in compensated divacancies is $\sim 2.16\,g\,cm^{-3}$ (assuming no local dilational effects). The density measurements of *Tanielian* et al. [10.35] suggest that the average microvoid in their films is slightly larger than a divacancy and that the changing hydrogen content is primarily due to a change in the concentration of these compensated microvoids.

When a–Si:H films are heated to temperatures greater than the deposition temperature (or $\gtrsim 350\,°C$), hydrogen gas evolves from the material [10.6–9] and the density of dangling bonds increases [10.8, 29]. Thus, the out-diffusion of hydrogen is a mechanism that can limit the operational life of a–Si:H solar cells.

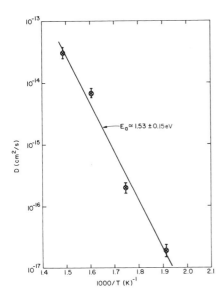

Fig. 10.2. The deuterium diffusion coefficient as a function of 1000/T

The diffusion of deuterium in a–Si:H films was studied by fabricating layered samples of hydrogenated and deuterated films and then analyzing the profiles with secondary ion mass spectrometry (SIMS) [10.10] before and after various heat treatments. Figure 10.2 shows a plot of the diffusion coefficient for deuterium as a function of $1000/T$, and $D(T) = 1.7\,10^{-2}\,\exp(-1.53 \pm 0.15$ eV$/kT)$ [cm^2 s^{-1}] [10.36]. The diffusion profiles were not influenced by fields of $\sim 3\,10^4$ V cm^{-1} or illumination levels of ~ 1 sun (~ 100 mW cm^{-2}) at temperatures of $\sim 300\,°$C. These results imply that degradation of a–Si:H solar cells due to hydrogen out-diffusion at $100\,°$C will not be significant until after more than 10^4 years.

10.3.2 Optical Properties

As shown in Fig. 10.3, the optical absorption coefficient α of a–Si:H increases as the substrate temperature increases, and the absorption coefficient can be more than an order of magnitude larger than that of crystalline Si over most of the visible light range (9). Thus, most of the solar radiation with $\lambda < 0.7\,\mu$m can be absorbed in an a–Si:H film $\sim 1.0\,\mu$m thick.

As discussed in Chap. 4, an optical band gap (E_{OPT}) for a–Si:H films can be determined by plotting $(\alpha h\nu)^{1/2}$ as a function of $h\nu$; this behavior is predicted for amorphous semiconductors if the band edges are parabolic and the matrix elements for the optical transitions are independent of energy [10.37]. The variation of E_{OPT} with substrate temperature is shown in Fig. 10.4 for a–Si:H films produced in dc proximity and rf capacitive glow discharges [10.23]. The increase in E_{OPT} with decreasing substrate temperature is due to the increasing

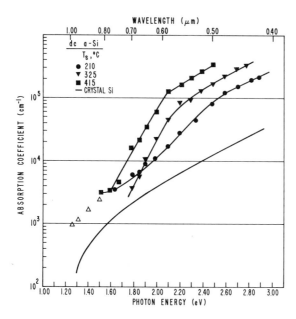

Fig. 10.3. Absorption coefficient as a function of photon energy for a-Si:H films produced in a dc proximity discharge. Data represented by the symbol △ were determined from thick films (~3 μm) deposited at 325 °C

hydrogen content of the films [10.9]. As discussed in Sect. 10.3.1, the hydrogen content can also be increased by varying other deposition parameters such as the discharge power or pressure, and in these cases E_{OPT} will also increase [10.31]. Recent photoelectron emission studies by *von Roedern* et al. [10.32] show that the addition of hydrogen to the amorphous silicon structure causes the top of the valence bond to move downward, and hence E_{OPT} increases.

The optical absorption data can be used to estimate the theoretical limit for the short-circuit current density (J_{sc}) of an a–Si:H solar cell. For an a–Si:H film deposited at a substrate temperature of ~320 °C, the maximum value of J_{sc} is ~20 mA cm^{-2} for a film 1.5 μm thick [10.5]. This estimate assumes that each absorbed photon produces an electron–hole pair and that all photo-generated carriers are collected. The calculated limit for J_{sc} will vary as the hydrogen content (or E_{OPT}) varies.

Optical absorption is also observed for photon energies less than E_{OPT} (see Fig. 10.3), but this absorption is not directly related to the hydrogen content of the a–Si:H films [10.9]. Instead, this optical absorption appears to be related to defect states (e.g., dangling bonds) in the energy gap and goes through a minimum at a substrate temperature of ~300 °C [10.38]. (Photoluminescence and spin density measurements also reveal a minimum in the density of defect states at a substrate temperature of ~300–350 °C; see Sect. 10.3.5.)

The refractive index of a–Si:H is relatively close to that of crystalline Si and only increases by ~5% as the substrate temperature increases from ~195 to ~420 °C (and the hydrogen content decreases from ~40 to ~10% at. %) [10.9]. This trend is consistent with a decreasing gap [10.39].

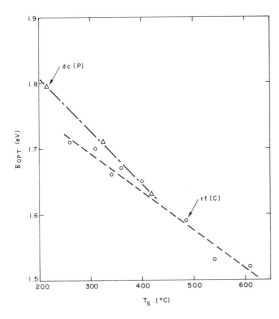

Fig. 10.4. Optical bandgap as a function of substrate temperature for a-Si:H films produced in dc proximity and rf capacitive discharges

10.3.3 Resistivity and Photoconductivity

The resistivity and photoconductivities of a–Si:H films are important in determining the performance of a–Si:H solar cells (see Sect. 10.5). Moreover, studies of these properties give information about the carrier lifetimes and the distributions, densities, and types of states in the gap. Both the resistivity and the photoconductivity of undoped a–Si:H films were studied by *Chittick* et al. in 1969 [10.11], and subsequently, several more detailed investigations have been performed in both undoped [10.40, 4] and doped a–Si:H films [10.22, 27, 42–44]. However, recently *Staebler* and *Wronski* [10.45] reported that large reversible conductivity changes could occur in a–Si:H films depending on their thermal history and their exposure to light. Since large changes in both the dark resistivity and the photoconductivity can occur even by exposure to room light, the usefulness of measurements made in the past is limited.

Subsequent studies have taken the effect of these conductivity changes into account [10.9, 46–48]. A reproducible, "annealed" state can be obtained by heating a–Si:H to 200 °C for ∼ 30 min and limiting the exposure to light. While continued exposure to room light causes gradual conductivity changes, it has been found that the properties tend to stabilize into a "light-soaked" state after ∼ 4 h of exposure to 200 mW cm⁻² of white light (see Fig. 10.5). The temperature dependence of the dark conductivity is shown in Fig. 10.6 for an a–Si:H film in the annealed state (curve *A*) and in the light-soaked state (curve *B*).

The dark resistivity of undoped a–Si:H in the annealed state can vary from $\gtrsim 10^{11}\,\Omega\,cm$ at a substrate temperature (T_s) of 100 °C to $\gtrsim 10^5\,\Omega\,cm$ for

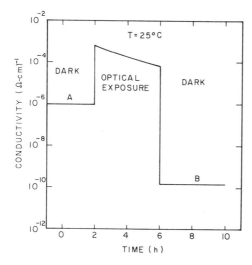

Fig. 10.5. Conductivity as a function of time before, during and after exposure to ∼200 mW cm⁻² of light

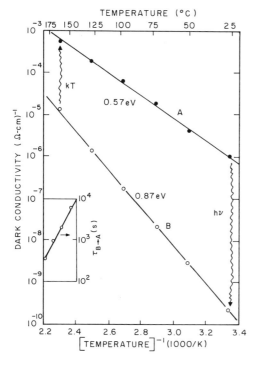

Fig. 10.6. Dark conductivity as a function of temperature for a sample (A) in the annealed state and (B) in the light-soaked state. The relaxation time shown in the inset has an activation energy of ∼1.5 eV

$T_s \simeq 550\,°C$ [10.11, 14]. The undoped films become more strongly n type as the substrate temperature increases. The resistivity and photoconductivity also depend on other deposition conditions such as discharge power and type of discharge [10.9]. This behavior indicates that the density of gap states is strongly influenced by the deposition conditions.

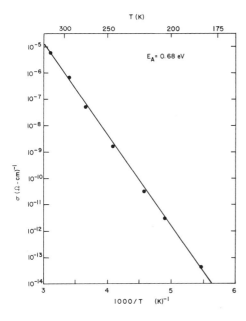

Fig. 10.7. Dark conductivity of a-Si:H as a function of 1000/T

The temperature dependence of the dark conductivity can be described by the expression

$$\sigma = \sigma_0 \exp(-E_a/kT), \tag{10.2}$$

where values of E_a vary from ~ 0.2 to $\sim 0.8\,\text{eV}$ for undoped a–Si:H [10.11, 14]. The conductivity has also been analyzed in terms of the expression for free carrier conduction [10.14, 43],

$$\sigma = q\mu_c n = q\mu_c N_c \exp\left(-\frac{E_c - E_f(T)}{kT}\right), \tag{10.3}$$

where μ_c is the extended state mobility, n is the density of free electrons, N_c is the effective density of states in the conduction band, and $E_c - E_f(T)$ is the position of the Fermi level with respect to the edge of the conduction band. Usually one assumes that $E_f(T) = E_f(0) + \gamma T$ so that $E_a = E_c - E_f(0)$.

The behavior of the resistivity and photoconductivity of undoped a–Si:H is consistent with free carrier transport over a wide temperature range. Figure 10.7 shows that a single activation energy of 0.68 eV can be used to fit the dark conductivity of an a–Si:H film over a temperature range from ~ 180 to $\sim 330\,\text{K}$. The conductivity data were obtained from a Schottky diode operating in the forward-bias regime. This result differs from previously published work [10.14] where a transition in carrier transport occurred at $\sim 250\,\text{K}$; below 250 K, the transport appeared to be due to hopping through localized states in

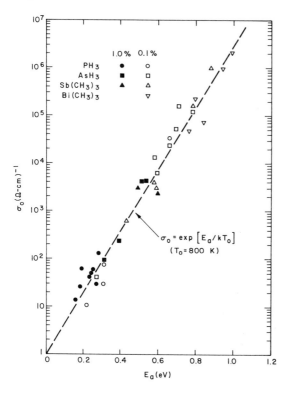

Fig. 10.8. A plot of the pre-exponential term, σ_0, as a function of the activation energy for a number of doped a-Si:H films

the gap. The discrepancy may be due to differences in the discharge conditions since, as noted earlier, the properties of a–Si:H depend strongly on deposition conditions.

An interesting observation is that the dark conductivity obeys the *Meyer–Neldel* rule [10.49], where the preexponential term σ_0 in (10.2) is given by

$$\sigma_0 = \sigma_{00}\exp(E_a/kT_0), \tag{10.4}$$

where σ_{00} and kT_0 are constants. This behavior is illustrated in Fig. 10.8 where $\ln\sigma_0$ is plotted as a function of E_a for a number of doped a–Si:H films. Undoped a–Si:H films also obey the Meyer–Neldel rule, and when exposed to light the changes in the dark conductivity follow (10.4) [10.47]. This behavior is not well understood, but may be due to the movement of the Fermi level through an exponential tail-state distribution as the density of a donor state changes. Other possible explanations involve a statistical temperature shift of the Fermi level or a gradual transition between two different conduction mechanisms [10.43].

The conduction in a–Si:H becomes nonohmic when the average electric field exceeds $\sim 10^4\ \mathrm{V\ cm^{-1}}$. A study of the high-field conduction as a function of temperature and film thickness indicates that Poole–Frenkel conduction

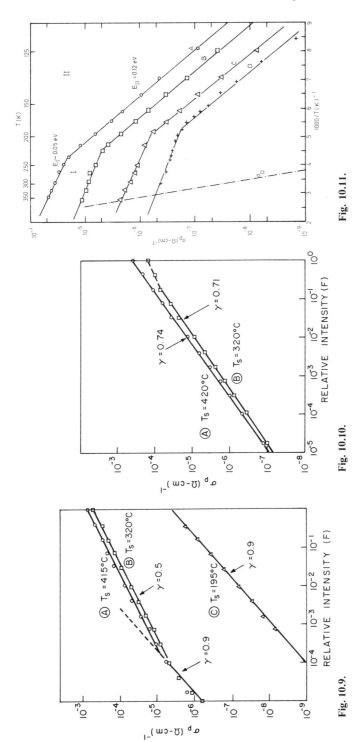

Fig. 10.9.

Fig. 10.10.

Fig. 10.11.

Fig. 10.9. Photoconductivity of an a-Si:H film produced in a rf electrodeless discharge as a function of illumination for various deposition temperatures

Fig. 10.10. Photoconductivity of an a-Si:H film produced in a dc proximity discharge as a function of illumination for various deposition temperatures

Fig. 10.11. Photoconductivity as a function of $1000/T$ for an a-Si:H film produced in a dc proximity discharge ($T_s \simeq 320\,°C$). The photon flux at $\lambda = 0.61\,\mu m$ is A: 2×10^{15}, $B = 2 \times 10^{14}$, C: 2×10^{13}, D: $2 \times 10^{12}\,cm^{-2}\,s^{-1}$. The field was $\sim 10^3\,V\,cm^{-1}$. Also shown is the dark conductivity in the annealed state

[10.50] is the likely mechanism [10.51]. This field ionization of carriers out of traps may be responsible for the high collection efficiency of carriers photo-generated in the space-charge region of a–Si : H solar cells (see Sect. 10.5.4).

As discussed in the preceding chapter by Spear and LeComber, a–Si:H films can be doped either n- or p-type, and resistivities as low as $10^2 \,\Omega$ cm can be obtained. Doped a–Si : H layers are utilized in the solar cell structures to make junctions and ohmic or low resistance contacts between metal electrodes and undoped a–Si : H (see Sect. 10.4). Such contacts are important not only in solar cells but also in photoconductivity studies where injecting contacts are required [10.9, 46].

The photoconductivity of undoped a–Si : H depends strongly on deposition conditions and exhibits a wide range of recombination kinetics similar to those found in crystalline photoconductors [10.52]. The recombination kinetics reflect the variety of distributions of deep centers in a–Si : H and cannot always be explained by a simple combination of nonmolecular and bimolecular processes.

Examples of the photoconductivity found in annealed, undoped a–Si : H produced by rf and dc discharges are shown in Figs. 10.9, 10, respectively [10.9]. The photoconductivity σ_p has a dependence on the illumination F, which is of the form

$$\sigma_p \propto F^\gamma, \tag{10.5}$$

where γ is a constant. Values of γ close to unity reflect a monomolecular type of recombination, whereas $\gamma = 0.5$ reflects a bimolecular type. However, values of $\gamma = 0.7$–0.75 over four orders of magnitude represent a recombination which reflects an exponential or quasi-exponential distribution of recombination centers. These photoconductivity characteristics are consistent with the treatment of recombination and trapping kinetics extensively discussed by *Rose* [10.52].

Recently, this treatment has been applied to the studies of the photoconductivity of both annealed and light-soaked states allowing extensive information to be obtained about the densities, energies, and types of centers in the gap [10.53].

Ohmic contacts have allowed volume-generated photocurrents to be studied not only over a wide range of illuminations, but also over a wide range of temperatures, including low temperatures where the quasi-Fermi levels approach the free carrier bands. The temperature dependence of the photoconductivity of undoped a–Si : H films indicates that free carrier transport occurs even at temperatures as low as 120 K. The temperature dependence of the photoconductivity for various illuminations is shown in Fig. 10.11 for the same film ($T_s = 320\,°C$) characterized in Fig. 10.10. In Fig. 10.11, two temperature regimes, I and II, are clearly distinguishable, which exhibit activation energies $E_I \sim 0.05$ eV and $E_{II} = 0.12$ eV with values of $\gamma = 0.7$ and $\gamma = 0.5$, respectively. The transition from region I to region II occurs at temperatures which depend on

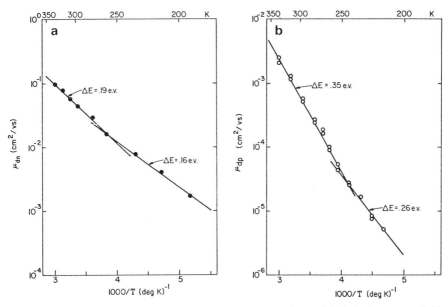

Fig. 10.12a, b. The drift mobilities of (**a**) electrons in *n*-type a-Si:H and (**b**) holes in *p*-type a-Si:H, both as a function of 1000/T

the intensity of illumination consistent with a change in the recombination kinetics [10.52]. In this model, one expects a transition to large activation energies at lower temperatures, and similar results have been found in a variety of rf- and dc-produced a–Si:H films. These results are in contrast to those reported by *Spear* et al. [10.41] on rf-produced films in which transitions to *smaller* activation energies were observed at temperatures below ~250 K; their results were interpreted as a transition from free carrier transport to hopping transport in localized states [10.41]. The absence of hopping conduction at the lower temperatures is significant not only in the characterization of a–Si:H, but also indicates that efficient solar cell operation can be achieved at low temperatures (see Sect. 10.5.5).

10.3.4 Carrier Mobilities and Lifetimes

Although the electronic properties of undoped a–Si:H have been extensively studied, direct measurements of the microscopic free carrier mobilities have not yet been performed. The drift mobility of electrons (μ_{dn}) has been measured by *Le Comber* et al. [10.14, 54] in rf-produced a–Si:H films, and recently *Moore* [10.55] has measured the drift mobilities of electrons and holes (μ_{dp}) in dc-produced a–Si:H films. The drift mobilities of both carriers, studied between ~400 and 200 K, are found to be trap controlled. At room temperature the values of μ_{dn} are $(2–5) \times 10^{-2}$ cm^2 V^{-1} s^{-1}, and those of μ_{dp} are $(5–6) \times 10^{-4}$

$cm^2 V^{-1} s^{-1}$ with activation energies of 0.19 and 0.35 eV respectively for temperatures $\gtrsim 250$ K [10.55]. The results obtained by *Moore* for the drift mobilities as a function of temperature are shown in Fig. 10.12. In both cases the mobilities are those of the majority carriers so that the trapping states can be related to the photoconductivity measurements such as discussed in Sect. 10.3.3.

Although the data in Fig. 10.12 show transitions to lower activation energies at the lower temperatures, these energies are not so low as to indicate hopping conduction. However, the data do indicate the presence of large densities of states ~ 0.2 eV below the conduction band and ~ 0.35 eV above the valence band. Due to uncertainties in the density of trapping states and in the effective density of band states, it is difficult to estimate the free carrier mobilities, but values on the order of ~ 1 $cm^2 V^{-1} s^{-1}$ seem reasonable [10.14, 54].

As indicated by photoconductivity results discussed in Sect. 10.3.3, a wide range of electron lifetimes is present in undoped a–Si:H films. These lifetimes are directly related to the various centers present in the gap which depend on the fabrication conditions and the conductivity state of the films.

The product of the free carrier mobility and the lifetime ($\mu\tau$) can be obtained directly from the photoconductivity measurements, and assuming $\mu_c \sim 1$ $cm^2 V^{-1} s^{-1}$, electron lifetimes between 10^{-3} and 10^{-7} s have been observed in undoped a–Si:H films [10.9]. As shown by the data in Figs. 10.9, 10, the lifetime depends on the illumination level; this reflects the displacement of the electron quasi-Fermi level and the introduction of new recombination centers into the carrier recombination kinetics [10.9, 53]. Moreover, the light-induced conductivity changes discussed in Sect. 10.3.3 reflect corresponding changes in the free carrier lifetimes [10.45, 47].

At present, the minority carrier lifetimes have not been measured in a–Si:H. However, an analysis of the operation of a–Si:H solar cells indicates that the hole diffusion lengths up to ~ 0.2 μm [10.56] are present. Thus if the extended state hole mobility is ~ 1 $cm^2 V^{-1} s^{-1}$, hole lifetimes up to $\sim 2 \times 10^{-8}$ s can be obtained.

10.3.5 Density of Gap States

The energies, densities, and types of states present in the gap of a–Si:H are of paramount importance in the operation of solar cells. Not only do they determine the free carrier lifetimes, but they also affect the junction characteristics. The net balance between the donor and acceptor states determines the space charge in the junction region and hence the distance over which an electric field is present.

Estimates of the densities and distributions of states in a–Si:H films have been made using field-effect measurements on a variety of films [10.57, 58]. However, the effects of surface states are ignored in the interpretation of these results. *Knights* and *Biegelsen* [10.59] have estimated a density of surface states

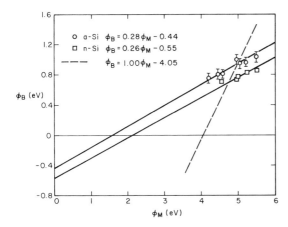

Fig. 10.13. Schottky barrier height vs metal work function

equal to $\sim 10^{13}\,\text{cm}^{-2}\,\text{eV}^{-1}$ from their electron spin resonance measurements on a–Si : H films of various thicknesses. *Wronski* and *Carlson* [10.60] found a similar density of surface states from a study of the Schottky barriers of various metal on a–Si : H. These results are shown in Fig. 10.13 where the dependence of the barrier heights reflects similar densities and distributions for both a–Si : H and crystalline Si. These densities of $\sim 10^{13}\,\text{cm}^{-2}\,\text{eV}^{-1}$ are equivalent to a density of $10^{18}\,\text{cm}^{-3}$ bulk states in a thickness of $\sim 1000\,\text{Å}$. Thus, surface states of $\sim 10^{13}\,\text{cm}^{-2}\,\text{eV}^{-1}$ can easily mask bulk densities which are $< 10^{18}\,\text{cm}^{-3}$. Bulk densities of $\sim 10^{15}$–$10^{16}\,\text{cm}^{-3}$ are indicated by the measurements of space-charge densities in a–Si : H Schottky-barrier cells [10.56] as well as by electron-spin-resonance measurements [10.29]. These results suggest that undoped a–Si : H films are closely compensated and that the density of deep centers is in the range of 10^{15}–$10^{16}\,\text{cm}^{-3}$.

Field-effect measurements showed that the density of gap states decreased with increasing substrate temperature [10.57, 58]. Recent measurements of the photoluminescence [10.8], the electron spin density [10.29], and the infrared absorption [10.38] show that the density of gap states goes through a minimum at a substrate temperature of $\sim 330\,^{\circ}\text{C}$. The density of defects increases at high substrate temperatures due to the outdiffusion of hydrogen [10.8, 36].

Recently, *Wronski* and *Daniel* [10.53] have used both steady-state and transient photocurrents to obtain information about the densities and energy distributions of states in the gap of undoped a–Si : H. Figures 10.14, 15 show the response times (τ_0) and the lifetimes (τ_n) for an rf-produced film in the annealed and light-soaked states, respectively, as a function of the volume generation of electron–hole pairs (f). The data in Fig. 10.14 are consistent with trap-dominated photoconductivity where shallow traps are dominant. The data for the a–Si : H film in the light-soaked state (see Fig. 10.15) are consistent with a large density of states near midgap. Thus, the exposure of a–Si : H to light appears to create deep centers that reduces the electron lifetime [10.47, 48].

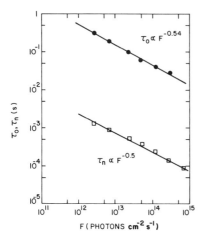

Fig. 10.14. The response time (τ_0) and the electron lifetime (τ_n) as a function of the volume generation of pairs (f) for an rf-produced film ($T_s \simeq 320\,°C$) in the annealed state

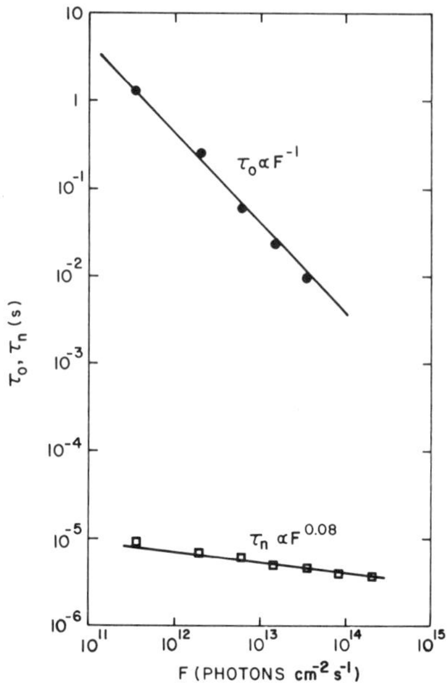

Fig. 10.15. The response time (τ_0) and the electron lifetime (τ_n) as a function of the volume generation of pairs (f) for an rf- produced film in the light-soaked state

The long electron lifetimes, of up to $\sim 10^{-3}$ s, found in undoped a–Si : H can be attributed to the presence of "sensitizing" centers near midgap [10.52] which have electron capture cross sections of $\sim 10^{-19}\,cm^2$ [10.53]. Although there appears to be a quasi-continuous distribution of states above midgap which have significantly greater electron capture cross sections, their densities do not exceed $\sim 10^{17}\,cm^{-3}\,eV^{-1}$ at least for energies up to $\lesssim 0.3\,eV$ below the conduction band. Also, no evidence is found for a large peak in the distribution of states at $\sim 0.4\,eV$ below the conduction band such as was inferred from the field-effect measurements of *Madan* et al. [10.58]; *Döhler* and *Hirose* [10.61] also found no evidence for this peak. The density of states $\sim 0.2\,eV$ below conduction band (Sect. 10.3.3) is found to be ~ 0.1 of the effective density of states in the conduction band, N_c. Since values for N_c of $\sim 10^{20}\,cm^{-3}$ have been obtained directly from the saturation current densities of a–Si : H Schottky barriers [10.62], the density of this level is $\sim 10^{19}\,cm^{-3}$. This density is very large, but the role of this level is that of an electron trap [10.53].

10.4 Solar Cell Structures

In this section we consider the various solar-cell structures that have been used to make thin-film cells with hydrogenated amorphous silicon. These structures include Schottky-barrier cells, *p–n* and *p–i–n* junctions, and heterojunctions [10.63] (see Fig. 10.16).

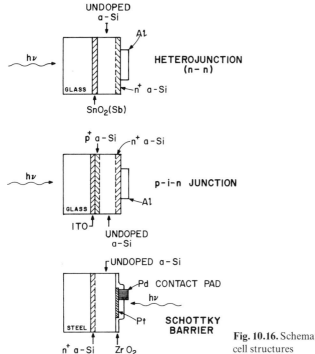

Fig. 10.16. Schematic diagrams of various solar-cell structures

10.4.1 Schottky-Barrier Cells

The simplest Schottky-barrier cells were fabricated by depositing undoped a–Si : H ($\sim 1\,\mu$m) on substrates such as Fe and Mo and then evaporating a thin film of a high-work-function metal such as Pt ($\sim 50\,$Å) on top of the undoped a–Si : H. However, such cells usually exhibited poor fill factors due to a poor contact to the substrate. The inclusion of a thin phosphorus-doped layer ($\gtrsim 0.1\,\mu$m) next to the substrate improved the reproducibility and performance of the cells (see Fig. 10.16). This doped layer was typically made from a glow discharge in SiH_4 containing $\sim 1\,\%\,PH_3$. Other n^+ contacting layers were made by evaporating Sb or CsN_3 into the SiH_4 discharge or by adding $\sim 0.1\,\%\,AsH_3$ to the SiH_4 discharge. The substrate temperatures for the doped and undoped layers were generally in the range of 300–400 °C.

Schottky barrier cells have also been fabricated by evaporating relatively low work function metals such as Al or Cr on p-type a–Si : H. In these cells, the substrate was contacted by a thin p^+ layer made from a discharge in SiH_4 containing $\sim 1\,\%\,B_2H_6$. The p-type a–Si : H ($\sim 1\,\mu$m) was made in a discharge with the volume fraction B_2H_6/SiH_4 equal to $\gtrsim 10^{-4}$.

If a thin nascent oxide is present on the a–Si : H prior to the evaporation of the metal barrier then an MIS-type cell is formed [10.64, 65]. An oxide layer,

~20–30 Å thick, can be formed by heating the a–Si:H film for fifteen minutes at 350 °C in air. While these devices exhibit large open-circuit voltages, they degrade in the presence of moisture; the degradation can be reversed by heating for several minutes at 200 °C in vacuum. This behavior is probably due to the presence of a large density of OH groups in the oxide layer.

Schottky-barrier or MIS cells without an antireflection coating only transmit ~40–50% of the incident light into the a–Si:H. The average transmission can be increased to ~80–90% by means of antireflection coatings (~450 Å) such as ZrO_2, TiO_2, Si_3N_4, and ITO (indium tin oxide). For large-area Schottky-barrier or MIS cells ($\gtrsim 0.1$ cm^2), current collection grids are necessary to keep the series resistance low.

10.4.2 p–n and p–i–n Junctions

A p–n junction cell may be fabricated by depositing a phosphorus-doped layer on top of a boron-doped layer or vice versa. However, the doping levels must be kept relatively low in at least one layer (i.e., B_2H_6/SiH_4 or $PH_3/SiH_4 \gtrsim 10^{-4}$) to prevent a significant loss in the minority carrier lifetime (see Sect. 10.5.3). Moreover, thin contacting layers of more heavily doped a–Si:H are usually required to obtain low resistance contacts to both front and back electrodes.

The p–i–n structure utilizes an undoped layer, ~1 μm thick, and thin p^+ and n^+ layers adjacent to the front and back electrodes (see Fig. 10.16). The undoped layer is actually not intrinsic but slightly n type [10.14]. The p^+ and n^+ layers are ~100–300 Å thick, and are usually made in a SiH_4 discharge containing ~1% of either B_2H_6 or PH_3. Since the resistivity of doped a–Si:H is $\gtrsim 10^2\,\Omega$ cm, the sheet resistance is $\gtrsim 10^8\,\Omega/\square$ for a layer 100 Å thick. Thus, a semitransparent conductive layer such as ITO or a thin metal film must be used as a contacting electrode on the illuminated side of the cell. Generally better contacts to the p^+ or n^+ layers have been obtained with thin metal films rather than with ITO. Apparently, a thin resistive oxide layer forms at the ITO/doped a–Si:H interface [10.63]. High work function metals such as Pt and Pd make good contacts to p^+ layers while metals such as Mo and Cr make good contacts to n^+ layers.

10.4.3 Heterojunctions

The first heterojunction cells were made by depositing undoped a–Si:H on glass substrates coated with ITO or SnO_2 (Sb doped). A thin n^+ layer was deposited on top of the undoped layer to provide an ohmic contact to an evaporated Al electrode (see Fig. 10.16). Other heterojunction structures were fabricated by sputtering ITO onto undoped a–Si:H with a thin n^+ layer next to the substrate. p–n heterojunctions were made by depositing p-type a–Si:H on ITO or on SnO_2, and also by depositing undoped a–Si:H (slightly n type) on

p-type GaAs or on p-type crystalline Si. These p–n heterojunctions exhibited poor photovoltaic properties with conversion efficiencies less than 1 %.

10.5 Photovoltaic Characteristics

The a–Si:H solar cell represents a new type of thin-film solar cell. Conversion efficiencies as high as 6 % have been obtained in a highly disordered material that has a large resistivity and a relatively low free carrier mobility. Having discussed the a–Si:H solar-cell structures in the last section, we now consider the electrical characteristics of these devices.

10.5.1 Current–Voltage Behavior

The current density of a diode in the dark obeys the relation

$$J = J_0 \left[\exp\left(\frac{qV}{nkT}\right) - 1 \right],$$

(10.6)

where J_0 is the saturation current density and n is the diode quality factor. Some representative, dark current–voltage data are shown in Fig. 10.17 for Schottky barriers on a–Si:H and crystalline Si as well as for a p–i–n structure in a–Si:H. For Schottky-barrier diodes, the values of n are between 1.0 and 1.2 which indicates the absence of generation and recombination in the junctions [10.62]. This is not the case for the p–i–n structure where the diode quality factor is ~ 2.3 [10.28].

The dark current–voltage characteristics of the Schottky-barrier diodes have been interpreted in terms of the diffusion theory of metal–semiconductor rectification [10.62]. In this theory, the saturation current density is given by

$$J_0 = q\mu_c N_c E_s \exp(-\phi_B/kT),$$

(10.7)

where μ_c is the electron mobility in the conduction ban, N_c is the effective density of states in the conduction bond, E_s is the electric field at the surface of the semiconductor, and ϕ_B is the Schottky-barrier height. Measurements of J_0 as a function of temperature allow a determination of ϕ_B, and values as large as 1.1 eV have beeen measured for Pt barriers on a–Si:H [10.62]. These measurements of ϕ_B are in good agreement with those obtained from photovoltage and capacitance measurements [10.60, 65].

Illuminated I–V characteristics are shown in Fig. 10.18 for a Pt Schottky barrier cell with a ZrO_2 antireflection coating [10.63]. This cell exhibited a conversion efficiency of 5.5 % in sunlight of 65 mW cm^{-2}, the highest efficiency obtained to date is 6.0 % in a 1.5 mm^2 Schottky-barrier cell. The best values of J_{sc}, V_{oc}, and FF are 14.5 mA cm^{-2}, 895 mV, and 0.674, respectively, for AM1

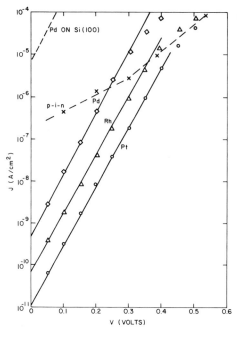

► **Fig. 10.17.** Dark current density as a function of forward bias voltage for Pd, Rh, and Pt Schottky barriers on a-Si:H. Also, shown are data for a Pd Schottky barrier on crystalline Si and a p–i–n junction in a-Si:H

Fig. 10.18. Illuminated I–V characteristics of a Pt Schottky-barrier cell in sunlight of 65 mW cm^{-2}

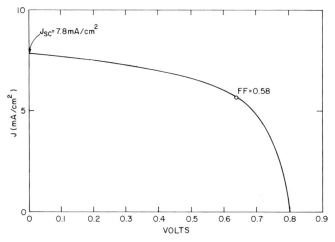

sunlight ($\sim 100\,\mathrm{mW\,cm^{-2}}$). If these values could all be obtained in the same cell, then the conversion efficiency would be $\sim 8.8\,\%$. The best efficiency obtained so far with a p–i–n cell is $\sim 3.3\,\%$ due to smaller values of J_{sc} and FF than those obtained with Schottky-barrier cells [10.66].

Since the open-circuit voltage is related to the short-circuit current density by

$$V_{\mathrm{oc}} = \frac{n'kT}{q}\ln\left(\frac{J_{\mathrm{sc}}}{J_0}+1\right), \tag{10.8}$$

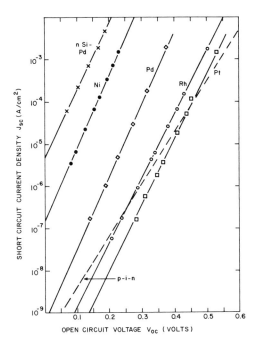

Fig. 10.19. The short-circuit density as a function of the open-circuit voltage for several Schottky barriers on a-Si:H. Also shown are data for a Pd barrier on crystalline Si and $p–i–n$ junction in a-Si:H

a plot of J_{sc} versus V_{oc} allows a determinator of n', the diode quality factor in the light (see Fig. 10.19). Generally, n' is ∼1.0–1.1 for Schottky barrier cells and ∼1.5 for $p–i–n$ cells. While the data in Fig. 10.17 show that the dark current densities become resistance limited at ∼10^{-5} A cm^{-2}, the data in Fig. 10.19 show that J_{sc} is not resistance limited even at current densities of ∼10^{-2} A cm^{-2}. This is due to the large photoconductive effect mentioned in Sect. 10.3.3.

10.5.2 Effects of Substrate Temperature

The photovoltaic properties of a–Si : H solar cells are strongly influenced by the substrate temperature (T_s) during deposition. For substrate temperatures ≳200 °C, the devices exhibit very poor photovoltaic properties due to large defect densities [10.29]. Consequently, other electronic properties such as photoconductivity [10.9] and photoluminescence [10.67] are also adversely affected by the short recombination lifetimes. As discussed in Sect. 10.3.1 the defects formed in a–Si:H films at substrate temperatures ≳200 °C may be associated with dihydride or trihydride groups.

Poor photovoltaic properties also result if the substrate temperature is >400 °C. In this case, the defects are dangling bonds left behind as hydrogen diffuses out of the film [10.36]. As shown in Fig. 10.20, J_{sc} for a Schottky-barrier cell decreases rapidly as T_s increases above 400 °C; the FF also drops rapidly as the effective series resistance of the cell increases [10.68].

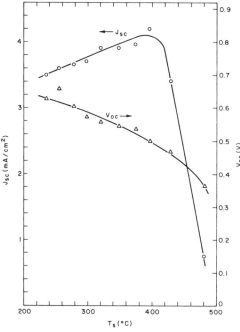

Fig. 10.20. The short-circuit current density and the open-circuit voltage of a Pt Schottky-barrier cell as a function of substrate temperature.

Efficient a–Si:H solar cells can only be made in the range of substrate temperatures between 200 and 400 °C. This range encompasses the minimum defect density as measured by photoluminescence [10.67], spin density [10.29] and infrared absorption [10.38]. As T_s increases from 200 to 400 °C, J_{sc} increases and V_{oc} generally decreases (see Fig. 10.20). J_{sc} increases with T_s because E_{OPT} decreases (and α increases; see Figs. 10.3, 4), and also because the carrier recombination lifetime improves [10.9, 23]). For a Schottky-barrier cell, V_{oc} decreases as T_s increases; this is due to the dependence of ϕ_B on E_{OPT} [10.20]. However, in the case of MIS or p–i–n cells, the dependence of V_{oc} on T_s appears to be more complicated.

10.5.3 Effects of Impurities

The photovoltaic properties of a–Si:H solar cells are not strongly affected by most impurities normally found in the glow-discharge environment. The effect of impurities in a–Si:H has been studied by adding various impurity gases to the SiH$_4$ discharge during the fabrication of Pt Schottky-barrier cells [10.69]. Table 10.1 shows the effect of various impurities on the photovoltaic properties measured in light comparable to AM1 sunlight ($\sim 100\,\mathrm{mW\,cm^{-2}}$). The a–Si:H films were deposited at $T_s \simeq 330$ °C, and the Schottky barriers were formed by evaporating $\sim 50\,\text{Å}$ of Pt in 2 mm^2 dots (no antireflection coatings). Although

Table 10.1. Photovoltaic properties for various impurity gases in the SiH_4 discharge

Impurity gas (Vol. % of total atmosphere)	Type of discharge	V_{oc} [mV]	J_{sc} [mA cm^{-2}]	Fill factor	Impurity content of film
None (Control)	dc(P)	700	6.0	0.54	(H/Si $\simeq$ 0.20)[a]
2.3 % H_2O	dc(P)	400	0.6	0.23	O/Si $\simeq$ 0.037
2.3 % H_2S	dc(P)	425	3.0	0.21	—
1.0 % GeH_4	rf(E)	370	3.0	0.27	Ge/Si $\simeq$ 0.025
10 % CH_4	rf(E)	662	4.0	0.53	C/Si $\simeq$ 0.07
30 % CH_4	rf(E)	230	0.02	0.18	C/Si $\simeq$ 0.21
10 % N_2	rf(C)	595	6.0	0.55	N/Si $\simeq$ 0.008
0.06 % PH_3	dc(P)	130	1.5	0.42	P/Si $\simeq$ 0.0004
50 % SiH_2Cl_2, 50 % H_2	dc(C)	321	0.05	0.25	Cl/Si $\simeq$ 0.07

[a] All films contained $\sim$ 10–30 at. % hydrogen

cells were fabricated in various types of discharges, the control samples (no impurity doping) exhibited variations of < 20 % in the measured photovoltaic properties. The impurity content of the a–Si : H films was determined by Auger electron spectroscopy.

Gases such as N_2 and CH_4 have little effect on the photovoltaic properties even when they constitute $\sim$ 10 % of the discharge atmosphere (see Table 10.1). However, a few percent of H_2O, H_2S, or GeH_4 cause a significant reduction in all the photovoltaic parameters. A small amount of PH_3 ($\sim$ 0.06 %) in the SiH_4 discharge causes a severe decreases in V_{oc} and J_{sc}, but only a relatively small decrease in the fill factor. A film made from a discharge in SiH_2Cl_2 and H_2 also exhibited reduced values of the photovoltaic parameters.

As shown in Table 10.1, all of the a–Si : H films contained significant amounts of the impurities. These results show that a–Si : H can accomodate relatively large concentrations of many impurities and still exhibit a significant photovoltaic effect. The observation that impurity gases such as GeH_4 (1 %) and CH_4 (30 %) cause degradation of the photovoltaic properties indicate that defects other than isoelectronic centers are being created in these films. These defects could be clusters of dopant atoms or atoms with one or more hydrogen atoms bonded to them.

Other impurity gases such as H_2O probably create defects similar to those found in crystalline Si such as A centers which are located $\sim$ 0.2 eV below the conduction band [10.70]. Similarly, PH_3 apparently gives rise to shallow donors in a–Si : H [10.12]. However, the decrease observed in J_{sc} (see Table 10.1) indicates that phosphorus doping also creates deeper levels. These recombination centers may be due to phosphorus–hydrogen complexes or clusters. Moreover, many of the dopant species may be bonded in sites that satisfy the normal valence state of the impurity [10.71] possibly as a result of bond compensation by hydrogen. These centers may act only as shallow neutral traps.

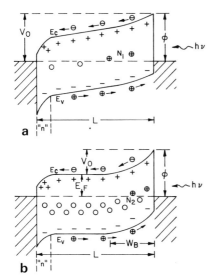

Fig. 10.21a, b. Schematic energy-band diagrams of a Schottky-barrier cell under short-circuit conditions; (**a**) a fully depleted cell and (**b**) a partially depleted cell

10.5.4 Collection Efficiency of Photogenerated Carriers

Solar cells are ambipolar devices whose efficient operation depends on the extraction of *both* photogenerated carriers from the semiconductor. Also, solar cells only utilize the electric fields that are generated by the junctions to collect the photogenerated carriers. In contrast to the a–Si:H solar cell, previous applications of photosensitive amorphous materials (in xerography or in image pick-up tubes) required externally applied voltages and relied primarily on the properties of one photogenerated carrier.

The a–Si:H solar cell exhibits built-in potentials of the order of 1 volt [10.28], and since the film thickness is $\sim 1\,\mu m$, fields of $\sim 10^4\,V\,cm^{-1}$ are present. The large optical absorption of a–Si:H (see Sect. 10.3.2) causes an appreciable density of electron–hole pairs to be generated in the region of the junctions; these photogenerated pairs are separated by the electric field in the junction and are then collected by the cell electrodes. The photogeneration of carriers in the junction region becomes particularly important for cells with minority carrier diffusion lengths significantly less than the film thickness [10.56].

In this section we consider the collection efficiency of a–Si:H solar cells where the collection efficiency is defined as the percentage of electron–hole pairs collected in the external circuit per incident photon. The collection efficiencies were measured under relatively low levels of illumination to avoid series resistance limitation (the series resistance can be large for short-wavelength light since the light is absorbed only near the surface and the dark resistivity of the bulk a–Si:H is large; see Sect. 10.5.5).

Schematic energy band diagrams of two a–Si:H Schottky-barrier structures are shown in Fig. 10.21. The two undoped a–Si:H films have different densities of deep ionized centers, due to different fabrication conditions, but the density of shallow, compensated donor and acceptor states is considered to be the same. Metal Schottky-barrier junctions are used as an illustration here because the electronic and optical properties of the cells are not influenced by doping (a thin n^+ layer exists only at the back contact). The cells are shown under short-circuit conditions at a low level of illumination so that there is very small splitting of the electron and hole quasi-Fermi levels. The usual semiconductor energy band diagram can be used here because a–Si : H Schottky barriers have been characterized over a wide temperature range in terms of the same parameters as crystalline semiconductor Schottky barriers [10.62, 65].

The Schottky-barrier height ϕ_B results in charge separation at the metal semiconductor interface which generates a built-in potential V_0, given by

$$V_0 = \phi_B - (E_c - E_f) - \frac{kT}{q},$$ (10.9)

where E_c is the edge of the conduction band and E_f is the Fermi level. The space-charge or depletion region W_B is given by

$$W_B = \left(\frac{\varepsilon}{2\pi q} \frac{V_0}{N} \right)^{1/2},$$ (10.10)

where N is the space-charge density and ε is the dielectric constant.

In Fig. 10.21a, the space-charge density N_1 is sufficiently low to allow the depletion width to extend through the entire film thickness L. (This is achieved in a 1-µm film when N_1 is less than $\sim 10^{15}$ cm^{-3}.) In Fig. 10.21b, on the other hand, N_2 is much higher and a quasi-neutral region, $L - W_B$, is present. The difference between the position of the Fermi level with respect to the conduction band at the ohmic contact and in the bulk produces an effect which also results in an electric field; however, this is not considered here.

To minimize the dependence of the collection efficiency on the photogenerated-hole transport, it is desirable to have as wide a depletion width as possible, preferably one that will lead to the type of situation that is shown in Fig. 10.21a. In this case all the carriers are generated in an electric field which sweeps out the electrons and holes in opposite directions. In a field configuration, such as shown in Fig. 10.21b, this applies only to the carriers generated in the region W_B. The holes generated in the region $L - W_B$ must diffuse to the space-charge region before they can take advantage of the electric field. In single-crystal Si cells, the diffusion length of the minority carriers is so large that the cell currents are generally not affected by changes in W_B. However, in a–Si : H the diffusion lengths are much smaller, and the effect of W_B on the collection of photogenerated carriers must be considered.

The collection efficiency characteristics of photogenerated carriers in a fully depleted structure, such as in Fig. 10.21a, are shown in Fig. 10.22. This structure consisted of a Pt Schottky barrier ($\sim 20\%$ transmitting) on a layer of

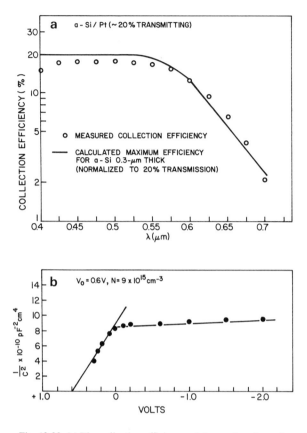

Fig. 10.22. (a) The collection efficiency of J_{sc} as a function of wavelength and (b) a plot of $1/C^2$ as a function of voltage for a Pt Schottky-barrier cell

undoped a–Si:H ($\sim 0.3\,\mu m$ thick) with a thin n^+ layer ($\sim 300\,\text{Å}$) at the back contact. Figure 10.22a shows the short-circuit currents collection efficiency as a function of wavelength for photon fluxes of 10^{13}–$10^{14}\,cm^2 s^{-1}$. Also shown is the efficiency calculated from optical absorption data for a 0.3 µm thick a-Si:H film where the incident flux of photons is reduced 80% by the metal film. There is excellent agreement between the measured and calculated efficiencies. Figure 10.22b shows the results of $C–V$ measurements carried out at a frequency of 100 Hz. The data indicate a space-charge density of $9 \times 10^{15}\,cm^{-3}$ and a built-in potential of 0.6 V. The capacitance at zero bias corresponds to a depletion width of 0.3 µm. The measured collection efficiency is also independent of reverse bias as expected for a saturated photocurrent where all the photogenerated carriers are collected under the short-circuit condition. Also, in the case discussed here, the collection efficiency is independent of the hole diffusion length (L_p) since there is no field free region.

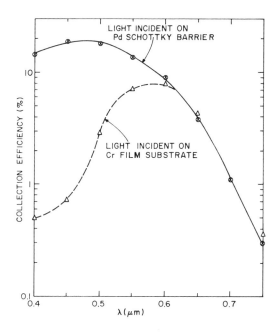

Fig. 10.23. The collection efficiency as a function of wavelength for front and back illuminations of a Pd Schottky-barrier cell

In the case of thick cell structures ($\gtrsim 1\,\mu$m) the field distribution is similar to that shown in Fig. 10.21b. The advantage of generating carriers in the region of the Schottky barrier is demonstrated in Fig. 10.23 where the collection efficiency is shown as a function of wavelength for front and back illuminations. The Pd Schottky barrier cell was fabricated on a glass substrate coated with a semitransparent Cr film; the n^+ layer contacting the Cr was $\sim 300\,\text{Å}$ thick. Thus, this structure could be illuminated either through the Pd barrier or through the Cr film (10.63).

The collection efficiency obtained for light incident on this Pd barrier is similar to that obtained in Fig. 10.22a even though the film thickness in the present case ($\sim 0.8\,\mu$m) is much greater than $0.3\,\mu$m. The similarity of the data is due to the fact that the junction region of the Pd Schottky barrier cell was $\sim 0.3\,\mu$m so that all the carriers photogenerated in this region are collected just as in the previous case. However, carriers photogenerated was the back of this cell are not efficiently collected. When the cell is back illuminated, the collection efficiency is low for short wavelengths indicating that holes generated near the Cr film have difficulty in reaching the Pd electrode.

Cells formed from undoped a-Si:H in which L_p is significantly less than $L - W_B$ (and also W_B) exhibit front-wall collection efficiencies which are strongly dependent on the bias. In the presence of a bias potential V, (10.10) is modified by replacing V_0 by $V_0 - V$, and thus W_B decreases with increasing forward bias. If $L_p \ll W_B$, the photocurrent will decrease as the forward bias is increased. In such a case the photocurrent consists mainly of carriers generated in the junction region with a negligible contribution from the field-free region.

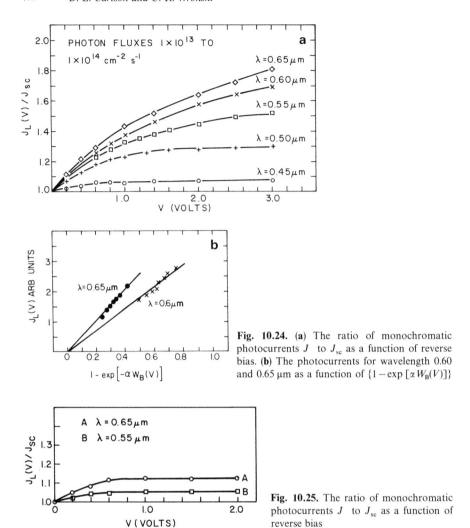

Fig. 10.24. (a) The ratio of monochromatic photocurrents J to J_{sc} as a function of reverse bias. **(b)** The photocurrents for wavelength 0.60 and 0.65 μm as a function of $\{1-\exp\,[\alpha\,W_B(V)]\}$

Fig. 10.25. The ratio of monochromatic photocurrents J to J_{sc} as a function of reverse bias

The photocurrent characteristics for a Pt Schottky barrier cell ($\sim 1.3\,\mu$m thick) where $L_p \ll W_B$ are shown in Fig. 10.24. The ratio of monochromatic photocurrent to the short-circuit current is shown as a function of reverse bias for several wavelengths of light [Fig. 10.24a]. Capacitance measurements showed that the space charge density was $\sim 10^{16}\,\text{cm}^{-3}$ and that the depletion width changed from 0.23 to 0.45 μm as the bias increased from 0 to -3 V.

The increase of the normalized photocurrents with reverse bias demonstrate the effect of changes in $W_B(V)$ and the effect of the absorption coefficients of the different wavelengths. In the range shown, the photocurrents saturate for the shorter wavelengths, and exhibit a continual increase for the longer wavelengths. Such behavior also indicates that most of the short circuit current

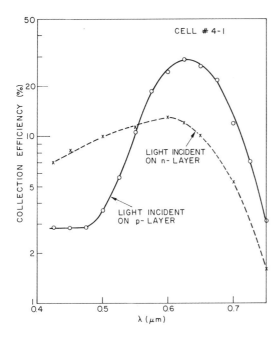

Fig. 10.26. The collection efficiency as a function of wavelength for front and back illuminations of a p–i–n cell

under white light illumination is generated in the field region. For monochromatic illumination at wavelength, λ, this photocurrent, $J_W(\lambda)$, for the region $W_B(V)$ is given by

$$J_W(\lambda) = F(\lambda)\{1 - \exp[-a(\lambda)W_B(V)]\}, \tag{10.11}$$

where $F(\lambda)$ is the photon flux entering the a–Si:H and $\alpha(\lambda)$ is the absorption coefficient. This equation is used to construct Fig. 10.24b, where the photocurrents for the wavelengths 0.6 and 0.65 μm at the different biases V are plotted against $\{1 - \exp[-\alpha(\lambda)W_B(V)]\}$. The values of $\alpha(\lambda)$ used were 3×10^4 and 1.2×10^4 cm^{-1} for $\lambda = 0.6$ and 0.65 μm, respectively, and were obtained from measurements on similar films. The extrapolated values to zero photocurrent indicate that in this film the values of L_p are not only significantly less than $L - W_B(V)$ but also less than the values of $W_B(V)$ (0.23–0.45 μm).

The presence of L_p values comparable to W_B allows holes generated beyond the junction region to be collected [10.56, 71]. This not only extends the spectral response (such as shown in Figs. 10.22, 23) but also reduces the bias dependence of the photocurrents generated by white light.

An example of such a cell is shown in Fig. 10.25 where the ratio of $J_L(V)/J_{sc}$ for $\lambda = 0.55$ and 0.65 μm is plotted against the reverse bias voltage V. Such cells, in which there is a significant contribution from the region $(L - W_B)$, operate efficiently under AM1 illumination and exhibit high fill factors as will be discussed in Sect. 10.5.5.

The ability to collect all the carriers photogenerated in the field region of an a–Si:H Schottky barrier cell indicates that the hole lifetime is greater than their transit time across this region. Also, the high collection efficiency obtained with blue light absorbed near the Schottky barrier reflects low recombination at the metal/undoped a–Si:H interface.

p–i–n solar cell structures exhibit relatively poor collection efficiencies at short wavelengths as shown in Fig. 10.26. The cell was formed by depositing a thin p^+ layer (~1 at.% boron, ~500 Å thick) on ITO-coated glass followed by ~1 μm of undoped a–Si:H and then a thin n^+ layer (~1 at.% phosphorous, ~200 Å thick) [10.5]. When the cell is illuminated through the p^+ layer, the low collection efficiency at short wavelengths indicates that significant recombination occurs for carriers photogenerated in the boron-doped region. When light is incident on the n^+ layer, the collection efficiency is also reduced at short wavelengths, but the recombination in the thinner n^+ layer does not appear to be as serious as in the p^+ layer. Apparently, the density of deep recombination centers is greater in doped a–Si:H than in undoped a–Si:H (see Sect. 10.5.3). Similar effects have been observed in highly doped, diffused junctions in single crystal solar cells [10.19].

10.5.5 Analysis of a–Si:H Solar-Cell Operation

In this section, we discuss the major parameters that influence the operation of a–Si:H solar cells and relate these parameters to both the relevant a–Si:H properties discussed in Sect. 10.3 and the conditions for efficient solar-cell operation discussed in Sect. 10.1.2. We must emphasize that since the electronic and optical properties of a–Si:H are strongly dependent on deposition conditions a wide variety of cell characteristics can be obtained. Furthermore, the performance of a–Si:H cells is usually limited by a combination of different parameters so that exact characterization is difficult [10.72].

As discussed in Sect. 10.4, the junctions utilized in a–Si:H cells have included p–i–n homojunctions, heterojunctions, and metal Schottky barriers. In most of these cases, the bulk of the cell (~1 μm thick) consists of undoped a–Si:H. Since the requirements in terms of the electrical properties of the a–Si:H are basically the same, irrespective of the junction used, the operation of the cell is discussed here in terms of the Schottky barrier structure. However, where appropriate, any significant difference arising from the use of other structures will be considered. The schematic energy band diagrams of the a–Si:H Schottky-barrier structures are shown in Fig. 10.21 and reference to them will be made in the course of this discussion.

Efficient photovoltaic energy conversion, discussed in Sect. 10.1.2, requires the generation of large-short-circuit currents, J_{sc}, and open-circuit voltages, V_{oc}, and also the ability of the cell to supply power to a load. In the case of single crystal Si cells, the dependence of J_{sc}, V_{oc}, and the fill factor on the different semiconductor and junction properties has been extensively studied and characterized [10.19]. However, in the case of a–Si:H solar cells, other parameters such as the photoconductivity must be considered in explaining the operation.

Before considering these effects we will discuss the dependence of the cell characteristics on the junction in a–Si:H cells since the properties of these junctions are similar to those found in crystalline semiconductors [10.65, 73].

Both the open-circuit voltages and short-circuit currents depend on the junctions characteristics. For large values of V_{oc}, the reverse saturation current J_0 must be low. For large value of J_{sc} at AM1 illuminations the junction must be able to transport and efficiently collect the large number of photogenerated carriers. The generation–recombination currents which affect the current–voltage characteristics of junctions in the dark (see Sect. 10.5.1) also affect the photovoltaic properties. The photovoltaic characteristics of the junctions and the ability of these junctions to transport photogenerated carriers are reflected in the dependence of the short-circuit currents J_{sc} and the open-circuit voltage V_{oc} on the intensity of illumination F. By going to AM1 intensities the junction performance under solar cell operation conditions can be evaluated.

In most a–Si:H cells, J_{sc} exhibits a linear dependence on F over many orders of magnitude up to AM1 sunlight illuminations [10.56]. Such behavior indicates that collection efficiencies can be obtained which are essentially independent of the density of carriers photogenerated in the cells. These results also indicate that V_{oc} can be directly correlated with J_0 and the effects of generation–recombination in the junctions. The sublinear dependence of the photocurrents discussed in Sect. 10.3.3 reflect the recombination kinetics of electrons in the quasi-neutral region of the cell. The intensity dependence is determined by the changes that occur in electron lifetime as the electron quasi-Fermi level is swept toward the conduction band. In contrast, the photocurrent collected in the external circuit of a solar cell is the sum of the extracted electron and hole currents. These currents are equal and are characterized by the carrier with the shorter lifetime and/or mobility (in a–Si:H, the holes) [10.5].

As discussed in Sect. 10.5.1, the diode quality factor for the dark current–voltage characteristics (n) can be determined from (10.6) while the diode quality factor under illumination (n') can be determined from (10.8). In single crystal cells these diode quality factors are the same. However, they can differ if the recombination of free carriers is different in the two cases. In the case of an illuminated cell, n' deviates from unity when the generation–recombination currents in the junction are comparable to J_{sc}. This situation appears to be occurring in a–Si:H p–i–n cells but not in Schottky-barrier cells where $n = n' \simeq 1.0$–1.1 (see Sect. 10.5.1).

The temperature dependence of J_{sc} and V_{oc} for a Pd Schottky-barrier cell is shown in Fig. 10.27, the Pd film was $\sim 25\%$ transmitting and was illuminated with $\sim 100\,\mathrm{mW\,cm^{-2}}$ of white light. As the temperature is decreased from 350 to 160 K, J_{sc} decreases only slightly apparently due to the change in the optical gap with temperature [10.74]. This result indicates that there is no significant change in the collection efficiency and the transport of carriers across the junction in this temperature range. The decrease in J_{sc} below $\sim 160\,\mathrm{K}$ can be

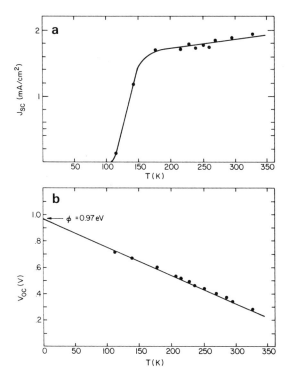

Fig. 10.27. (a) The short-circuit current density and (b) the open-circuit voltage as a function of temperature

attributed to the increased series resistance of the bulk a–Si:H which will be discussed later.

The temperature dependence of V_{oc} shown in Fig. 10.27b is in agreement with (10.7, 8) which can be combined to yield

$$V_{oc} \simeq \frac{n'kT}{8}[\ln J_{sc} - \ln(q\mu_c N_c E_s)] + n'\phi_B. \tag{10.12}$$

Since $n' \simeq 1$, the intercept of $V_{oc}(T)$ is ϕ_B at $T=0$; the value of $\phi_B \simeq 0.97$ eV is in excellent agreement with the value obtained at room temperature by other techniques [10.60]. These results show that both carriers are collected without recombination in the junction to temperatures well below 200 K and also that there is no transition to hopping transport through gap states at $T \sim 250$ K as has been reported in other a–Si:H films [10.14]. A transition to hopping transport would cause a change in the slope of the V_{oc} against T plot and would yield an effectively reduced barrier height from the V_{oc} intercept at $T=0$.

Deviations from the junction characteristics just discussed are obtained when a thin insulating film, sufficient to affect the free carrier transport, is introduced at the metal/undoped a–Si:H interface. Such films can change the diode and photovoltaic characteristics in a fashion similar to that observed on single crystal structures when an ~ 15–30 Å oxide film is introduced at the

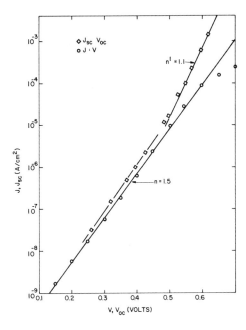

Fig. 10.28. Dark current density as a function of forward bias, and short-circuit current density as a function of open-circuit voltage for an MIS junction

metal/semiconductor interface. Such films do not limit the collection of photogenerated carriers but result in diode characteristics exhibiting n values significantly greater than unity. Similar structures can also be obtained with a–Si:H either by growing thin oxide films (~ 15 min at $350\,°C$ in air) or by evaporation of the metal contact in a residual pressure of oxygen [6.63]. An example of the dark I–V and J_{sc}–V_{oc} characteristics that can be obtained with such metal/thin oxide/a–Si:H structures is shown in Fig. 10.28. The structure was similar to a Pd Schottky barrier cell, but a thin oxide was introduced at the metal/a–Si:H interface. The dark I–V characteristics exhibit $n = 1.5$ with a normal series resistance limit at the high current densities due to the bulk of the a–Si:H. Even though J_{sc} increases linearly with intensity over the entire range of intensities used, the J_{sc}–V_{oc} plot exhibits two values of n'. At low levels of illumination, and $J_{sc} \gtrsim 10^{-5}$ A cm^{-2}, $n' \simeq n = 1.5$ but at the higher illumination $n' = 1.1$. This transition of n' to a value close to unity, without a corresponding transition in n, results from the fact that in forward bias the electrons flow from the a–Si:H into the Schottky barrier metal, whereas in the case of short circuit currents it is the holes. Consequently the kinetics of carrier trapping and recombination at the oxide/a–Si:H interface are different for the two cases and a transition to $n' \simeq 1$ can occur when the field across the thin insulator becomes independent of the current flow across the junction.

Several possible mechanisms have been proposed for the characteristics of metal/thin oxide/single crystal contacts [10.64, 75, 76] and some of these can be applied to the a–Si:H structures. It should be pointed out here also that these mechanisms are quite different from the recombination in the barrier region of the a–Si:H which also yield results similar to those shown in Fig. 10.28. Such barrier recombination is observed in intimate metal/*doped* a–Si:H Schottky barriers. This is a strong indication that doping introduces deep recombination centers in addition to ionized donors and acceptors, in agreement with other results obtained with doped a–Si:H (see Sect. 10.5.3) and the characteristics of *p–i–n* cell structures (See Sect. 10.5.1, 4).

As shown in Fig. 10.19, the J_{sc}–V_{oc} plots exhibit exponential characteristics to illuminations of ~1 sun and J_{sc} ~ 10 mA cm^{-2}. All a–Si:H cells exhibit this behavior and it reflects the fact that the internal resistances of the cells under sunlight illumination are significantly smaller than in the dark. This is an extremely important effect, associated with the photoconductivity of a–Si:H, which makes the undoped a–Si:H structures photoconductive solar cells [10.56, 65].

For a cell to have negligible power loss due to its internal resistance, the resistance in series with the current generating region, R_S, must be low. In single crystal silicon cells operating under AM1 illumination this is achieved if R_S is ~0.1 Ω for a cell having an area of 1 cm^2. Since the a–Si:H cells have large open-circuit voltages and lower short-circuit currents under AM1 illumination, the requirements on R_S are less stringent, but still require R_S to be $\gtrsim 1$ Ω. In cells where the photogenerated carriers are only collected from a fraction of their thickness, the resistance of the remaining fraction of the a–Si:H film constitutes a series resistance to the electron current. This resistance is added to the usual front and back contact resistances when the total series resistance (R_S) is being considered. Because of the high intrinsic resistivity of a–Si:H the lateral resistance of the films necessitates the use of transparent conductivity contacts on the illuminated surface of the cell. These contacts must have a sufficiently high sheet conductance so as to satisfy the requirement of $R_S \gtrsim 1$ Ω. To minimize the contact resistance and to ensure optimum operation of the cell requires the presence of ohmic back contacts.

In the presence of such front and back contacts, R_S is determined by the quasi-neutral region and is given by $R_S = (L - W_B)/\sigma$ where σ is the conductivity of the a–Si:H. Under far forward bias, the diode currents become limited by this series resistance and in this regime are given by [10.77]

$$J \cong \frac{(V - V_0)}{R_S} \simeq \frac{(V_0 - V)\sigma}{L}, \tag{10.13}$$

since under these conditions $W_B \ll L \simeq 1$ μm. The dark conductivities of undoped a–Si:H can range from 10^{-5} to 10^{-11} Ω^{-1} cm^{-1} depending on the substrate temperature and result in forward bias diode characteristics with the onset of R_S limited current densities at ~10^{-3} down to ~10^{-9} A cm^{-2} respectively. However, these large series resistances are greatly reduced by the volume photoconductivity of the undoped a–Si:H discussed in Sect. 10.3.3. This effect is shown in Fig. 10.29 for an undoped a–Si:H, Pd Schottky-barrier structure. The far forward-bias currents in the dark and under illumination are shown as

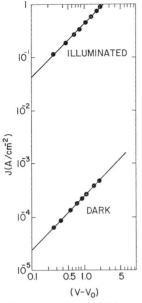

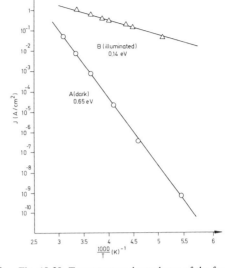

Fig. 10.29. Far forward bias current densities of a Pd Schottky-barrier cell in the dark and under illumination as a function of $V - V_0$

Fig. 10.30. Temperature dependence of the far forward bias current densities for a Pd Schottky-barrier cell in the dark and under illumination

a function of the applied voltage, V, minus V_0. ($V_0 = 0.45$ in the dark and 0.55 under illumination.) The cell was illuminated with $\sim 100\,\mathrm{mW\,cm^{-2}}$ of white light filtered to pass wavelengths $> 0.625\,\mu\mathrm{m}$ so as to generate volume absorption of the light. Figure 10.29 shows a reduction of R_s from $\sim 4 \times 10^2\,\Omega$ to $2.5\,\Omega$ due to the illumination which corresponds to a change in the conductivity of the bulk from $\sim 2.5 \times 10^{-7}\,\Omega^{-1}\mathrm{cm}^{-1}$ to $4 \times 10^{-5}\,\Omega^{-1}\mathrm{cm}^{-1}$.

These conductivities and their temperature dependences are in good agreement with measurements carried out on similar a–Si:H films using coplanar structures with two ohmic contacts [10.46, 65, 73]. The temperature dependence of these far forward bias currents obtained with a similar Pd/Schottky-barrier structure are shown in Fig. 10.30 where the forward biases applied were 1.5 V in the dark and 3 V under illumination. The activation energies, 0.65 and 0.14 eV for the dark currents and photocurrents respectively, are in good agreement with the activation energies of the dark conductivity and the bulk photoconductivity of a–Si:H discussed in Sect. 10.3.3. The photoconductivity of undoped a–Si:H allows short-circuit current densities of $\sim 10\,\mathrm{mA\,cm^{-2}}$ to be collected without bulk series resistance limitations as indicated by the results shown in Figs. 10.19, 29, 30. The low activation energies of these photoconductivities (Sect. 10.3.3 and Fig. 10.20) ensure that this condition is maintained even at temperatures well below 200 K as seen in Fig. 10.27. In contrast, the large activation energy of the dark conductivity, such as shown in Fig. 10.30, results in a rapid increase in R_s upon cooling leading to a complete domination of the forward-bias diode current character-istics by R_S below room temperature [10.65].

There is an additional effect due to the photoconductivity of a–Si : H which should be considered in solar cell operation [10.56, 65]. The volume generated photoconductivity of a–Si : H under AM1 illumination results in a quasi-Fermi level which can be up to ~0.2 eV closer to the conduction band than in the dark. Such a displacement of the electron quasi-Fermi level causes a change from the energy band diagram of a cell structure represented in Fig. 10.21b for a low level of illumination. The new density of electrons results in a steady state condition which increases V_0 by an amount equal to the displacement of the Fermi level so that (10.9) can still be satisfied. The increase in V_0, which has to be included in (10.9), has the effect of extending the space-charge region W_B. This subtle effect is mentioned here because it counteracts the decreases in W_B, discussed below, which have profound effects on the cell operation.

A cell under load is subject to a forward bias that not only generates a current flow in the opposite direction to the photogenerated current, but also decreases the voltage drop across the space charge region [thus decreasing $W_B(V)$]. Consequently, even a fully depleted cell, such as represented by Fig. 10.21a under short-circuit conditions, can have a field distribution such as shown in Fig. 10.21b under load. As discussed earlier, the collection efficiency of carriers generated in the barrier region can approach 100%, but holes generated in the quasi-neutral region, $L - W_B(V)$, must rely on diffusion to be collected. An extensive characterization of minority carrier diffusion has not yet been performed due to experimental difficulties [10.55]. However, hole dif-fusion lengths up to a few tenths of a micrometer have been estimated from the characterization of a–Si : H photovoltaic structures [10.56]. These small dif-fusion lengths are consistent with photoconductivity [10.78] and hole drift mobility measurements [10.55] on lightly doped, p-type a–Si : H. The diffusion of holes can contribute greatly to the operation of cells in which $L_p \sim W_B(V)$.

A detailed theoretical treatment of the illuminated $I–V$ characteristics in the various a–Si : H solar cell structures has to take into account the large variations in the electronic and optical properties as discussed in the previous sections. In addition, it is complicated by the commonly observed dependencies of the collection efficiencies on bias such as discussed in Sect. 10.5.4. Since the photocurrents obtained with white light correspond to the sum of the contributions from all wavelengths, they are strongly dependent on the spectral distribution of the illumination, the values of $\alpha(\lambda)$, the values of $W_B(V)$, the diffusion length of holes and the field distributions in the cells. Because of the variation in these properties for different cells it is very difficult to carry out a quantitative correlation between observed and calculated $I–V$ characteristics without making numerous assumptions. Consequently, this is not undertaken here but instead the general trend of the dependence of the characteristics on the principal parameters is discussed.

When $L_p \ll [L - W_B(V)]$ and $L_p \ll W_B(V)$, the photocurrents exhibit a strong dependence on bias as shown in Fig. 10.24a. Carriers generated by short-

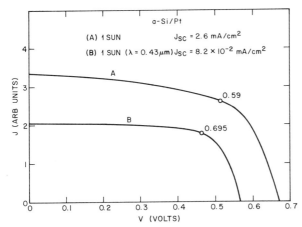

Fig. 10.31. Illuminated I–V characteristics of a Pt Schottky-barrier cell. Curve A was obtained with illumination comparable to AM1 sunlight and curve B with the same illumination but with a narrow passband blue filter present

wavelength light are efficiently collected due to the large values of $\alpha(\gtrsim 10^5\,\text{cm}^{-1})$ even when $W_B(V)$ is decreased to $\sim 0.1\,\mu\text{m}$ under load. However, the collection of carriers generated by longer wavelength light depends strongly on $W_B(V)$, and this variation in the photocurrents with bias limits the values of the fill factor to between 0.4 and 0.5 for white light illumination.

If the cell is fully depleted even under load, then the photovoltaic characteristics are relatively independent of L_p. The illuminated I–V characteristics for such a cell are shown in Fig. 10.31. (The collection efficiency and C–V characteristics are shown in Fig. 10.22 for the same cell.) This cell exhibited only a small change in the zero and reverse bias capacitance when illuminated with $\sim 100\,\text{mW cm}^{-2}$ of white light ($\sim$AM1).

The fill factors for this structure are shown in Fig. 10.31 for both white light illumination and illumination having a narrow passband filter at $\lambda = 0.43\,\mu\text{m}$. The respective values of J_{sc} are $2.6\,\text{mA cm}^{-2}$ and $8.2 \times 10^{-2}\,\text{mA cm}^{-2}$ and the fill factors are 0.59 and 0.695. The cell exhibited near-ideal diode characteristics in the dark and had a J_0 of $5 \times 10^{-13}\,\text{A cm}^{-2}$. In this structure the white light fill factor is limited by a series resistance at the back contact similar to that considered in single crystal cells. This was indicated by the improvement in fill factor as the intensity of white light illumination was decreased. However, the changes obtained with white light indicated that at the optimum load not all the holes generated at the back of the cell were collected. With white light adjusted to provide a J_{sc} of $\sim 10 \times 10^{-2}\,\text{mA cm}^{-2}$, the fill factor was not as good as the one shown for the blue light. A larger fill factor is obtained with the blue light because all the carriers are generated in a region ($\sim 0.1\,\mu\text{m}$ thick) adjacent to the metal. This limits the generation of holes to a high field region even under load and, at the lower photocurrent levels ($\sim 10^{-1}\,\text{mA cm}^{-2}$), the extraction is not R_s limited. Since the collection efficiency for blue light is

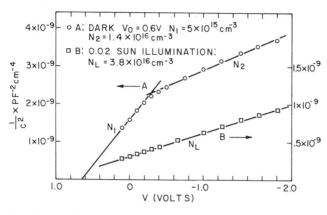

Fig. 10.32. $1/C^2$ as a function of voltage for a Pt Schottky-barrier solar cell in the dark and white light illumination corresponding to ~ 0.02 AM1

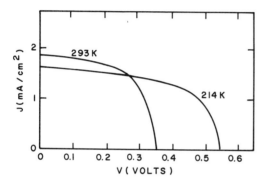

Fig. 10.33. Illuminated I–V characteristics of a Pd Schottky-barrier cell at 293 and 214 K

independent of bias the measured fill factor of 0.695 is in reasonable agreement with that expected from the forward bias diode characteristics.

Cells with high efficiencies have been fabricated in which L_p is not negligible with respect to $L - W_B(V)$. High efficiencies (4–6%) were obtained even though there was a significant increase in space charge density under AM1 illumination due to the trapping of photogenerated holes. Figure 10.32 shows the increase in space charge density as the illumination was increased to ~ 0.02 suns (~ 0.02 AM1) for a Pt Schottky-barrier cell with a ZrO_2 antireflection coating. The capacitance measurements at higher illuminations could not be carried out because high dissipation factors could not be balanced out on the system used. The barrier region in the dark exhibits two slopes which correspond to space-charge densities $N_1 = 5 \times 10^{15}$ cm^{-3} and $N_2 = 1.6 \times 10^{16}$ cm^{-3}. The two values N_1 and N_2 can be attributed to a density of electron traps, $\sim 10^{16}$ cm^{-3}, below the Fermi level in the bulk. Under illumination the space charge density of the barrier region increases to $N_L = 3.8 \times 10^{16}$ cm^{-3} and the zero bias depletion width decreases to 0.18 µm from the dark value of 0.41 µm.

Despite the decrease in the depletion width that occurs upon increased illumination there was no corresponding decrease in the collection efficiency of the cell; the short-circuit current was proportional to intensity up to the illuminations corresponding to AM1. Furthermore, both white light and monochromatic photocurrents did not increase significantly under reverse bias (see Sect. 10.5.4). These results indicate that there is a significant contribution from carriers generated outside the field region which compensates for the large change in the depletion width as the illumination is increased. Consequently, L_p must be comparable to the change in $W_B(V)$ or $L_p \sim 0.2\,\mu\text{m}$. Also, similar values of the fill factors were obtained with comparable currents generated by either red or blue light. In poor cells with low values of L_p the fill factors for red light are significantly lower than blue light.

When $L_p \sim W_B(V)$, the white light photocurrents are relatively insensitive to bias, and fill factors $\gtrsim 0.6$ are obtained as shown in Fig. 10.33 for a Pd Schottky barrier structure. These illuminated $I-V$ characteristics were obtained at 293 and 214 K with $\sim 100\,\text{mW cm}^{-2}$ of white light (the Pd film was $\sim 25\%$ transmitting). These characteristics are in agreement with the temperature dependence of the junction and bulk a–Si:H properties shown in Figs. 10.27, 30. J_{sc} changes only slightly between 293 and 214 K while V_{oc} exhibits a significant increase. The fill factor remains essentially constant as a result of the efficient photocarrier collection and the photoconductivity of the bulk.

10.6 Conclusions

This chapter has reviewed the recent development of the a–Si:H solar cell. While the study of the material is less than a decade old, the discovery of the photovoltaic effect in a–Si:H occurred only in 1974. Many of the properties of the material and the devices are not well understood. In particular, our knowledge concerning the defect states in a–Si:H is rudimentary at best, and our understanding of the relationship between the discharge kinetics and film properties is even less well developed. As discussed in Sect. 10.5.5, the minority carrier transport properties must be improved before efficient a–Si:H solar cells become a reality.

While much more research must be done before practical devices appear on the market, it is encouraging that moderate conversion efficiencies (6%) have been obtained in a relatively short time and that preliminary testing indicates that the material is relatively stable (see Sect. 10.3.1).

The single greatest advantage of this new emerging technology is the promise of low-cost fabrication of large area solar arrays. Since a–Si:H solar cells can be fabricated on inexpensive substrates such as glass and steel sheet, the total material costs are on the order of a few dollars per square foot. Moreover, glow-discharge deposition can be readily scaled up to continuous, in-line processing of large sheets so that processing costs should be quite low.

Thus, it is possible that in the not too distant future, thin films of an amorphous semiconductor may be providing the people of the world with inexpensive, nonpolluting power from an inexhaustible source.

References

10.1   D. M. Chapin, C. S. Fuller, G. L. Pearson: J. Appl. Phys. **25**, 676 (1954)
10.2   J. Lindmeyer, C. Y. Wrigley: "Development of a 20% efficient solar cell" (NSF-43090, National Science Foundation, Washington, D.C. 1975)
10.3   L. W. James, R. L. Moon: Proc. 11th IEEE Photovoltaic Specialists Conf., Scottsdale, Arizona, May 6–8 (1975) pp. 402–408
10.4   D. E. Carlson: U.S. Patent No. 4,064,521 (1977)
10.5   D. E. Carlson, C. R. Wronski: Appl. Phys. Lett. **28**, 671 (1976)
10.6   A. Triska, D. Dennison, H. Fritzsche: Bull. Am. Phys. Soc. **20**, 392 (1975)
10.7   M. H. Brodsky, M. A. Frisch, J. F. Ziegler, W. A. Lanford: Appl. Phys. Lett. **30**, 561 (1977)
10.8   J. I. Pankove, D. E. Carlson: Appl. Phys. Lett. **31**, 450 (1977)
10.9   P. J. Zanzucchi, C. R. Wronski, D. E. Carlson: J. Appl. Phys. **48**, 5227 (1977)
10.10 G. J. Clark, C. W. White, D. D. Allred, B. R. Appleton, C. W. Magee, D. E. Carlson: Appl. Phys. Lett. **31**, 582 (1977)
10.11 R. C. Chittick, J. H. Alexander, H. F. Sterling: J. Electrochem. Soc. **116**, 77 (1969)
10.12 W. E. Spear, P. G. LeComber: Solid State Commun. **17**, 1193 (1975)
10.13 W. E. Spear, P. G. LeComber, S. Kinmond, M. H. Brodsky: Appl. Phys. Lett. **28**, 105 (1976)
10.14 P. G. LeComber, A. Madan, W. E. Spear: J. Non-Cryst. Solids **11**, 219 (1972)
10.15 G. A. N. Connell, J. R. Pawlik: Phys. Rev. B**13**, 787 (1976)
10.16 W. Paul, A. J. Lewis, G. A. N. Connell, J. D. Monstakas: Solid State Commun. **20**, 969 (1976)
10.17 A. J. Lewis, G. A. N. Connell, W. Paul, J. R. Pawlik, R. J. Temkin: Proc. Int. Conf. Tetrahedrally Bonded Amorphous Semicond., ed. by M. H. Brodsky, S. Kirkpatrick, D. Weaire (American Institute of Physics, New York 1974) p. 27
10.18 A. L. Robinson: Science **197**, 1069 (1977)
10.19 D. J. Morel, A. K. Jhosh, F. Feng, G. L. Stogryn, P. E. Purwin, R. S. Shaw, C. Fishman: Appl. Phys. Lett. **32**, 495 (1978)
10.20 J. Lindmeyer: COMSAT Tech. Rev. **2**, 105 (1972)
10.21 S. M. Sze: *Physics of Semiconductor Devices* (Wiley-Interscience, New York 1963)
10.22 J. C. Knights: Philos. Mag. **34**, 663 (1976)
10.23 J. J. Hanak, P. J. Zanzucchi, D. E. Carlson, C. R. Wronski, J. I. Pankove: Proc. 7th Intern. Vac. Congr. 3rd Intern. Conf. Solid Surfaces, Vienna, Austria, Sept. 12–16, ed. by R. Dobrozensky, S. Rudenauer, F. P. Viehböck, R. Breth (1977) p. 1947
10.24 M. H. Brodsky: Thin Solid Films **40**, L23 (1977)
10.25 J. C. Knights, J. M. Hayes, J. C. Mikkelsen, Jr.: Phys. Rev. Lett. **39**, 712 (1977)
10.26 S. R. Herd, P. Chaudhari, M. H. Brodsky: J. Non-Cryst. Solids **7**, 309 (1972)
10.27 W. Beyer, R. Fischer: Appl. Phys. Lett. **31**, 850 (1977)
10.28 D. E. Carlson, C. R. Wronski: J. Electron. Mater. **6**, 95 (1977)
10.29 H. Fritzsche, C. C. Tsai, P. Persans: Solid State Technol. **21**, 55 (1978)
10.30 M. H. Brodsky, M. Cardona, J. J. Cuomo: Phys. Rev. B**16**, 3556 (1977)
10.31 D. E. Carlson, C. W. Magee, A. R. Triano: J. Electron. Soc. **126**, 688 (1979)
10.32 B. von Roedern, L. Ley, M. Cardona: Phys. Rev. Lett. **39**, 1576 (1977)
10.33 A. Barna, P. B. Barna, G. Rodroczi, L. Toth, P. Thomas: Phys. status solidi (a) **41**, 81 (1977)
10.34 V. A. Singh, C. Weigel, J. W. Corbett, L. M. Roth: Phys. status solidi (b) **81**, 637 (1977)
10.35 M. H. Tanielian, H. Fritzsche, C. C. Tsai: Bull. Am. Phys. Sos. **22**, 336 (1977)
10.36 D. E. Carlson, C. W. Magee: Appl. Phys. Lett. **33**, 81 (1978)
10.37 E. A. Davis: *Amorphous Semiconductors*, ed. by P. G. LeComber, J. Mort (Academic Press, New York 1973) p. 450

10.38 P.J.Zanzucchi: Personal communication
10.39 J.C.Phillips: *Bonds and Bands in Semiconductors* (Academic Press, New York 1973)
10.40 R.J.Loveland, W.E.Spear, A.Al-Sharboty: J. Non-Cryst. Solids **13**, 55 (1973/74)
10.41 W.E.Spear, R.J.Loveland, A.Al-Sharboty: J. Non-Cryst. Solids **15**, 410 (1974)
10.42 W.E.Spear, P.G.LeComber: Philos. Mag. **33**, 935 (1976)
10.43 W.Rehm, R.Fischer, J.Stuke, H.Wagner: Phys. status solidi (b) **79**, 539 (1977)
10.44 D.A.Anderson, W.E.Spear: Philos. Mag. **36**, 695 (1977)
10.45 D.L.Staebler, C.R.Wronski: Appl. Phys. Lett. **31**, 292 (1977)
10.46 C.R.Wronski, D.E.Carlson: In *Proc. 7th Int. Conf. Amorphous and Liquid Semiconductors*, ed. by W.E.Spear (CICL, University of Edinburgh 1977) p. 452
10.47 D.L.Staebler, C.R.Wronski: Extended Abstracts of the Fall Meeting of the Electrochemical Society, Atlanta, Georgia, Oct. 9–14 (1977) pp. 805–806
10.48 D.L.Stabler, C.R.Wronski: To be published
10.49 W.Meyer, H.Neldel: Z. Tech. Phys. **18**, 588 (1937)
10.50 J.G.Simmons: Phys. Rev. **155**, 657 (1967)
10.51 C.R.Wronski, D.E.Carlson: To be published
10.52 A.Rose: *Concepts in Photoconductivity and Allied Problems* (Interscience, New York 1963)
10.53 C.R.Wronski, R.E.Daniel: To be published
10.54 P.G.LeComber, W.E.Spear: Phys. Rev. Lett. **25**, 509 (1970)
10.55 A.R.Moore: Appl. Phys. Lett. **31**, 762 (1977)
10.56 C.R.Wronski: IEEE Trans. ED-**24**, 351 (1977)
10.57 W.E.Spear, P.G.LeComber: J. Non-Cryst. Solids **8–10**, 727 (1972)
10.58 A.Madan, P.G.LeComber, W.E.Spear: J. Non-Cryst. Solids **20**, 239 (1976)
10.59 J.C.Knights, D.K.Biegelsen: Solid State Commun. **22**, 133 (1977)
10.60 C.R.Wronski, D.E.Carlson: Solid State Commun. **23**, 421 (1977)
10.61 G.H.Döhler, M.Hirose: In *Proc. 7th Int. Conf. Amorphous and Liquid Semiconductors*, ed. by W.E.Spear (CICL, University of Edinburgh 1977) p. 373
10.62 C.R.Wronski, D.E.Carlson, R.E.Daniel: Appl. Phys. Lett. **29**, 602 (1976)
10.63 D.E.Carlson: IEEE Trans. ED-**24**, 449 (1977)
10.64 M.A.Green, R.B.Godfrey: Appl. Phys. Lett. **29**, 610 (1976)
10.65 C.R.Wronski: Jpn. J. Appl. Phys., IV, Suppl. **17-1**, 299 (1978)
10.66 J.J.Hanak, V.Korsun: Proc. 13th IEEE Photovoltaic Specialists Conf. (1978)
10.67 D.Engemann, R.Fisher: *Proc. 12th Int. Conf. Physics of Semiconductors*, ed. by M.H.Pilkuhn (Teubner, Stuttgart 1974) p. 1042
10.68 J.J.Hanak: Data supplied by RCA Laboratories, Princeton, N.J.
10.69 D.E.Carlson: Tech. Digest of 1977 IEEE Intern. Electron Dev. Mtg., Washington, DC (IEEE, N.Y. 1977) p. 214
10.70 G.D.Watkins, J.W.Corbett: Phys. Rev. **121**, 1001 (1961)
10.71 N.F.Mott, E.A.Davis: *Electron Processes in Noncrystalline Materials* (Clarendon Press, Oxford 1971)
10.72 C.R.Wronski: Proc. 13th Photovoltaic Specialists Conf., (IEEE, New York 1978) p. 744
10.73 C.R.Wronski, D.E.Carlson, R.E.Daniel, A.R.Triano: Tech. Digest of 1976 IEEE Intern Electron Dev. Mtg., Washington, DC (IEEE, N.Y. 1976) p. 75
10.74 J.I.Pankove, D.E.Carlson: Appl. Phys. Lett. **29**, 620 (1976)
10.75 J.Schewchum, M.A.Green, F.D.King: Solid State Electron. **17**, 551 (1974)
10.76 S.J.Fonash: J. Appl. Phys. **47**, 3597 (1976)
10.77 A.Goodman: Surface Sci. **1**, 54 (1964)
10.78 C.R.Wronski, A.R.Moore: To be published

Additional References with Titles

Of Particular Interest are Four Recent Volumes:

G.S.Cargill III, P.Chaudhari (eds.): *Proceedings of the Topical Conference on Atomic Scale Structure of Amorphous Solids*, Yorktown Heights, 1978; [J. Non-Cryst. Solids **31** 1978)]; Republished as book (North-Holland, Amsterdam 1979)

P.H.Gaskell, J.D.MacKenzie (eds.): *Electronic Properties and Structure of Amorphous Solids: Mott Festschrift*, J. Non-Cryst. Solids **32** (1979)

N.F. Mott, E.A.Davis: *Electronic Processes in Non-Crystalline Materials*, 2nd ed. (Clarendon Press, Oxford 1979)

W.Paul, M.Kastner (eds.): *Proceeding of the 8th International Conference on Amorphous and Liquid Semiconductors*, Cambridge, 1979 (to appear as a volume of J. Non-Cryst. Solids)

Chapter 2

P.W.Anderson, E.Abrahams, T.V.Ramakrishnan: Possible explanation of nonlinear conductivity in thin-film metal wires. Phys. Rev. Lett. **43**, 718 (1979)

P.A.Lee: Real space scaling studies of localization. Phys. Rev. Lett. **42**, 1492 (1979)

J.Robertson: Theory of doping of amorphous tetrahedrally bonded semiconductors. Philos. Mag. B **40**, 31 (1979)

D.Thouless: The effect of inelastic electron scattering on the conductivity of very thin wires. (Submitted for Publication)

Chapter 3

J.Orenstein, M.Kastner: Time-resolved optical absorption and mobility of localized charge carriers in a-As_2Se_3. Phys. Rev. Lett. **43**, 161 (1979)

R.A.Street, G.Lucovsky: Ionicity effects on defects in chalcogenide alloys. Solid State Commun. **31**, 289 (1979)

Chapter 4

D.E.Ackley, J.Tauc, W.Paul: Picosecond relaxation of optically induced absorption in amorphous semiconductors. Phys. Rev. Lett. **43**, 715 (1979)

E.C.Freeman, W.Paul: Optical constants of rf sputtered hydrogenated amorphous silicon. Phys. Rev. B **20**, 715 (1979)

R.L.Fork, C.V.Shank, A.M.Glass, A.Migus, M.A.Bosch, J.Shah: Picosecond dynamics of optically induced absorption in the band gap of As_2S_3. Phys. Rev. Lett. **43**, 394 (1979)

P.O'Connor, J.Tauc: Photoinduced Optical Absorption in Amorphous Si_xGe_{1-x}:H. Phys. Rev. Lett. **43**, 311 (1979)

Chapter 5

S.H.Holmberg, R.R.Shanks, V.A.Bluhm: Chalcogenide memory materials. J. Electro. Mater. **8**, 333 (1979)

J.A.McInnes, P.N.Butcher: Numerical calculations of dc Hopping conductivity, Philos. Mag. **39**, 1 (1979)

G. Pfister, K. S. Liang, M. Morgan, P. C. Taylor, E. J. Frieble, S. G. Bishop: Hole transport, photo-luminescence, and photoinduced spin resonance in thallium-doped amorphous As_2Se_3. Phys. Rev. Lett. **41**, 1318 (1979)

Chapter 6

M. A. Bosch, J. Shah: Time-resolved photoluminescence spectroscopy in amorphous As_2S_3. Phys. Rev. Lett. **42**, 118 (1979)

W. Rehm, R. Fischer: Fast radiationless recombination in amorphous silicon. Phys. Stat. Solidi (b) (in press)

T. M. Searle, T. S. Nashashibi, I. G. Austin, R. Devonshire, G. Lockwood: Radiative and non-radiative tunneling in glow-discharge and sputtered amorphous silicon. Philos. Mag. 1979 (in press)

J. Shah, M. A. Bosch: Band-to-band luminescence in amorphous solids: Implications for the nature of electronic band states. Phys. Rev. Lett. **42**, 1420 (1979)

R. A. Street, D. K. Biegelsen, J. Stuke: Defects in bombarded amorphous silicon. Philos. Mag. (in press)

R. A. Street, J. G. Knights, D. K. Biegelsen: Luminescence studies of plasma-deposited hydrogenated silicon. Phys. Rev. B **18**, 1880 (1978)

C. Tsang, R. A. Street: Recombination in plasma-deposited amorphous Si : H; luminescence decay. Phys. Rev. B **19**, 3027 (1979)

Chapter 7

R. Bachus, B. Movaghar, L. Schweitzer, U. Voget-Grote: Influence of the exchange interaction on the ESR linewidth in amorphous silicon. (Submitted for publication)

B. Movaghar, B. Ries, L. Schweitzer: Influence of magnetic resonance on the luminescence of amorphous silicon. Philos. Mag. (in press)

B. Movaghar, B. Ries, L. Schweitzer: Theory of the resonant and non-resonant photoconductivity change in amorphous silicon. Philos. Mag. (in press)

U. Voget-Grote, W. Kümmerle, R. Fischer, J. Stuke: The influence of spin defects on recombination and electronic transport in amorphous silicon. Philos. Mag. (in press)

Chapter 8

G. Lucovsky: Spectroscopic evidence for valence-alternation-pair defect states in vitreous SiO_2. Philos. Mag. B **39**, 513 (1979)

W. Paul: On the interpretation of the vibrational spectra of amorphous silicon-hydrogen alloys: a dissent. Philos. Mag. (submitted 1979)

J. C. Phillips: Structure of amorphous $(Ge,Si)_{1-x}Y_x$ alloys. Phys. Rev. Lett. **42**, 1151 (1979)

Chapter 9

W. Beyer, H. Overhof: Transport properties of doped amorphous silicon. Solid State Commun. **31**, 1 (1979)

D. I. Jones, R. A. Gibson, P. G. LeComber, W. E. Spear: Hydrogen content, electrical properties and stability of glow discharge amorphous silicon. Solar Energy Materials **1**, Nov. (1979)

S. Kalbitzer, G. Müller, P. G. LeComber, W. E. Spear: The effects of ion implanation on the electrical properties of amorphous silicon. Philos. Mag, (in press)

P. G. LeComber, W. E. Spear, D. Allan: Transport studies in doped amorphous silicon. J. Non-Cryst. Solids **32**, 1 (1979)

P. G. LeComber, W. E. Spear, A. Ghaith: Amorphous-silicon field-effect device and possible application. Electron. Lett. **15**, 179 (1979)

A. Madan, S. R. Ovshinsky, E. Benn: Electrical and optical properties of amorphous Si : F : H alloys. Philos. Mag. **40** (1979)

B. v. Roedern, L. Ley, M. Cardona: Spectroscopic determination of the position of the fermi level in doped amorphous hydrogenated silicon. Solid State Commun. **29**, 415 (1979)

B. v. Roedern, L. Ley, M. Cardona, F. W. Smith: Photoemission studies on in situ prepared hydrogenated amorphous silicon films. (Submitted for publication)

A. J. Snell, K. D. Mackenzie, P. G. LeComber, W. E. Spear: The interpretation of capacitance and conductance measurements on metal-amorphous silicon barriers. Philos. Mag. B**40**, 1 (1979)

W. E. Spear, D. Allan, P. G. LeComber, A. Ghaith: A new approach to the interpretation of transport results in a-Si. Philos. Mag. B**39** (in press)

W. E. Spear, P. G. LeComber, S. Kalbitzer, G. Müller: Doping of amorphous silicon by alkali-ion implantation. Philos. Mag. B**39**, 159 (1979)

R. H. Williams, R. R. Varma, W. E. Spear, P. G. LeComber: The fermi level position on doped amorphous silicon. J. Phys. C. **12**, L209 (1979)

Chapter 10

Y. Hamakawa, H. Okamoto, Y. Nitta: A new type of amorphous silicon photovoltaic cell generating more than 2 V. Appl. Phys. Lett. **39**, 187 (1979)

J. J. Hanak: Monolithic solar cell panel of amorphous silicon. Sol. Energy **23** (1979) (in press)

T. D. Moutakas: Sputtered Hydrogenated Amorphous Silicon. J. Electron. Mater. **8**, 391 (1979)

Subject Index

Applied Physics

A monthly journal

Board of Editors
S. Amelinckx, Mol; **V. P. Chebotayev,** Novosibirsk;
R. Gomer, Chicago, IL; **P. Hautojärvi,** Espoo;
H. Ibach, Jülich; **V. S. Letokhov,** Moskau;
H. K. V. Lotsch, Heidelberg; **H. J. Queisser,** Stuttgart;
F. P. Schäfer, Göttingen; **K. Shimoda,** Tokyo;
R. Ulrich, Stuttgart; **W. T. Welford,** London;
H. P. J. Wijn, Eindhoven

Coverage

application-oriented experimental and theoretical
physics:

Solid-State Physics	*Quantum Electronics*
Surface Science	*Laser Spectroscopy*
Solar Energy Physics	*Photophysical Chemistry*
Microwave Acoustics	*Optical Physics*
Electrophysics	*Optical Communication*

Special Features

rapid publication (3–4 month)
no page charge for concise reports
prepublication of titles and abstracts
microfiche edition available as well

Languages
mostly English

Articles

original reports, and short communications
review and/or tutorial papers

Manuscripts

to Springer-Verlag (Attn. H. Lotsch), P.O. Box 105 280
D-6900 Heidelberg 1, F. R. Germany

Place North-America orders with:
Springer-Verlag New York Inc., 175 Fifth Avenue,
New York, N.Y. 10010, USA

Springer-Verlag
Berlin
Heidelberg
New York

Dynamics of Solids and Liquids by Neutron Scattering

Editors: S. W. Lovesey, T. Springer
1977. 156 figures, 15 tables. XI, 379 pages
(Topics in Current Physics, Volume 3)
ISBN 3-540-08156-9

Contents:
S. W. Lovesey: Introduction. – *H. G. Smith, N. Wakabayashi:* Phonons. – *B. Dorner, R. Comès:* Phonons and Structural Phase Transformations. – *J. W. White:* Dynamics of Molecular Crystals, Polymers, and Adsorbed Species. – *T. Springer:* Molecular Rotations and Diffusion in Solids, in Particular Hydrogen in Metals. – *R. D. Mountain:* Collective Modes in Classical Monoatomic Liquids. – *S. W. Lovesey, J. M. Loveluck:* Magnetic Scattering.

E. N. Economou
Green's Functions in Quantum Physics

1979. 49 figures, 2 tables. IX, 251 pages
(Springer Series in Solid-State Sciences, Volume 7)
ISBN 3-540-09154-8

Contents:
Green's Functions in Mathematical Physics. – Green's Functions in One-Body Quantum Problems. – Green's Functions in Many-Body Systems. – Appendices.

O. Madelung
Introduction to Solid-State Theory

Translated from the German by B. C. Taylor
1978. 144 figures. XI, 486 pages
(Springer Series in Solid-State Sciences, Volume 2)
ISBN 3-540-08516-5

Contents:
Fundamentals. – The One-Electron Approximation. – Elementary Excitations. – Electron-Phonon Interaction: Transport Phenomena. – Electron-Electron Interaction by Exchange of Virtual Phonons: Superconductivity. – Interaction with Photons: Optics. – Phonon-Phonon Interaction: Thermal Properties. – Local Description of Solid-State Properties. – Localized States. – Disorder. – Appendix: The Occupation Number Representation.

Neutron Diffraction

Editor: H. Dachs
1978. 138 figures, 32 tables. XIII, 357 pages
(Topics in Current Physics, Volume 6)
ISBN 3-540-08710-9

Contents:
H. Dachs: Principles of Neutron Diffraction. – *J. B. Hayter:* Polarized Neutrons. – *P. Coppens:* Combining X-Ray and Neutron Diffraction: The Study of Charge Density Distributions in Solids. – *W. Prandl:* The Determination of Magnetic Structures. – *W. G. Schmatz:* Disordered Structures. – *P.-A. Lindgård:* Phase-Transitions and Critical Phenomena. – *G. Zaccaï:* Application of Neutron Diffraction to Biological Problems. – *P. Chieux:* Liquid Structure Investigation by Neutron Scattering. – *H. Rauch, D. Petrachek:* Dynamical Neutron Diffraction and Its Application.

Springer-Verlag
Berlin
Heidelberg
New York